International Symposium on Quasifission Process in Heavy Ion Reactions 2010

Journal of Physics: Conference Series Volume 282

Messina, Italy
8 – 9 November 2010

Editors:

Giorgio Giardina
Avazbek K. Nasirov
Sara Pirrone

ISBN: 978-1-61782-663-4
ISSN: 1742-6588

Printed from e-media with permission by:

Curran Associates, Inc.
57 Morehouse Lane
Red Hook, NY 12571

Some format issues inherent in the e-media version may also appear in this print version.

Printed by Curran Associates, Inc. (2011)

For permission requests, please contact the Institute of Physics
at the address below.

Institute of Physics
Dirac House, Temple Back
Bristol BS1 6BE UK

Phone: 44 1 17 929 7481
Fax: 44 1 17 920 0979

techtracking@iop.org

Additional copies of this publication are available from:

Curran Associates, Inc.
57 Morehouse Lane
Red Hook, NY 12571 USA
Phone: 845-758-0400
Fax: 845-758-2634
Email: curran@proceedings.com
Web: www.proceedings.com

TABLE OF CONTENTS

INVITED TALKS

ORAL CONTRIBUTIONS

PREFACE

Editors:
Giorgio Giardina (University of Messina)
Avazbek K Nasirov (Joint Institute for Nuclear Research)
Sara Pirrone (Istituto Nazionale di Fisica Nucleare Sezione di Catania)

Organizing Committee:
Chairman: G Giardina (Messina – Italy)
Co-Chairman: A K Nasirov (Dubna – Russia)
Co-Chairman: S Pirrone (Catania – Italy)
Scientific Secretary: G Mandaglio (Messina – Italy) and A Di Pietro (Catania – Italy)

Organizing Institutions:
Fondazione University of Messina Bonino-Pulejo (Messina)

Topics:

- Capture and Deep Inelastic Collision
- Quasifission
- Complete Fusion
- Fast Fission
- Fusion-Fission
- Ternary Fission
- Synthesis of Heavy and Superheavy Elements
- Pre-equilibrium emission

Local Organizing Committee:
G Fazio (Messina), G Giardina (Messina), G Mandaglio (Messina), M Manganaro (Messina), M Romaniuk (Messina), C Saccà (Messina), A Di Pietro (Catania), S Pirrone (Catania), A Nasirov (Dubna)

Sponsored by:
University of Messina, Fondazione Bonino-Pulejo (Messina), INFN Sezione di Catania, INFN Laboratori Nazionali del Sud Catania.

Website: http:/nucleo.unime.it/symp2010/

Dates

Issue 1 (2011)

Peer review statement

All papers published in this volume of *Journal of Physics: Conference Series* have been peer reviewed through processes administered by the proceedings Editors. Reviews were conducted by expert referees to the professional and scientific standards expected of a proceedings journal published by IOP Publishing.

Journal of Physics: Conference Series **282** (2011) 012001 doi:10.1088/1742-6596/282/1/012001

Peculiarities of sub-barrier reactions with heavy ions

V.V.Sargsyan[1], **G.G.Adamian**[1,2], **N.V.Antonenko**[1], **W.Scheid**[3], **and H.Q.Zhang**[4]

[1]Joint Institute for Nuclear Research, 141980 Dubna, Russia
[2]Institute of Nuclear Physics, 702132 Tashkent, Uzbekistan
[3]Institut für Theoretische Physik der Justus–Liebig–Universität, D–35392 Giessen, Germany
[4]China Institute of Atomic Energy, Post Office Box 275, Beijing 102413, China

E-mail: adamian@theor.jinr.ru

Abstract. With the quantum diffusion approach the behavior of capture cross sections and mean-square angular momenta of captured systems are revealed in the reactions with deformed and spherical nuclei at sub-barrier energies. The calculated results are in a good agreement with existing experimental data. With decreasing bombarding energy under the barrier the external turning point of the nucleus-nucleus potential leaves the region of short-range nuclear interaction and action of friction. Because of this change of the regime of interaction, an unexpected enhancement of the capture cross section is expected at bombarding energies far below the Coulomb barrier. This effect is shown its worth in the dependence of mean-square angular momentum of captured system on the bombarding energy. From the comparison of calculated capture cross sections and experimental capture (fusion) cross sections, the importance of quasifission near the entrance channel is shown for the actinide-based reactions and reactions with medium-heavy nuclei at sub-barrier energies.

1. Introduction

The measurement of excitation functions down to the extreme sub-barrier energy region is important for studying the long range behavior of nucleus-nucleus interaction, and very little data exist on the fusion, fission and capture cross sections at extreme sub-barrier energies [1–27]. The experimental data obtained are of interest for solving astrophysical problems related to nuclear synthesis. Indications for an enhancement of the S-factor, $S = E_{\rm c.m.}\sigma \exp(2\pi\eta)$ [28, 29], where $\eta(E_{\rm c.m.}) = Z_1 Z_2 e^2\sqrt{\mu/(2\hbar^2 E_{\rm c.m.})}$ is the Sommerfeld parameter, at energies $E_{\rm c.m.}$ below the Coulomb barrier have been found in Refs. [10, 15, 18]. Its origin is still under discussion.

From the comparison of capture cross sections and fusion cross sections one can show a significant role of the quasifission channel in the reactions with various medium-light and heavy nuclei at sub-barrier energies. The competition between the complete fusion and quasifission can strongly reduce the value of the fusion cross section and, respectively, the value of the evaporation residue cross section [30–32]. This effect is especially crucial in the production of superheavy nuclei. It worth remembering that first evidences of hindrance for compound nucleus formation in the reactions with massive nuclei at low energies near the Coulomb barrier were observed at GSI already long time ago [33].

To clarify the behavior of capture cross sections at sub-barrier energies, a further development of the theoretical methods is required [32, 34–36]. The conventional coupled-channel approach with realistic set of parameters is not able to describe the capture cross sections either below

Journal of Physics: Conference Series **282** (2011) 012001 doi:10.1088/1742-6596/282/1/012001

or above the Coulomb barrier [18]. The use of a quite shallow nucleus-nucleus potential [37] with an adjusted repulsive core considerably improves the agreement between the calculated and experimental data. Besides the coupling with collective excitations, the dissipation, which is simulated by an imaginary potential in Ref. [37] or by damping in each channel in Ref. [38], seems to be important. The quantum diffusion approach based on the quantum master-equation for the reduced density matrix has been suggested in Ref. [39]. This model takes into consideration the fluctuation and dissipation effects in collisions of heavy ions which model the coupling with various channels. In the present paper the model [39] is applied.

2. Model

2.1. The nucleus-nucleus potential

In the model [39] the potential describing the interaction of two nuclei is taken in the form [40]

$$V(R, Z_i, A_i, \theta_i, J) = V_C(R, Z_i, A_i, \theta_i, J) + V_N(R, Z_i, A_i, \theta_i, J) + \frac{\hbar^2 J(J+1)}{2\mu R^2}, \qquad (1)$$

where V_N, V_C, and the last summand stand for the nuclear, the Coulomb, and the centrifugal potentials, respectively. The nuclei are proposed to be spherical or deformed. The potential depends on the distance R between the center of mass of two interacting nuclei, mass A_i and charge Z_i of nuclei ($i = 1, 2$), the orientation angles θ_i of the deformed (with the quadrupole deformation parameters β_i) nuclei and the angular momentum J. The static quadrupole deformation parameters are taken from Ref. [41] for the even-even deformed nuclei. For the nuclear part of the nucleus-nucleus potential, we use the double-folding formalism, in the form

$$V_N = \int \rho_1(\mathbf{r_1})\rho_2(\mathbf{R} - \mathbf{r_2})F(\mathbf{r_1} - \mathbf{r_2})d\mathbf{r_1}d\mathbf{r_2}, \qquad (2)$$

where $F(\mathbf{r_1} - \mathbf{r_2}) = C_0[F_{\rm in}\frac{\rho_0(\mathbf{r_1})}{\rho_{00}} + F_{\rm ex}(1 - \frac{\rho_0(\mathbf{r_1})}{\rho_{00}})]\delta(\mathbf{r_1} - \mathbf{r_2})$ is the density-dependent effective nucleon-nucleon interaction and $\rho_0(\mathbf{r}) = \rho_1(\mathbf{r}) + \rho_2(\mathbf{R} - \mathbf{r})$, $F_{\rm in,ex} = f_{\rm in,ex} + f'_{\rm in,ex}\frac{(N_1-Z_1)(N_2-Z_2)}{(N_1+Z_1)(N_2+Z_2)}$. Here, $\rho_i(\mathbf{r_i})$ and N_i are the nucleon densities and neutron numbers of the light and the heavy nuclei of the dinuclear system, respectively. Our calculations are performed with the following set of parameters: C_0 = 300 MeV fm^3, $f_{\rm in}$ = 0.09, $f_{\rm ex}$ = -2.59, $f'_{\rm in}$ = 0.42, $f'_{\rm ex}$ = 0.54 and ρ_{00} = 0.17 fm^{-3} [40]. The densities of the nuclei are taken in the two-parameter symmetrized Woods-Saxon form with the nuclear radius parameter r_0=1.15 fm (for the nuclei with $A_i \geq 16$) and the diffuseness parameter a depending on the charge and mass numbers of the nucleus [40]. We use a= 0.53 fm for the light magic nucleus ^{16}O, a= 0.55 fm for the intermediate nuclei (^{30}Si, ^{32,34,36}S, ^{48}Ca, ^{50}Ti, ^{64}Ni, ^{208}Pb), and a= 0.56 fm for the actinides. For the ^{4}He, nucleus r_0=1.02 fm and a=0.48 fm.

The Coulomb interaction of two deformed nuclei has the following form:

$$V_C(R, Z_i, A_i, \theta_i, J) = \frac{Z_1 Z_2 e^2}{R} + \left(\frac{9}{20\pi}\right)^{1/2} \frac{Z_1 Z_2 e^2}{R^3} \sum_{i=1,2} R_i^2 \beta_i \left[1 + \frac{2}{7}\left(\frac{5}{\pi}\right)^{1/2} \beta_i\right] P_2(\cos\theta_i), \quad (3)$$

where $P_2(\cos\theta_i)$ is the Legendre polynomial.

In Fig. 1 there is shown the nucleus-nucleus potential V for the ^{16}O + ^{238}U reaction (for simplicity, ^{238}U is assumed to be spherical) which has a pocket. With increasing centrifugal part of the potential the pocket depth becomes smaller, while the position of the pocket minimum moves towards the barrier at $R = R_b \approx R_1 + R_2 + 2$ fm, where $R_i = 1.15A_i^{1/3}$ are the radii of colliding nuclei. This pocket is washed out at large angular momenta $J > 65$. Thus, only a limited part of angular momenta contributes to the capture process.

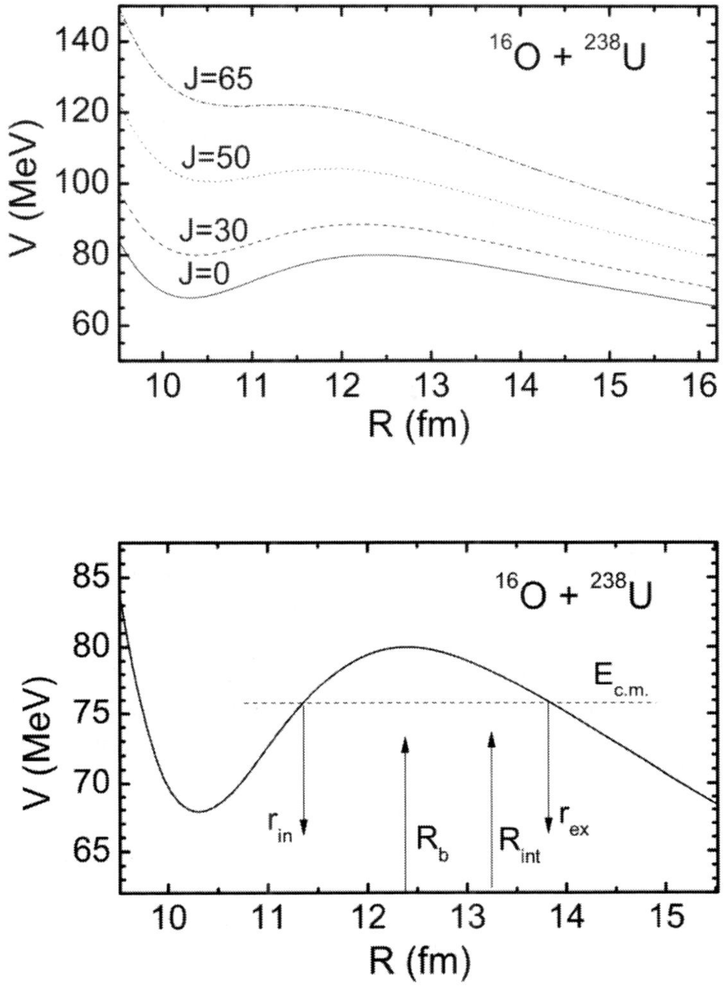

Figure 1. (Upper part) The nucleus-nucleus potentials calculated at $J = 0$ (solid curve), 30 (dashed curve), 50 (dotted curve), and 65 (dash-dotted curve) for the ^{16}O + ^{238}U reaction. The interacting nuclei are assumed to be spherical in the calculations. (Lower part) The position R_b of the Coulomb barrier, radius of interaction R_{int}, and external $r_{\rm ex}$ and internal $r_{\rm in}$ turning points for some values of $E_{\rm c.m.}$ are indicated at the nucleus-nucleus potential (J=0) for the same reaction.

2.2. Capture cross section

The capture cross section is a sum of partial capture cross sections $\sigma_{\rm cap}(E_{\rm c.m.}, J)$ [39]

$$\sigma_{cap}(E_{\rm c.m.}) = \frac{\pi\hbar^2}{2\mu E_{\rm c.m.}} \sum_J (2J+1) \int_0^{\pi/2} d\theta_1 \sin\theta_1 \int_0^{\pi/2} d\theta_2 \sin\theta_2 P_{\rm cap}(E_{\rm c.m.}, J, \theta_1, \theta_2), \tag{4}$$

where $\mu = m_0 A_1 A_2/(A_1 + A_2)$ is the reduced mass (m_0 is the nucleon mass), and the summation is over the possible values of angular momentum J at a given bombarding energy $E_{\rm c.m.}$. Knowing the potential of the interacting nuclei for each orientation, one can obtain the partial capture probability $P_{\rm cap}$ which is defined by the passing probability of the potential barrier in the relative distance R at a given J. The value of $P_{\rm cap}$ is obtained by integrating the propagator G from the

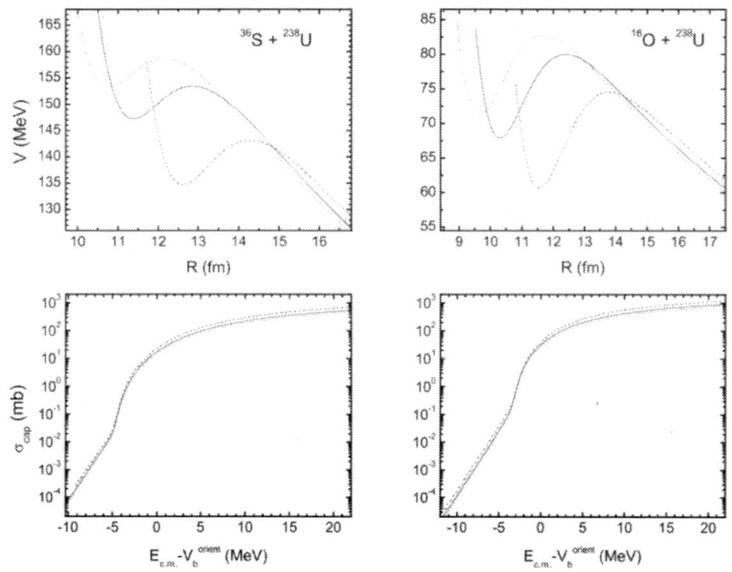

Figure 2. (Upper part) The nucleus-nucleus potentials calculated at J=0 for the reactions ^{36}S+^{238}U and ^{16}O+^{238}U. (Lower part) The dependence of the capture cross section for nuclei colliding with a fixed orientation on $E_{\rm c.m.} - V_b^{orient}$ where V_b^{orient} is the height of the Coulomb barrier for certain orientations. The results of calculations for the sphere-sphere (the interacting nuclei are spherical), sphere-pole and sphere-side configurations are shown by solid, dashed and dotted lines, respectively. The static quadrupole deformation parameters are: $\beta_2(^{238}\mathrm{U})$=0.286 and $\beta_1(^{16}\mathrm{O})$=$\beta_1(^{36}\mathrm{S})$=0.

initial state (R_0, P_0) at time $t = 0$ to the final state (R, P) at time t (P is a momentum) [39]:

$$P_{\rm cap} = \lim_{t\to\infty} \int_{-\infty}^{r_{\rm in}} dR \int_{-\infty}^{\infty} dP\; G(R,P,t|R_0,P_0,0) = \lim_{t\to\infty} \frac{1}{2}{\rm erfc}\left[\frac{-r_{\rm in} + \overline{R(t)}}{\sqrt{\Sigma_{RR}(t)}}\right]. \tag{5}$$

The second line in (5) is obtained by using the propagator $G = \pi^{-1}|\det \mathbf{\Sigma}^{-1}|^{1/2}\exp(-\mathbf{q}^T\mathbf{\Sigma}^{-1}\mathbf{q})$ ($\mathbf{q}^T = [q_R, q_P]$, $q_R(t) = R - \overline{R(t)}$, $q_P(t) = P - \overline{P(t)}$, $\overline{R(t=0)} = R_0$, $\overline{P(t=0)} = P_0$, $\Sigma_{kk'}(t) = 2\overline{q_k(t)q_{k'}(t)}$, $\Sigma_{kk'}(t=0) = 0$, $k, k' = R, P$) calculated for an inverted oscillator which approximates the nucleus-nucleus potential V in the variable R. The frequency ω of this oscillator with an internal turning point $r_{\rm in}$ (Fig. 1) is defined from the condition of equality of the classical actions of approximated and realistic potential barriers of the same hight at given J. It should be noted that the passage through the Coulomb barrier approximated by a parabola has been previously studied in Refs. [35, 42]. This approximation is well justified for the reactions and energy range, which are considered here. Finally, one can find the expression for the capture probability:

$$P_{\rm cap} = \frac{1}{2}{\rm erfc}\left[\left(\frac{\pi s_1(\gamma - s_1)}{2\mu(\omega_0^2 - s_1^2)}\right)^{1/2} \frac{\mu\omega_0^2 R_0/s_1 + P_0}{[\gamma \ln(\gamma/s_1)]^{1/2}}\right], \tag{6}$$

where γ is the internal-excitation width, $\omega_0^2 = \omega^2\{1 - \hbar\tilde{\lambda}\gamma/[\mu(s_1+\gamma)(s_2+\gamma)]\}$ is the renormalized frequency in the Markovian limit, the value of $\tilde{\lambda}$ is related to the strength of linear coupling in coordinates between collective and internal subsystems. The s_i are the real roots ($s_1 \geq 0 > s_2 \geq s_3$) of the following equation: $(s+\gamma)(s^2 - \omega_0^2) + \hbar\tilde{\lambda}\gamma s/\mu = 0$. The details

Journal of Physics: Conference Series **282** (2011) 012001 doi:10.1088/1742-6596/282/1/012001

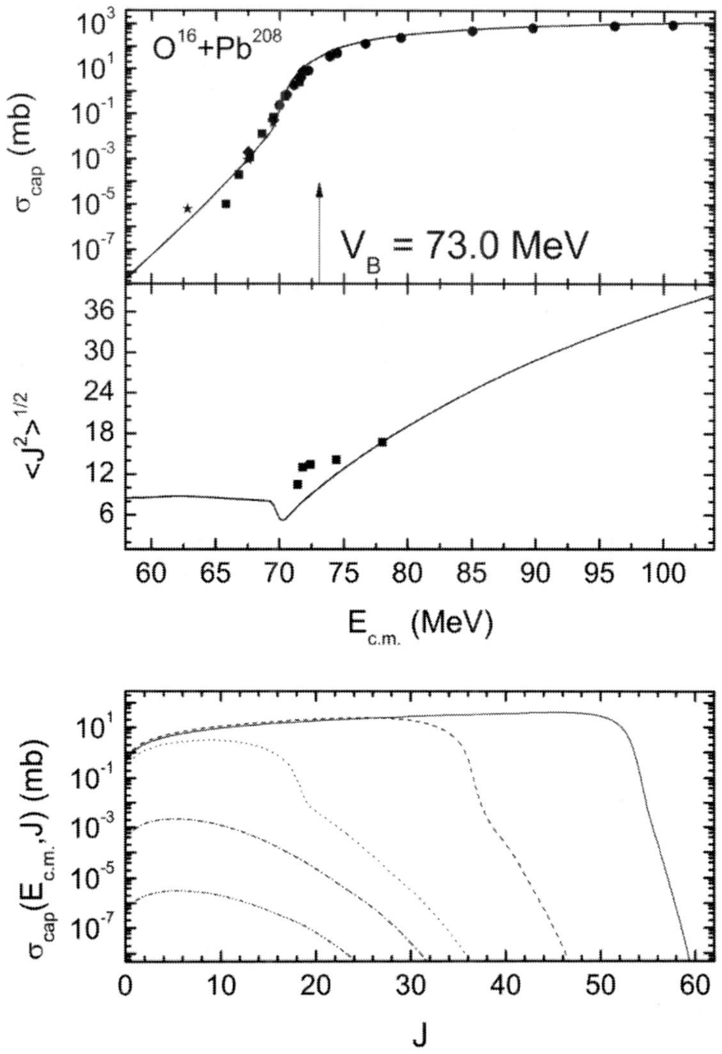

Figure 3. The calculated (lines) capture cross section (upper part), average angular momenta of captured system (middle part) versus $E_{\rm c.m.}$, and partial capture cross sections (lower part) versus J at $E_{\rm c.m.}$=65 (dash-dot-dotted line), 69.5 (dash-dotted line), 73 (dotted line), 83 (dashed line), and 100 (solid line) MeV for the ^{16}O+^{208}Pb reaction with the spherical nuclei. The experimental cross sections marked by closed squares, circles, rhombus, stars are from Refs. [5, 8, 11, 18], respectively. The experimental values of $\langle J^2\rangle$ (solid squares) are taken from Ref. [45]. The value of the Coulomb barrier V_b is indicated by arrow.

of the used formalism are presented in [39]. We have to mention that most of the quantum-mechanical, dissipative effects and non-Markovian effects accompanying the passage through the potential barrier are taken into consideration in our formalism [35, 39]. For example, the non-Markovian effects appear in the calculations through the internal-excitation width γ.

As shown in Fig. 1, the nuclear forces start to play a role at $R_{int} = R_b + 1.1$ fm where the nucleon density of colliding nuclei approximately reaches 10% of the saturation density. If the value of $r_{\rm ex}$ corresponding to the external turning point is larger than the interaction radius R_{int}, we take $R_0 = r_{\rm ex}$ and $P_0 = 0$ in Eq. (6). For $r_{\rm ex} < R_{int}$, it is naturally to start our treatment

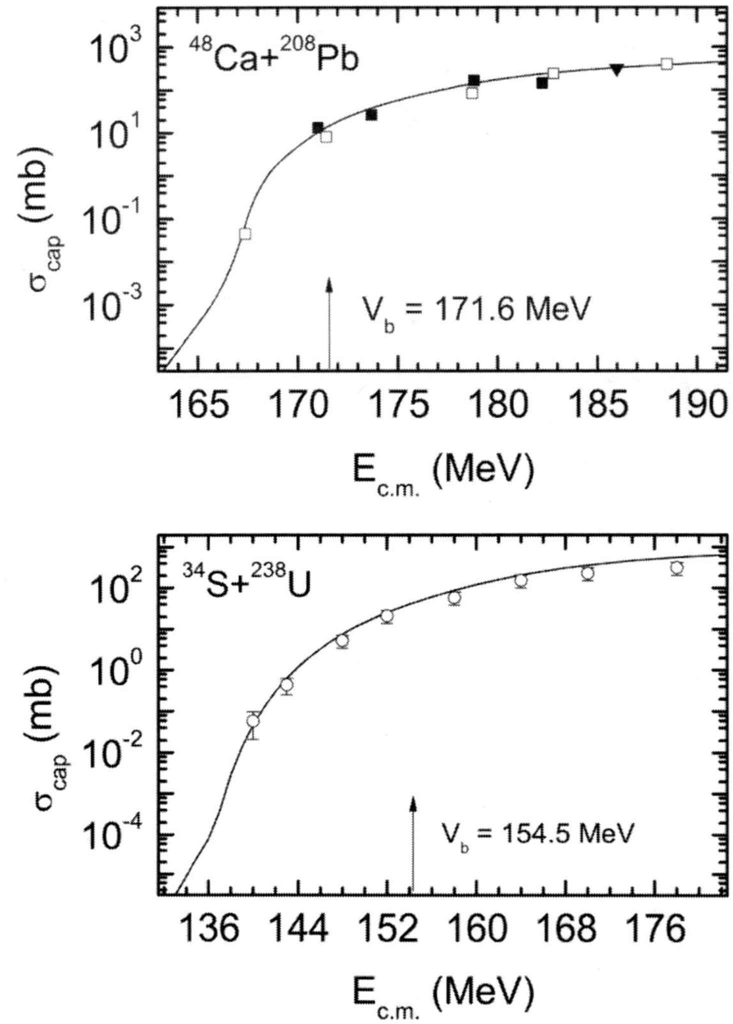

Figure 4. The calculated (solid lines) capture cross section versus $E_{c.m.}$ for the reactions ^{48}Ca+^{208}Pb (upper part) and ^{34}S+^{238}U (lower part). The experimental cross sections are taken from Refs. [46] (closed squares and triangle), [19] (open squares), and [26] (open circles). The value of the Coulomb barrier V_b is indicated by arrow. The static quadrupole deformation parameters are: $\beta_1(^{48}\text{Ca})=\beta_2(^{208}\text{Pb})=0$, $\beta_1(^{34}\text{S})=0.125$, and $\beta_2(^{238}\text{U})=0.286$.

with $R_0 = R_{int}$ and P_0 defined by the kinetic energy at $R = R_0$. In this case the friction hinders the classical motion to proceed towards smaller values of R. If $P_0 = 0$ at $R_0 > R_{int}$, the friction almost does not play a role in the transition through the barrier. So, at $R < R_{int}$ the relative motion may be more coupled with other degrees of freedom. At $R > R_{int}$ the relative motion is almost independent of the internal degrees of freedom. Thus, two regimes of interaction at sub-barrier energies differ by the action of the nuclear forces and the role of friction at $R = r_{ex}$.

3. Calculated results

Besides the parameters related to the nucleus-nucleus potential, two parameters $\hbar\gamma$=32 MeV ($\hbar\gamma$=15 MeV) and the friction coefficient $\hbar\lambda = -\hbar(s_1 + s_2)$=2 MeV are used for calculating the capture probability in reactions with deformed actinides (spherical lead nucleus). The most

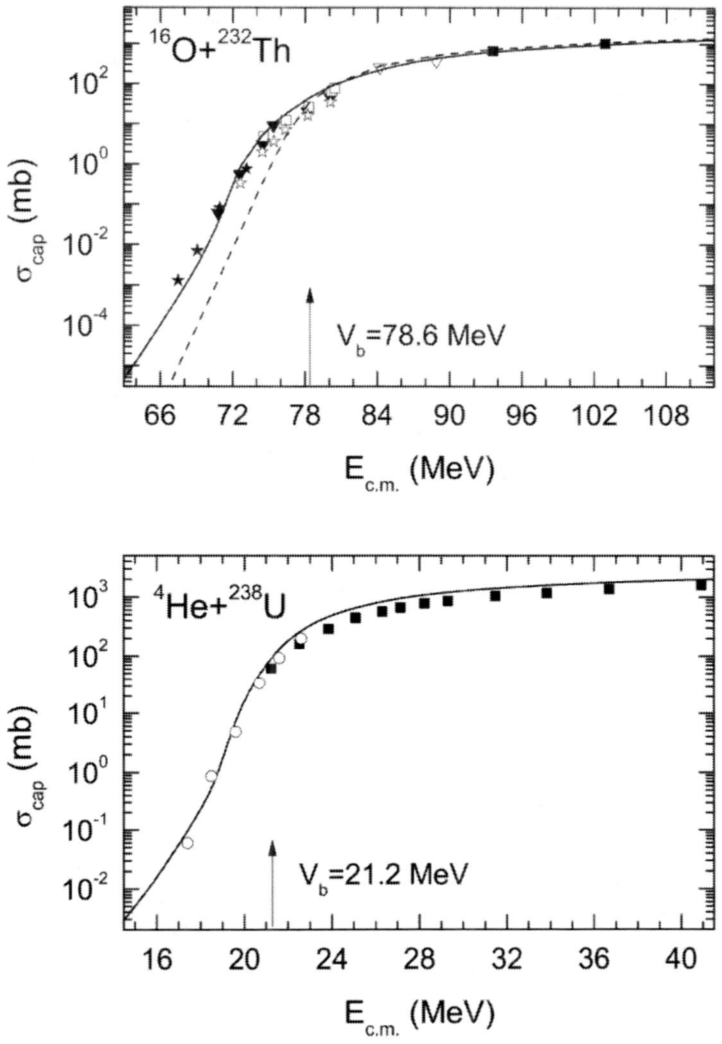

Figure 5. The same as in Fig. 4, but for the reactions for the reactions ^{16}O + ^{232}Th and ^{4}He + ^{238}U. The experimental data in the upper part are taken from Refs. [47] (open triangles), [2] (closed triangles), [48] (open squares), [49] (closed squares), [6] (open stars) and [7] (closed stars). The fission cross sections from Refs. [9] and [50] are shown in the lower part by open circles and solid squares, respectively. The value of the Coulomb barrier V_b for the spherical nuclei is indicated by arrow. The dashed curve represents the calculation with the Wong formula. The static quadrupole deformation parameters are: $\beta_2(^{232}\mathrm{Th})=0.261$ $\beta_2(^{238}\mathrm{U})=0.286$ and $\beta_1(^{4}\mathrm{He})=\beta_1(^{16}\mathrm{O})=0$.

realistic friction coefficients in the range of $\hbar\lambda \approx$ 1-2 MeV are suggested from the study of deep inelastic and fusion reactions [43]. These values are close to those calculated within the mean field approach [44]. All calculated results presented are obtained with the same set of parameters and are rather insensitive to a reasonable variation of them [39].

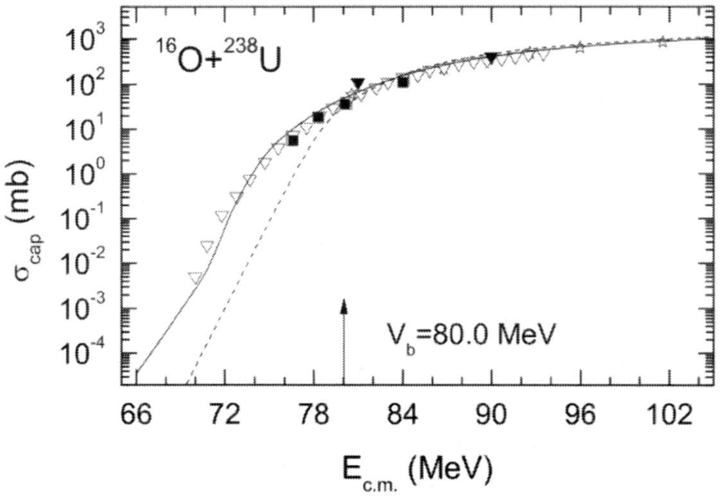

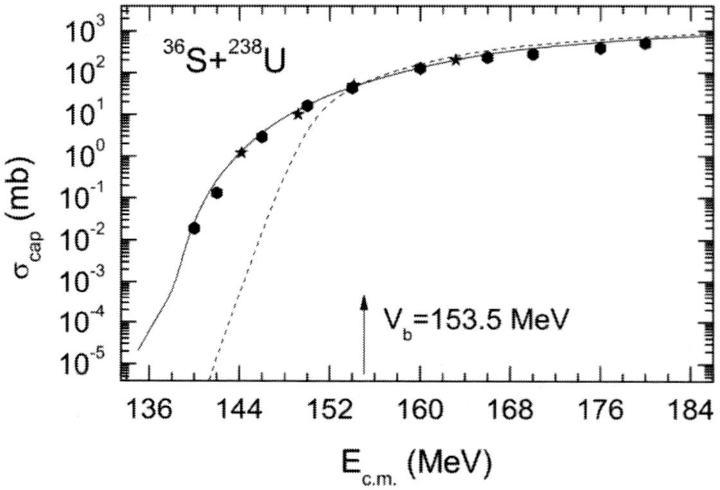

Figure 6. The same as in Fig. 4, but for the reactions $^{16}O + {}^{238}U$ and $^{36}S + {}^{238}U$. The experimental cross sections are taken from Refs. [13] (open triangles), [51] (closed triangles), [6] (open squares), [4] (closed squares), [50] (open stars), [24] (closed stars), and [20] (rhombuses). The dashed curve represents the calculation with the Wong formula. The static quadrupole deformation parameters are: $\beta_2(^{238}U)$=0.286 and $\beta_1(^{16}O)=\beta_1(^{36}S)$=0.

3.1. Effect of orientation

The influence of orientation of the deformed heavy nucleus on the capture process in the reactions $^{36}S + {}^{238}U$ and $^{16}O + {}^{238}U$ is studied in Fig. 2. We demonstrate that the capture cross section σ_{cap} at fixed orientation as a function of $E_{\rm c.m.} - V_b^{orient}$, where V_b^{orient} is the Coulomb barrier for this orientation, is almost independent of the orientation angle θ_2.

3.2. Comparison with experimental data and predictions

In Figs. 3 - 7 the calculated capture cross sections for the reactions ^{16}O,^{48}Ca+^{208}Pb, ^{16}O,^{32}S+^{232}Th, ^{4}He,^{16}O,^{30}Si,^{34,36}S+^{238}U are in a rather good agreement with the available experimental data. Because of the uncertainties in the definition of the deformation of the light

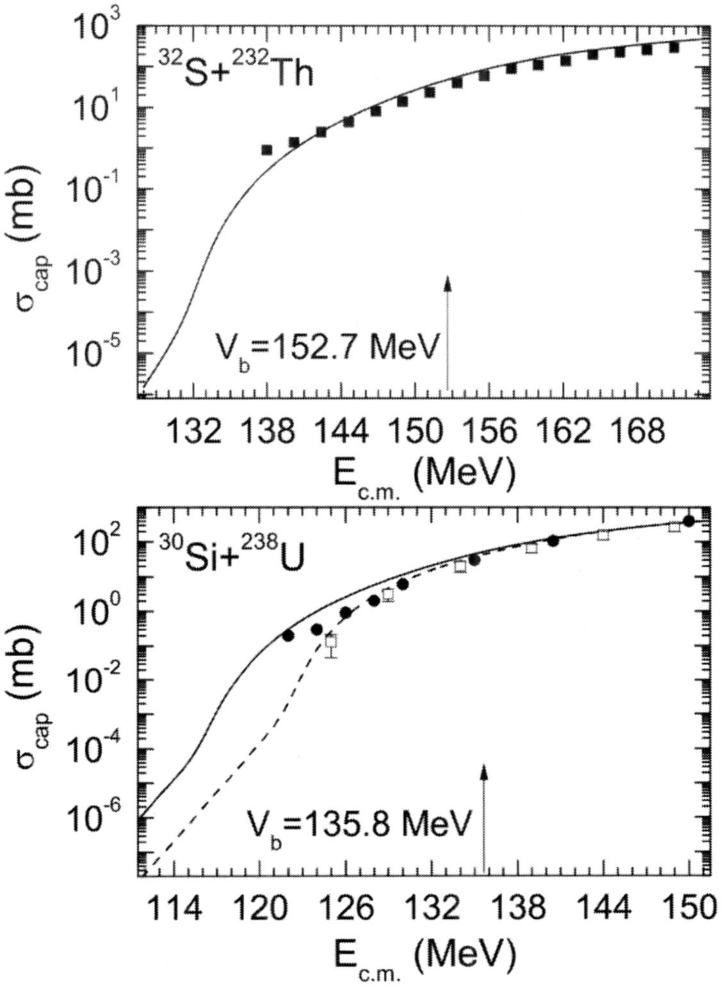

Figure 7. The same as in Fig. 4, but for the reactions ^{32}S + ^{232}Th and ^{30}Si + ^{238}U. The experimental data are taken from Refs. [22] (solid squares), [16] (solid circles), and [26] (open squares). The static quadrupole deformation parameters are: $\beta_2(^{238}U)=0.286$, $\beta_2(^{232}Th)=0.261$, $\beta_1(^{32}S)=0.312$, and $\beta_1(^{30}Si)=0.315$. For the ^{30}Si + ^{238}U reaction, the results of calculations with $\beta_1(^{30}Si)=0$ (the predictions of the mean-field and macroscopic-microscopic models) are presented by dashed line in the lower part of the figure.

nucleus and in the experimental data [16, 26], we show the calculated results for the $^{30}Si+^{238}U$ reaction with $\beta_1(^{30}Si)$ from Ref. [41] as well as with $\beta_1(^{30}Si)=0$ (lower part of Fig. 7). Note that $\beta_1(^{30}Si)=0$ for the ground state were predicted within the mean-field and the macroscopic-microscopic models.

For the ^{32}S + ^{208}Pb reaction, the calculated capture cross sections in Fig. 8 are consistent with the experimental data [12] if the value of Coulomb barrier V_b is increased by 1.5 MeV. The predicted capture cross sections for the reactions ^4He + ^{208}Pb and $^{34,36}S$ + ^{244}Pu,^{248}Cm are shown in Figs. 9 and 10, respectively.

As seen from Figs. 5 and 6 (dashed lines) the Wong formula [52] does not reproduce the capture cross section at $E_{c.m.} < V_b$ even taking into consideration the static quadrupole deformation of target-nucleus.

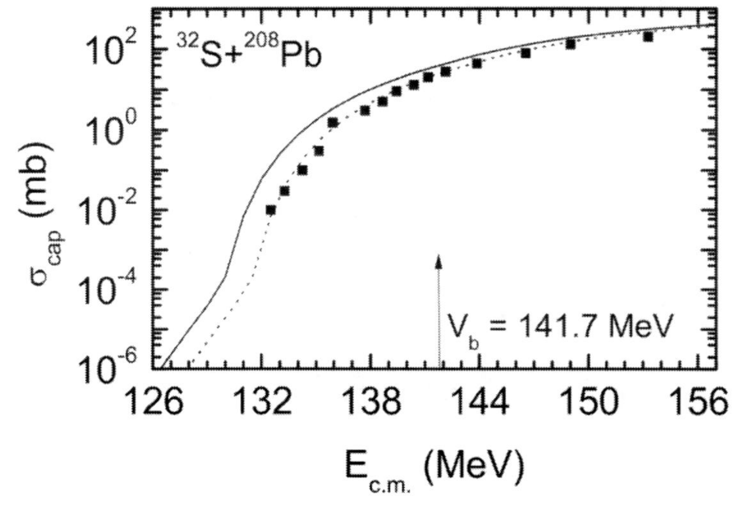

Figure 8. The same as in Fig. 4, but for the reaction ^{32}S(β_1=0.312) + ^{208}Pb(β_2=0). The capture cross sections calculated with the value of the Coulomb barrier V_b+1.5 MeV are shown by dotted line. The experimental data are taken from Ref. [12] (closed squares).

One can see in Figs. 3 - 10 that there is a sharp fall-off of the cross sections just under the Coulomb barrier corresponding to spherical nuclei. With decreasing $E_{\rm c.m.}$ up to about 3 - 10 MeV (when the projectile is spherical and the target is deformed) and 15 - 20 MeV (when both projectile and target are deformed nuclei) below the Coulomb barrier the regime of interaction is changed because at the external turning point the colliding nuclei do not reach the region of nuclear interaction where the friction plays a role. As result, at smaller $E_{\rm c.m.}$ the cross sections fall with a smaller rate. With larger values of R_{int} the change of fall rate occurs at smaller $E_{\rm c.m.}$. However, the uncertainty in the definition of R_{int} is rather small. Therefore, an effect of the change of fall rate of sub-barrier capture cross section should be in the data if we assume that the friction starts to act only when the colliding nuclei approach the barrier. Note that at energies of 10 - 20 MeV below the barrier the experimental data have still large uncertainties to make a firm experimental conclusion about this effect. The effect seems to be more pronounced in collisions of spherical nuclei, where the regime of interaction is changed at $E_{\rm c.m.}$ up to about 1 - 5 MeV below the Coulomb barrier [39]. The collisions of deformed nuclei occur at various mutual orientations affecting the value of R_{int}.

The calculated mean-square angular momenta

$$\langle J^2\rangle = \frac{\pi\hbar^2}{2\mu E_{\rm c.m.}\sigma_{cap}}\sum_J J(J+1)(2J+1)\int_0^{\pi/2} d\theta_1 \sin\theta_1 \int_0^{\pi/2} d\theta_2 \sin\theta_2 P_{\rm cap}(E_{\rm c.m.}, J, \theta_1, \theta_2) \quad (7)$$

of captured systems versus $E_{\rm c.m.}$ are presented in Figs. 3, 10, and 11 for the reactions ^{16}O,^{4}He+^{208}Pb and ^{16}O + ^{232}Th,^{238}U. At energies below the barrier $\langle J^2\rangle$ has a minimum. This minimum depends on the deformations of nuclei and on the factor $Z_1 \times Z_2$. For the reactions ^{16}O + ^{232}Th,^{238}U, these minima are about 7 - 8 MeV below the corresponding Coulomb barriers, respectively. The experimental data [45] indicate the presence of the minimum as well. On the left-hand side of this minimum the dependence of $\langle J^2\rangle$ on $E_{\rm c.m.}$ is rather weak. A similar weak dependence has been found in Refs. [53] in the extreme sub-barrier region. Note that the found behavior of $\langle J^2\rangle$, which is related to the change of the regime of interaction between the colliding nuclei, would affect the angular anisotropy of the products of fission-like fragments

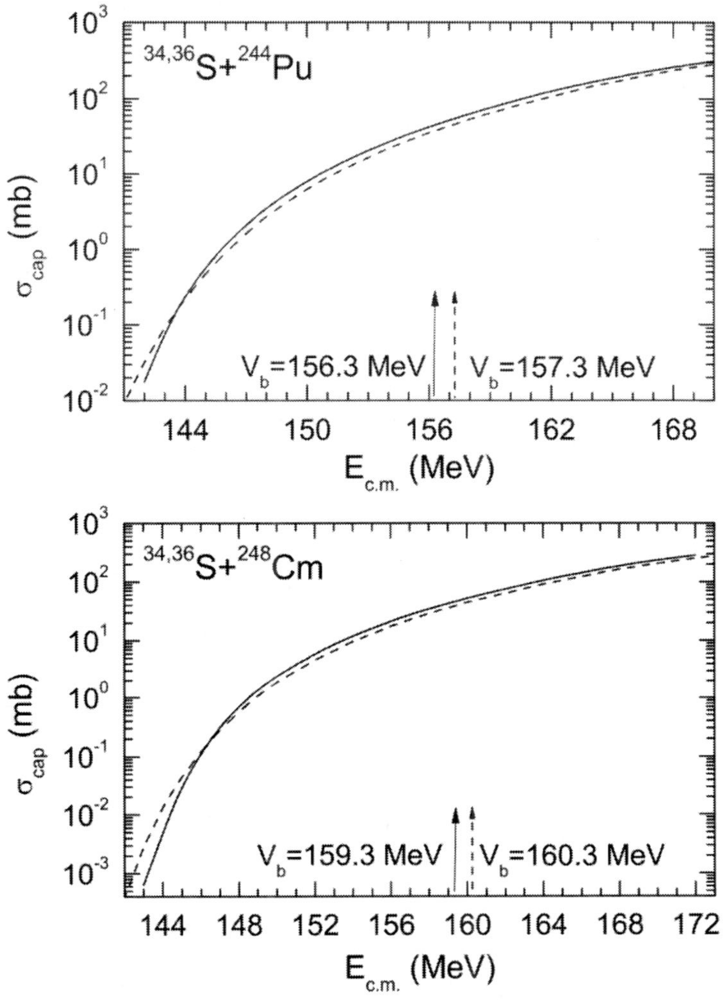

Figure 9. The predicted capture cross sections for the reactions ^{34}S + ^{244}Pu,^{248}Cm (dashed lines) and ^{36}S + ^{244}Pu,^{248}Cm (solid lines). The values of the Coulomb barriers for the reactions ^{34,36}S + ^{244}Pu,^{248}Cm are indicated by dotted and solid arrows, respectively. The static quadrupole deformation parameters are: $\beta_2(^{244}\mathrm{Pu})=0.293$, $\beta_2(^{248}\mathrm{Cm})=0.297$, $\beta_1(^{34}\mathrm{S})=0.125$, and $\beta_1(^{36}\mathrm{S})=0$.

following capture. Indeed, the values of $\langle J^2\rangle$ are extracted from data on angular distribution of fission-like fragments [25].

The agreement between calculated with Wong-type formula and experimental $\langle J^2\rangle$ is not good. At energies below the barrier $\langle J^2\rangle$ has no a minimum (see Fig. 11). However, for the considered reactions the saturation values of $\langle J^2\rangle$ are close to those obtained in our formalism.

3.3. Astrophysical factor, L-factor and barrier distribution

In Figs. 10 and 12 the calculated astrophysical S–factors versus $E_{\rm c.m.}$ are shown for the reactions ^{4}He+^{208}Pb and ^{16}O+^{238}U. The S-factor has a maximum which is seen in experiments [10,15,37]. After this maximum S-factor slightly decreases with decreasing $E_{\rm c.m.}$ and then starts to increase. This effect seems to be more pronounced in collisions of spherical nuclei. The same behavior has been revealed in Refs. [54] by extracting the S-factor from the experimental data.

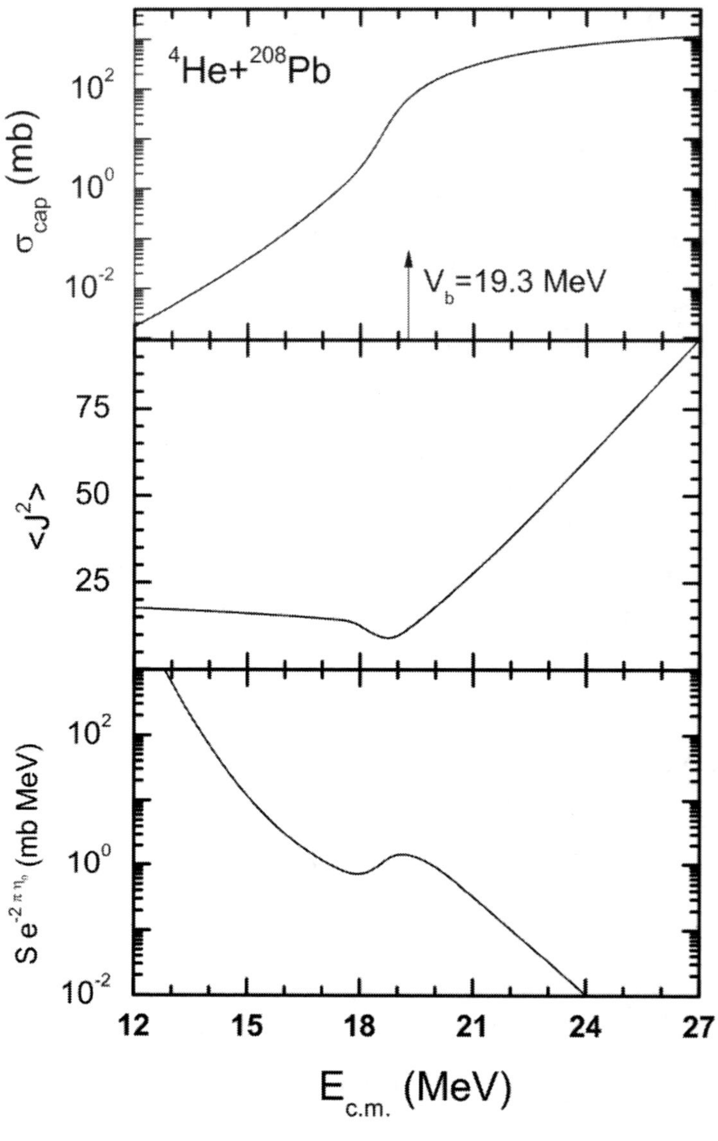

Figure 10. The calculated capture cross section (upper part), average angular momenta of compound nucleus (middle part) and astrophysical S-factor (lower part) with $\eta_0 = \eta(E_{\rm c.m.} = V_b)$ versus $E_{\rm c.m.}$ for the ^{4}He + ^{208}Pb reaction.

In Fig. 12, the so-called logarithmic derivative, $L(E_{\rm c.m.}) = d(\ln(E_{\rm c.m.}\sigma_{cap}))/dE_{\rm c.m.}$, and the barrier distribution $d^2(E_{\rm c.m.}\sigma_{cap})/dE^2_{\rm c.m.}$ are presented for the ^{16}O+^{238}U reaction. The logarithmic derivative strongly increases below the barrier and then has a maximum at $E_{\rm c.m.} \approx V_b^{orient}$(sphere-pole)-3 MeV (at $E_{\rm c.m.} \approx V_b$-3 MeV for the case of spherical nuclei). The maximum of L corresponds to the minimum of the S-factor.

In Fig. 12, the barrier distributions calculated with an energy increment 0.2 MeV have only one maximum at $E_{\rm c.m.} \approx V_b^{orient}$(sphere-sphere)= V_b as in the experiment [55]. With an increasing increment the barrier distribution is shifted to lower energies. Assuming a spherical target nucleus in the calculations, we obtain a more narrow barrier distribution.

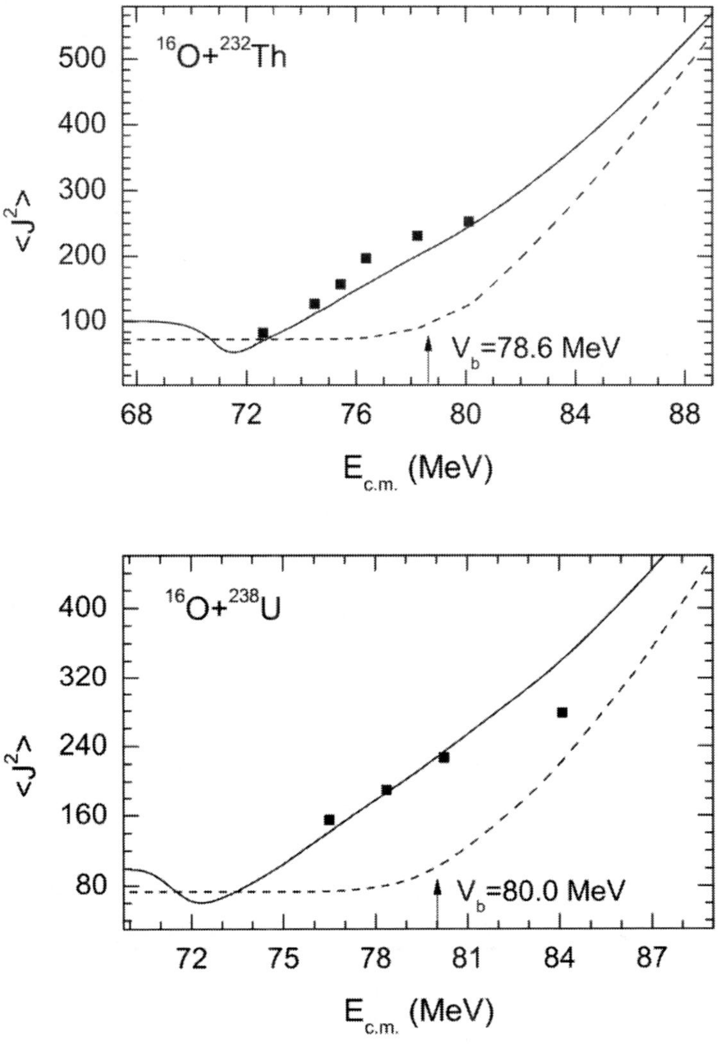

Figure 11. The calculated mean-square angular momenta versus $E_{c.m.}$ for the reactions ^{16}O + ^{232}Th,^{238}U are compared with experimental data [6]. The dashed curve represents the calculation by the Wong-type formula.

3.4. Capture cross sections in reactions with large fraction of quasifission

In the case of large values of $Z_1 \times Z_2$ the quasifission process competes with complete fusion at energies near barrier and can lead to a large hindrance for fusion, thus ruling the probability for producing superheavy elements in the actinide-based reactions [31,56]. Since the sum of the fusion cross section σ_{fus}, and the quasifission cross section σ_{qf} gives the capture cross section,

$$\sigma_{cap} = \sigma_{fus} + \sigma_{qf},$$

and $\sigma_{fus} \ll \sigma_{qf}$ for actinide-based reactions ^{48}Ca + ^{232}Th,^{238}U,^{244}Pu,^{246,248}Cm and ^{50}Ti + ^{244}Pu [31], we have

$$\sigma_{cap} \approx \sigma_{qf}.$$

In a wide mass-range near the entrance channel, the quasifission events overlap with the products of deep-inelastic collisions and can not be firmly distinguished. Because of this the

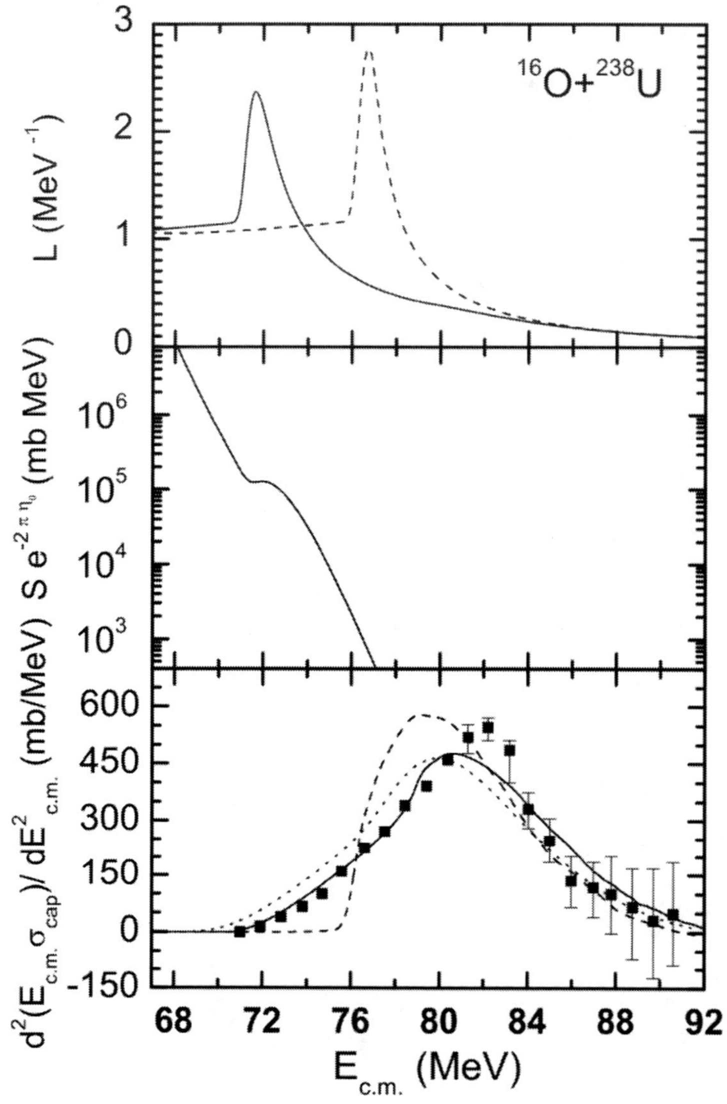

Figure 12. The calculated values of the astrophysical S-factor with $\eta_0 = \eta(E_{\rm c.m.} = V_b)$ (middle part), the logarithmic derivative L (upper part) and the fusion barrier distribution $d^2(E_{\rm c.m.}\sigma_{cap})/dE^2_{\rm c.m.}$ (lower part) for the ^{16}O+^{238}U reaction. The value of L calculated with the assumption of $\beta_1(^{16}{\rm O})=\beta_2(^{238}{\rm U})=0$ is shown by a dashed line. The solid and dotted lines show the values of $d^2(E_{\rm c.m.}\sigma_{cap})/dE^2_{\rm c.m.}$ calculated with the increments 0.2 and 1.2 MeV, respectively. The closed squares are the experimental data of Ref. [55].

mass region near the entrance channel is taken out in the experimental analyses of Refs. [57,58]. Thus, by comparing the calculated and experimental capture cross sections one can study the importance of quasifission near the entrance channel for the actinide-based reactions leading to superheavy nuclei.

The capture cross sections for the quasifission reactions [57–59] are shown in Figs. 13-15. One can observe a large deviations of the experimental data of Refs. [57,58] from the the calculated results. The possible reason is an underestimation of the quasifission yields measured in these reactions. Thus, the quasifission yields near the entrance channel are important. Note that

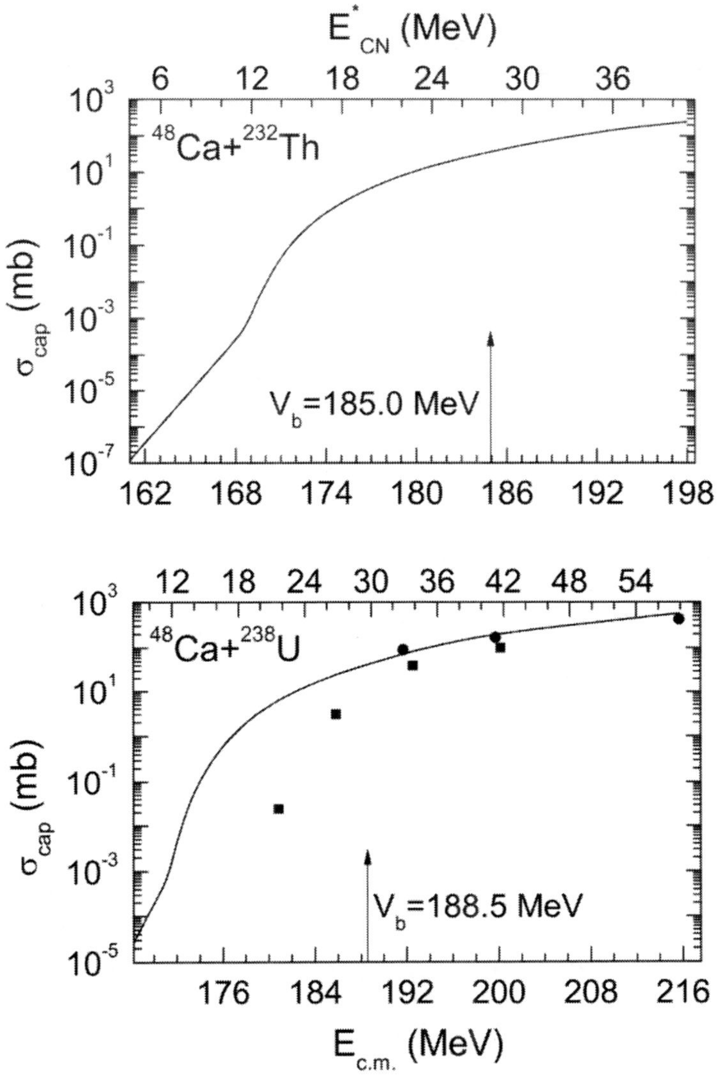

Figure 13. The same as in Fig. 4, but for the ^{48}Ca + ^{232}Th,^{238}U reactions. The excitation energies E^*_{CN} of the corresponding nuclei are indicated. The experimental data are taken from Refs. [57] (marked by squares) and [59] (marked by circles). The static quadrupole deformation parameters are: $\beta_2(^{238}\mathrm{U})$=0.286, $\beta_2(^{232}\mathrm{Th})$=0.261, and $\beta_1(^{48}\mathrm{Ca})$=0.

there are the experimental uncertainties in the bombarding energies.

One can see in Fig. 16 that the experimental and the theoretical cross sections become closer with increasing bombarding energy. This means that with increasing bombarding energy the quasifission yields near the entrance channel mass-region decrease with respect to the quasifission yields in other mass-regions. As seen in Fig. 16, the quasifission yields near the entrance channel mass-region increase with $Z_1 \times Z_2$.

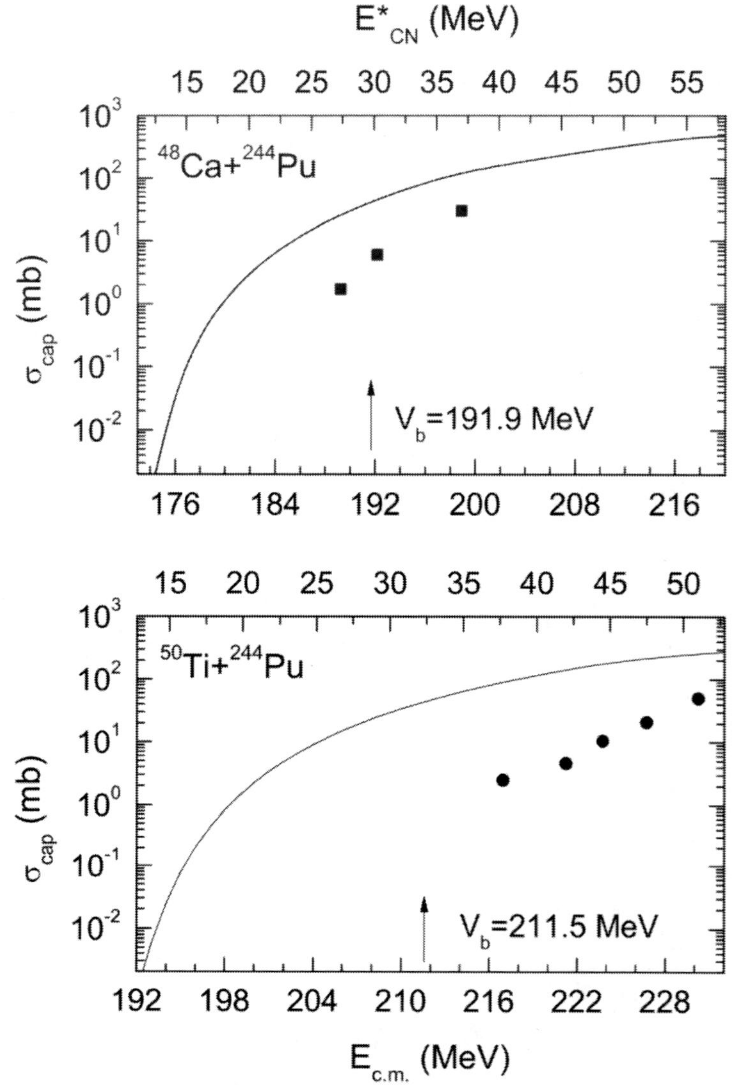

Figure 14. The same as in Fig. 13, but for the indicated ^{48}Ca,^{50}Ti + ^{244}Pu reactions. The experimental data are from Refs. [58] (squares) and [57] (circles). The static quadrupole deformation parameters are: $\beta_2(^{244}\mathrm{Pu})=0.293$, and $\beta_1(^{48}\mathrm{Ca})=\beta_1(^{50}\mathrm{Ti})=0$.

4. Origin of fusion hindrance in reactions with medium-mass nuclei at sub-barrier energies

In Figs. 17 and 18 the calculated capture cross section are presented for the reactions ^{36}S + ^{48}Ca,^{64}Ni and ^{64}Ni + ^{64}Ni. The values of V_b are adjusted to the experimental data for the fusion cross sections shown as well. For the systems mentioned above, the difference between the sub-barrier capture and fusion cross sections becomes larger with decreasing bombarding energy $E_{\mathrm{c.m.}}$. Assuming that the estimated capture and the measured fusion cross sections are correct, the small fusion cross section at energies well below the Coulomb barrier may indicate that other reaction channel is open and the system evolves by other reaction mechanism after the capture. The observed hindrance factor may be understood in term of quasifission whose cross section should be added to the one of fusion to obtain a meaningful comparison with

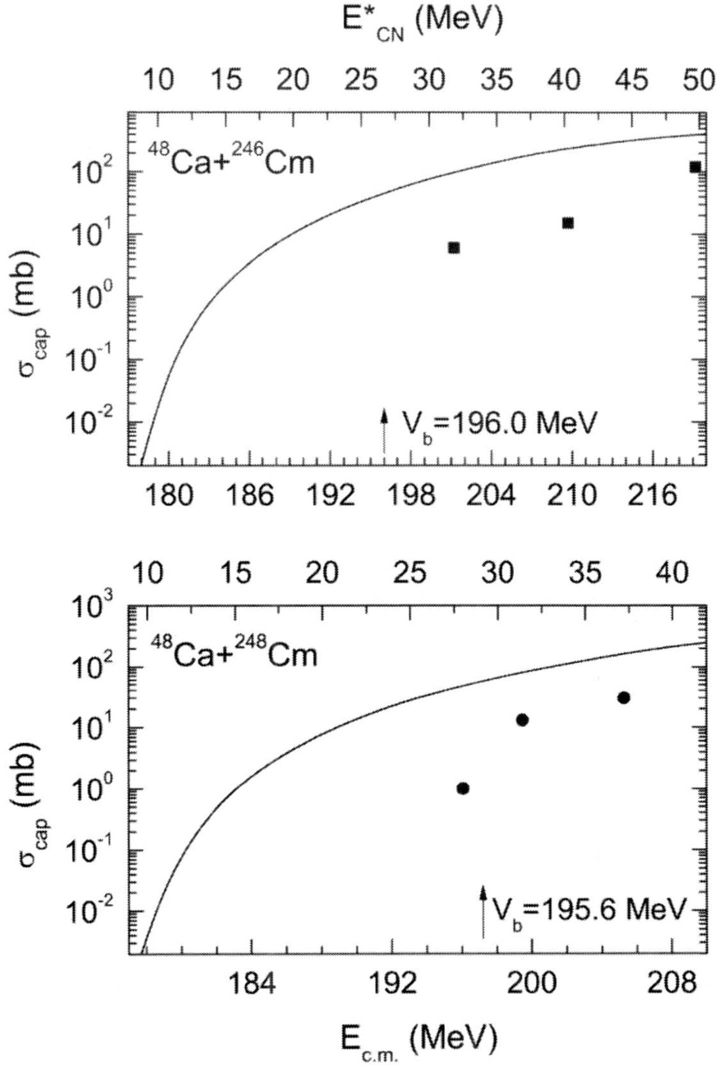

Figure 15. The same as in Fig. 13, but for the reactions ^{48}Ca + ^{246,248}Cm. The experimental data are from Refs. [57] (squares) and [58] (circles). The static quadrupole deformation parameters are: $\beta_2(^{246}\mathrm{Cm})$=0.298, $\beta_2(^{248}\mathrm{Cm})$=0.297, and $\beta_1(^{48}\mathrm{Ca})$=0.

the calculated capture cross section. The quasifission event corresponds to the formation of a nuclear-molecular state or dinuclear system with small excitation energy that separates (in the competition with the compound nucleus formation process) by the quantal tunneling through the Coulomb barrier in a binary event with mass and charge close to the entrance channel. In this sense the quasifission is the general phenomenon which takes place in the reactions with the massive [30–33], medium-mass and, probably, light nuclei.

From the present analysis an unambiguous signature of the role of quasifission (binary decay) channel could not be inferred. These reaction mode has to be studied in the future experiments: from the mass (charge) distribution measurements one can show the distinct components due to quasifission. The low-energy experimental data would probably provide straight information since the high-energy data may be shaded by competing reaction processes such as quasifission

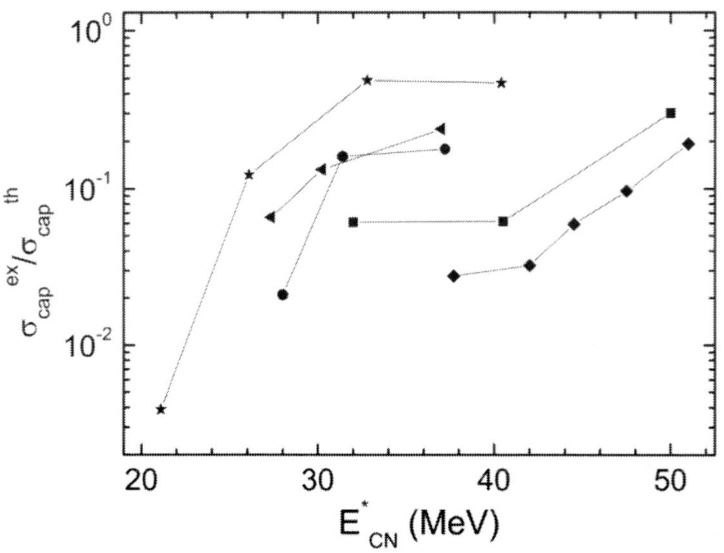

Figure 16. The ratio of theoretical and experimental capture cross sections versus the excitation energy $E_{\mathrm{c.m.}}$ of the compound nucleus for the reactions ^{48}Ca+^{238}U (closed stars), ^{48}Ca+^{244}Pu (closed triangles), ^{48}Ca+^{246}Cm (closed squares), ^{48}Ca+^{248}Cm (closed circles), and ^{50}Ti+^{244}Pu (closed rhombuses).

and deep inelastic collisions. More experimental and theoretical studies of sub-barrier fusion hindrance are needed to improve our understanding of quasifission process, which may be especially important in astrophysical fusion reactions. Note that the binary decay events were already observed experimentally in [61] for the ^{58}Ni + ^{124}Sn reaction at energies below the Coulomb barrier but assumed to be related to deep-inelastic scattering.

5. Summary

The quantum diffusion approach is applied to study the capture process in the reactions with deformed nuclei at sub-barrier energies. The available experimental data at energies above and below the Coulomb barrier are well described, showing that the static quadrupole deformations of the interacting nuclei are the main reasons for the capture cross section enhancement at sub-barrier energies. Since the deformations of the interacting nuclei mainly influence the slope of curve at $E_{\mathrm{c.m.}} < V_b$ and one can extract the ground state deformation of projectile or target from the experimental capture cross section data.

Due to a change of the regime of interaction (the turning-off of the nuclear forces and friction) at sub-barrier energies, the curve related to the capture cross section as a function of bombarding energy has smaller slope. This change is also reflected in the functions $\langle J^2\rangle$, $L(E_{\mathrm{c.m.}})$, and $S(E_{\mathrm{c.m.}})$. The mean-square angular momentum of captured system versus $E_{\mathrm{c.m.}}$ has a minimum and then saturates at sub-barrier energies. This behavior of $\langle J^2\rangle$ would increase the expected anisotropy of the angular distribution of the products of fission and quasifission following capture. The astrophysical factor has a maximum and a minimum at energies below the barrier. The maximum of L-factor corresponds to the minimum of the S-factor. One can suggest the experiments to check these predictions.

The importance of quasifission near the entrance channel is shown for the actinide-based reactions and reactions with medium-mass nuclei at sub-barrier energies.

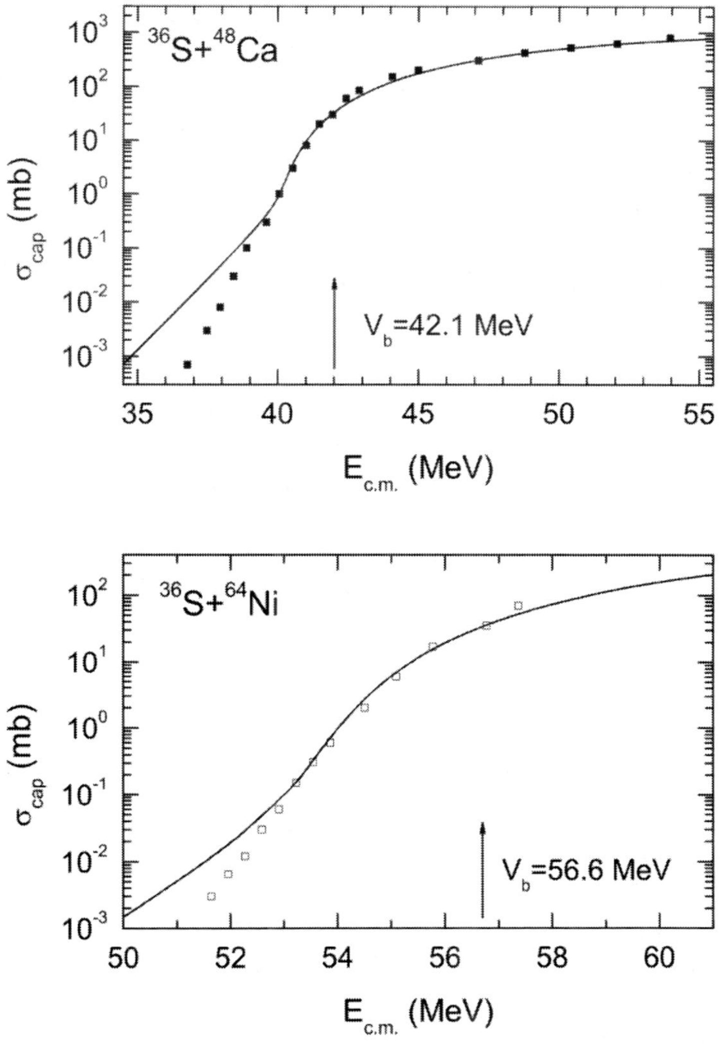

Figure 17. The calculated capture cross sections versus $E_{c.m.}$ for the indicated reactions. The experimental fusion cross sections marked by closed and open squares are taken from Refs. [21, 27], respectively. The values of the Coulomb barrier are indicated by arrows. The static quadrupole deformation parameters are: $\beta_1(^{36}S)=\beta_2(^{48}Ca)=0$ and $\beta_2(^{64}Ni)=0.087$ [60].

Acknowledgments
We thank H. Jia and D. Lacroix for fruitful discussions and suggestions. We are grateful to K. Nishio for providing us his experimental data. This work was supported by DFG, NSFC, and RFBR. The IN2P3-JINR, MTA-JINR and Polish-JINR Cooperation programs are gratefully acknowledged.

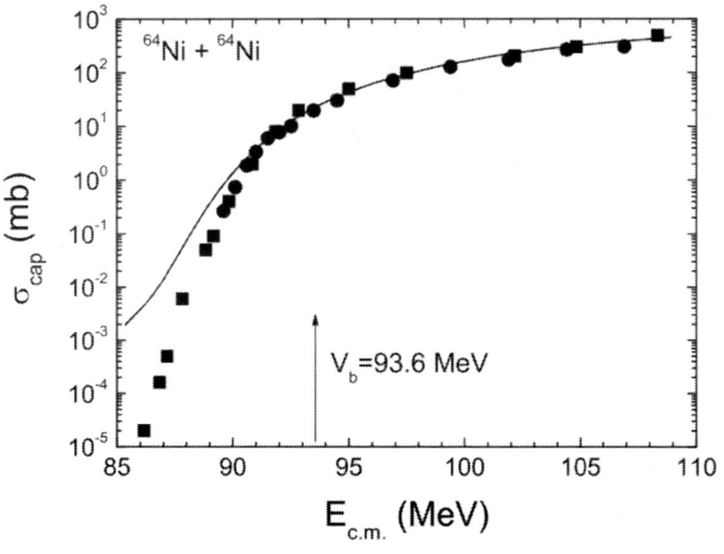

Figure 18. The calculated capture cross sections versus $E_{\rm c.m.}$ for the indicated reaction. The experimental fusion cross sections marked by closed squares and circles are taken from Refs. [1, 14], respectively. Here, $\beta_{1,2}(^{64}{\rm Ni})$=0.087.

[1] Beckerman M *et al.* (1982) *Phys. Rev.* C **25** 837
[2] Zhang H Q *et al.* (1990) *Phys. Rev.* C **42** 1086
[3] Scarlassara F *et al.* (1991) *Z. Phys.* A **338** 171
[4] Zhang H Q *et al.* (1994) *Phys. Rev.* C **49** 926
[5] Oganessian Yu Ts *et al.* (1996) *JINR Rapid Communications* **75** 123
[6] Liu Z *et al.* (1996) *Phys. Rev.* C **54** 761
[7] Nadkarni D M *et al.* (1999) *Phys. Rev.* C **59** R580
[8] Morton C R *et al.* (1999) *Phys. Rev.* C **60** 044608
[9] Trotta M *et al.* (2000) *Phys. Rev. Lett.* **84** 2342
[10] Jiang C L *et al.* (2002) *Phys. Rev. Lett.* **89** 052701
[11] Tretyakova S P *et al.* (2004) *Nucl. Phys.* A **734** E33; Tretyakova S P *et al.* (2004) *Nucl. Phys.* A **738** 487
[12] Dasgupta M and Hinde D J (2004) *Nucl. Phys.* A **734** 148
[13] Nishio K *et al.* (2004) *Phys. Rev. Lett.* **93** 162701
[14] Jiang C L *et al.* (2004) *Phys. Rev. Lett.* **93** 012701
[15] Jiang C L *et al.* (2005) *Phys. Rev.* C **71** 044613
[16] Nishio K *et al.* (2006) *Eur. Phys. J.* A **29** 281
[17] Loveland W *et al. Phys. Rev.* C 74 (2006) 044604; Vinodkumar A M *et al. Phys. Rev.* C 78 (2008) 054608
[18] Dasgupta M *et al.* (2007) *Phys. Rev. Lett.* **99** 192701
[19] Prokhorova E *et al.* (2008) *Nucl. Phys.* A **802** 45
[20] Nishio K *et al.* (2008) *Phys. Rev.* C. **77** 064607
[21] Stefanini A M *et al.* (2008) *Phys. Rev.* C **78** 044607
[22] Hinde D J *et al.* (2008) *Phys. Rev. Lett.* **101** 092701
[23] Shrivastava A *et al.* (2009) *Phys. Rev. Lett.* **103** 232702
[24] Itkis M G *et al.* (2010) *Nucl. Phys.* A **834** 374c
[25] Zhang H Q *et al.* (2010) *Phys. Rev.* C. **81** 034611
[26] Nishio K *et al.* (2010) *Phys. Rev.* C **82** 024611
[27] Montagnoli G *et al.* (2010) *Nucl. Phys.* A **834** 159c; (2010) *Phys. Rev.* C **82** 064609
[28] Langanke K and Barnes C A (1996) *Adv. Nucl. Phys.* **22** 173
[29] Aprahamian A, Langanke K and Wiescher M (2005) *Prog. Part. Nucl. Phys.* **54** 535
[30] Volkov V V (2004) *Particles and Nuclei* **35** 797
[31] Adamian G G, Antonenko N V and Scheid W (2003) *Phys. Rev.* C **68** 034601
[32] Giardina G *et al.* (2000) *Nucl. Phys.* A **671** 165; Nasirov A *et al.* (2005) *Nucl. Phys.* A **759** 342

[33] Keller J G *et al.* (1986) *Nucl. Phys.* A **452** 173
[34] Balantekin A B, Takigawa N (1998) *Rev. Mod. Phys.* **70** 77; Canto L E *et al.* (2006) *Phys. Rep.* **424** 1
[35] Sargsyan V V *et al.* (2009) *Phys. Rev.* C **80** 034606; (2009) *Phys. Rev.* C **80** 047603
[36] Hupin G and Lacroix D (2010) *Phys. Rev.* C **81** 014609; Ayik S, Yilmaz B and Lacroix D (2010) *Phys. Rev.* C **81** 034605
[37] Esbensen H and Jiang C L (2009) *Phys. Rev.* C **79** 064619; Misicu S and Esbensen H (2007) *Phys. Rev.* C **75** 034606; Esbensen H and Misicu S (2007) *Phys. Rev.* C **76** 054609
[38] Hagino K and Rowley N(2009) *AIP Conf. Proc.* **1098** 18
[39] Sargsyan V V *et al.* (2010) *Eur. Phys. J.* A **45** 125; (2011) *Eur. Phys. J.* A in print
[40] Adamian G G *et al.* (1996) *Int. J. Mod. Phys.* E **5** 191
[41] Raman S, Nestor C W, Jr and Tikkanen P. (2001) *At. Data Nucl. Data Tables* **78** 1
[42] Hofmann H (1997) *Phys. Rep.* **284** 137
[43] Adamian G G, Nasirov A K, Antonenko N V and Jolos R V (1994) *Phys. Part. Nucl.* **25** 583
[44] Washiyama K, Lacroix D and Ayik S (2009) *Phys. Rev.* C **79** 024609; C **79** 054606
[45] Vandenbosch R (1992) *Annu. Rev. Nucl. Part. Sci.* **42** 447
[46] Pacheco A J *et al.* (1992) *Phys. Rev.* C **45** 2861; Bock R *et al.* (1982) *Nucl. Phys.* A **388** 334
[47] Back B B *et al.* (1985) *Phys. Rev.* C. **32** 195
[48] Murakami T *et al.* (1986) *Phys. Rev.* C **34** 1353
[49] Kailas S *et al.* (1999) *Phys. Rev.* C **59** 2580
[50] Viola V E, Jr and Sikkeland T (1962) *Phys. Rev.* **128** 767
[51] Toke J *et al.* (1985) *Nucl. Phys.* A **440** 327
[52] Wong C Y (1973) *Phys. Rev. Lett.* **31** 766
[53] Balantekin A B, Bennett J R and Kuyucak S (1994) *Phys. Lett.* B **335** 295
[54] Langanke K and Koonin S E (1983) *Nucl. Phys.* A **410** 334; Redder A *et al.* (1987) *Nucl. Phys.*A **462** 385
[55] Hinde D J *et al.* (1995) *Phys. Rev. Lett.* **74** 1295
[56] Trotta M *et al.* (2005) *Eur. Phys. J.* A **25** 615
[57] Itkis M G *et al.* (2007) *Nucl. Phys.* A **787** 150c
[58] Itkis M G *et al.* (2004) *Nucl. Phys.* A **734** 136c
[59] Shen W Q *et al.* (1987) *Phys. Rev.* C **36** 115
[60] Möller P *et al.* (1995) *At. Data Nucl. Data Tables* **59** 185
[61] Wolfs F L H, Henning W, Rehm K E and Schiffer J P (1987) *Phys. Lett.* B **196** 113

Production of exotic nuclei in quasifission-type reactions

G.G.Adamian[1,2], **N.V.Antonenko**[1], **V.V.Sargsyan**[1], **W.Scheid**[3], **A.S.Zubov**[1]

[1]Joint Institute for Nuclear Research, 141980 Dubna, Russia
[2]Institute of Nuclear Physics, 702132 Tashkent, Uzbekistan
[3]Institut für Theoretische Physik, Justus-Liebig-Universität, 35392 Giessen, Germany

E-mail: antonenk@theor.jinr.ru

Abstract. The possibilities for producing neutron-rich isotopes ^{82,84,86}Zn and ^{86,88,90,92}Ge are demonstrated in the reactions ^{48}Ca+^{238}U, ^{244}Pu at incident energies near the Coulomb barrier. The production cross sections of new neutron-rich isotopes of nuclei with charge numbers Z=64–80 are estimated as well. The dynamics of the binary reaction is considered as the diffusive multinucleon transfer between the interacting nuclei in the collisions when the excitation energy of the produced exotic isotope is lower than the threshold for the neutron emission. In the quasifission reactions ^{48}Ca+^{244,246,248}Cm at beam energies close to the corresponding Coulomb barriers one can produce new isotopes of superheavies with $Z = 103-108$, which mainly undergo fission.

1. Introduction

Besides the reactions at intermediate energies [1–7], the multinucleon transfer reactions at low energies are actively discussed to produce exotic nuclei [8–14]. These binary reactions have been known for producing exotic nuclei for many years [15–24]. In Refs. [12] the possibility have been shown to produce the neutron-rich nuclei close to drip-line in the transfer-type reactions ^{48}Ca+^{232}Th,^{238}U,^{248}Cm at incident energies close to the Coulomb barrier. The ^{238}U(5.5 MeV/nucleon)+^{48}Ca reaction has been used to produce the odd and even neutron-rich Ca isotopes and study their low-lying states [13]. In Refs. [9–11,14] the neutron-rich nuclei with A=50–80 have been studied through multinucleon transfer reactions by bombarding ^{208}Pb and ^{238}U targets with beams of ^{48}Ca, ^{58,64}Ni, ^{70}Zn, and ^{82}Se.

The multinucleon transfer reactions and quasifission-type reactions are expected to be efficient to synthesize exotic nuclei. We are going to demonstrate the possibilities for producing neutron-rich isotopes ^{82,84,86}Zn and ^{86,88,90,92}Ge in the reactions ^{48}Ca+^{238}U, ^{244}Pu at incident energies near the Coulomb barrier. In these reactions the contributions of fast nonequilibrium processes of the entrance channel are expected to be negligible. The dynamics of the binary reaction is considered as the diffusive multinucleon transfer between the interacting nuclei in the collisions when the excitation energy of the produced exotic isotope is lower than the threshold for the neutron emission.

As the first step, we deal with the production of neutron-rich Zn and Ge isotopes with neutron number $N > 50$, which are the products of multinucleon transfer channel in the actinide-based reactions. Note that these reactions are intensively and successfully used to produce superheavy

Journal of Physics: Conference Series **282** (2011) 012002 doi:10.1088/1742-6596/282/1/012002

nuclei and products of multinucleon transfer process or quasifission which competes with the complete fusion process [25–28]. Within the same formalism, we will report for the future experiments the possibilities for producing new neutron-rich isotopes of nuclei with Z=64–80 as complementary to light fragments in the uranium-based multinucleon transfer reaction with a ^{48}Ca beam at incident energy $E_{\rm c.m.}$=189 MeV, the value of Coulomb barrier for the spherical nuclei. The possibilities for transfer-induced fission of new isotopes of heaviest nuclei with Z=103–108 will be discussed in the curium-based multinucleon transfer reactions with ^{48}Ca beam at incident energies close to the corresponding Coulomb barriers.

As shown in Refs. [12], the diffusive multinucleon transfer-type reactions can be described as an evolution of the dinuclear system (DNS) which is formed in the entrance channel of the reaction after dissipation of the kinetic energy and angular momentum of the relative motion [15, 16, 29–37]. The dynamics of the process is considered as a diffusion of the DNS in the charge and mass asymmetry coordinates, which are defined here by the charge and neutron numbers Z and N of light nucleus of the DNS. During the evolution in charge and mass asymmetry coordinates, the excited DNS can decay into two fragments in relative distance R between the centers of the DNS nuclei. So, within the DNS model the production of the exotic nucleus is treated as a three-step process. First, the initial DNS with light nucleus (Z_i, N_i) is formed in the peripheral collision for a short time. Second, the DNS with light exotic nucleus (Z, N) is produced by nucleon transfers. Then this DNS separates into two fragments.

2. Model

The cross section of the production of primary nucleus in the diffusive nucleon transfer reaction is written as a sum over all partial waves J

$$\begin{aligned} \sigma_{Z,N}(E_{\rm c.m.}) &= \sum_J \sigma_{Z,N}(E_{\rm c.m.}, J), \qquad (1) \\ \sigma_{Z,N}(E_{\rm c.m.}, J) &= \int_0^{\pi/2}\int_0^{\pi/2} d\cos\Theta_1 d\cos\Theta_2 \; \sigma_c(E_{\rm c.m.}, J, \Theta_i) Y_{Z,N}(E_{\rm c.m.}, J, \Theta_i). \end{aligned}$$

Here, the average over the orientations of statically deformed interacting nuclei (Θ_i (i=1,2) are the orientation angles with respect to the collision axis) is taken into consideration [38]. For the correct description of $\sigma_{Z,N}$, the partial capture cross section σ_c in the entrance channel and the formation-decay probability $Y_{Z,N}$ of the DNS configuration with charge and mass asymmetries given by Z and N should be properly calculated. The value of $\sigma_c(E_{\rm c.m.}, J, \Theta_i) = \frac{\pi\hbar^2}{2\mu E_{\rm c.m.}}(2J+1)T(E_{\rm c.m.}, J, \Theta_i)$ defines the transition of the colliding nuclei over the Coulomb barrier with the probability T and the formation of the DNS when the kinetic energy $E_{\rm c.m.}$ and angular momentum J of the relative motion are transformed into the excitation energy and angular momentum of the DNS. The capture (transition) probability $T(E_{\rm c.m.}, J, \Theta_i) = (1+\exp[2\pi(V_J(R_b, Z, N, \Theta_i) - E_{\rm c.m.})/\hbar\omega_J(Z, N, \Theta_i)])^{-1}$ is calculated with the Hill-Wheeler formula. The effective nucleus-nucleus potential [29]

$$V_J(R, Z, N, \Theta_i) = V_N(R, Z, N, \Theta_i) + V_C(R, Z, N, \Theta_i) + \hbar^2 J(J+1)/(2\Im) \qquad (2)$$

is calculated as a sum of nuclear V_N, Coulomb V_C and centrifugal interactions and approximated near the Coulomb barrier at $R = R_b$ by the inverted harmonic-oscillator potential with the barrier height $V_J(R_b, Z, N, \Theta_i)$ and frequency $\omega_J(Z, N, \Theta_i)$. In the entrance channel the moment of inertia is $\Im = \mu R^2$ but after the DNS formation it corresponds to the rigid body limit at sticking condition. The nuclear potential V_N is calculated with the double-folding model using a nuclear radius parameter r_0=1.15 fm and a diffuseness a=0.54 fm for ^{48}Ca and a=0.56 fm for the actinide targets [29]. The quadrupole deformation parameters of actinides are taken from Ref. [39].

The primary charge and mass yields of fragments can be expressed by the product of the formation probability $P_{Z,N}(t)$ of the DNS configuration with charge and mass asymmetries given by Z and N ($A = Z + N$ is the mass number of nucleus), and of the decay probability of this configuration in R represented by the one-dimensional Kramers rate $\Lambda^{qf}_{Z,N}$ [34]:

$$Y_{Z,N} = \Lambda^{qf}_{Z,N} \int\limits_0^{t_0} P_{Z,N}(t)dt. \tag{3}$$

Here, t_0 is the time of reaction which is determined as in Ref. [34] from the normalization condition $\sum\limits_{Z,N} Y_{Z,N} = 0.98$. The mass yield of transfer products is defined as follows

$$Y_A = \sum_Z Y_{Z,A-Z}. \tag{4}$$

Using the macroscopical method suggested in Ref. [34], one can find $P_{Z,N}$ from the solution of the system of master equations

$$\begin{aligned}\frac{d}{dt}P_{Z,N}(t) &= \Delta^{(-,0)}_{Z+1,N} P_{Z+1,N}(t) + \Delta^{(+,0)}_{Z-1,N} P_{Z-1,N}(t)\\ &+ \Delta^{(0,-)}_{Z,N+1} P_{Z,N+1}(t) + \Delta^{(0,+)}_{Z,N-1} P_{Z,N-1}(t)\\ &- (\Delta^{(-,0)}_{Z,N} + \Delta^{(+,0)}_{Z,N} + \Delta^{(0,-)}_{Z,N} + \Delta^{(0,+)}_{Z,N} + \Lambda^{qf}_{Z,N} + \Lambda^{fis}_{Z,N}) P_{Z,N}(t),\end{aligned} \tag{5}$$

with initial condition $P_{Z,N}(0) = \delta_{Z,Z_i}\delta_{N,N_i}$ and the transport coefficients which characterize the proton and neutron transfer rates from a heavy to a light nucleus ($\Delta^{(+,0)}_{Z,N}$, $\Delta^{(0,+)}_{Z,N}$) or in opposite direction ($\Delta^{(-,0)}_{Z,N}$, $\Delta^{(0,-)}_{Z,N}$). In Eqs. (5) we take only the transitions $Z \rightleftharpoons Z \pm 1$ and $N \rightleftharpoons N \pm 1$ into account in the spirit of the independent-particle model. Here, $\Lambda^{fis}_{Z,N}$ is the fission rate of the DNS heavy nucleus [34]. The solution of the master equations (5) with the decay terms and the microscopically calculated transport coefficients ensures a realistic description of the DNS evolution in charge and mass asymmetry [34].

As in Ref. [40], the excitation energy of the initial DNS should be enough to form the DNS with the certain exotic nucleus, i.e. to overcome the energy threshold $\Delta B_{Z,N,J}$ for this. The value of

$$\Delta B_{Z,N,J} = U(R_b, Z, N, J) - U(R_m, Z_i, N_i, J) \tag{6}$$

is defined using the DNS potential energy calculated as in Ref. [32]

$$U(R, Z, N, J) = B_L + B_H + V_J(R, Z, N, \Theta_i), \tag{7}$$

where B_L and B_H are the mass excesses of the light and heavy fragments, respectively. Here, the DNS potential energy is calculated at the touching distance $R = R_m \approx R_L(1+\beta_L Y_{20}(\Theta_L)) + R_H(1 + \beta_H Y_{20}(\Theta_H)) + 0.5$ fm (β_L and β_H are the quadrupole deformation parameters of the nuclei with radii R_L and R_H) and the position of the Coulomb barrier $R = R_b \approx R_m + 1.2$ fm for the systems ^{48}Ca+^{238}U,^{244}Pu. Note that the values of R_m and R_b depend on charge and mass asymmetries. As follows from Eqs. (6) and (7) the Q_{gg} values influence the production cross sections because of the binary character of the reaction.

The excitation energy of the initial DNS is $E^*(Z_i, N_i, J) = E_{\rm c.m.} - V_J(R_m, Z_i, N_i, \Theta_i)$. With this value the excitation energy of the DNS with exotic nucleus (Z, N) is $E^*(Z, N, J) = E^*(Z_i, N_i, J) - \Delta B_{Z,N,J}$. Assuming the situation of thermal equilibrium, the excitation energy

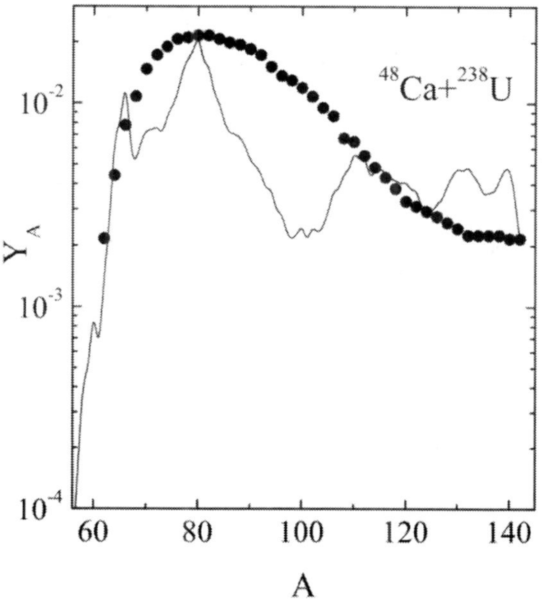

Figure 1. The calculated (solid lines) mass yield (in relative units) of the quasifission products as a function of the mass number of the light fragment for ^{48}Ca+^{238}U reaction at bombarding energy $E_{\rm c.m.}$=190.2 MeV. The experimental data [41] are shown by solid points.

of the nucleus with mass $A = Z + N$ in this DNS is $E^*_L(Z,N,J) = E^*(Z,N,J)A/A_{tot}$, where A_{tot} is the total mass number of the DNS. It is clear that the probability of formation of the DNS with exotic nucleus (Z, N) increases with $E^*(Z_i, N_i, J)$. However, the increase of $E^*(Z_i, N_i, J)$ is possible up to the moment when $E^*_L(Z, N, J)$ reaches the neutron separation energy $S_n(Z, N)$. Further increase of $E^*(Z_i, N_i, J)$ would lead to the strong loss of neutron-rich nuclei because of the neutron emission. If the primary nucleus is excited, one should take into consideration its survival probability W_{sur} in the deexcitation process to obtain the evaporation residue cross section as follows

$$\sigma^{ER}_{Z,N-x}(E_{\rm c.m.}) = \sum_J \sigma_{Z,N}(E_{\rm c.m.},J)W_{sur}(E_{\rm c.m.},J,x), \tag{8}$$

where x is the number of evaporated neutrons from the excited primary nucleus. W_{sur} is treated as in Ref. [40] and takes into consideration the competition with other deexcitation channels.

In order to test our method of calculation of $\sigma^{ER}_{Z,N}(E_{\rm c.m.})$ and $\sigma_{Z,N}(E_{\rm c.m.})$, we treat the production of Ti in the multinucleon transfer reactions ^{58}Ni($E_{\rm c.m.}$ = 256.8 MeV)+ ^{208}Pb [10] and ^{64}Ni($E_{\rm c.m.}$= 307.4 MeV)+^{238}U [9] at bombarding energies near the Coulomb barrier. In these reactions the available excitation energies supply 2 neutron evaporation from the primary Ti isotopes having the maximal yields. In the ^{58}Ni + ^{208}Pb reaction ^{50}Ti and ^{52}Ti are produced with the cross sections 1 and 0.2 mb [10], respectively, which are consistent with our calculated cross sections 0.6 and 0.35 mb, respectively. In the ^{64}Ni+^{238}U reaction the experimental [9] and theoretical production cross sections for ^{52}Ti are 0.5 and 1.6 mb, respectively. In the reaction ^{48}Ca($E_{\rm c.m.}$= 204 MeV)+^{248}Cm→^{40}S+(^{254}Fm+2n), the calculated $\sigma^{ER}_{Z,N}(E_{\rm c.m.})$ for ^{254}Fm is about 0.5 μb, which is close to the experimental result presented in Refs. [19]. In the ^{48}Ca($E_{\rm c.m.}$= 274.6 MeV) + ^{238}U reaction the experimental [14] and calculated ratios of secondary yields $Y(^{62}{\rm Fe})/Y(^{58}{\rm Cr})$ for the neutron-rich ^{62}Fe and ^{58}Cr isotopes are about 0.2 and 0.3, respectively.

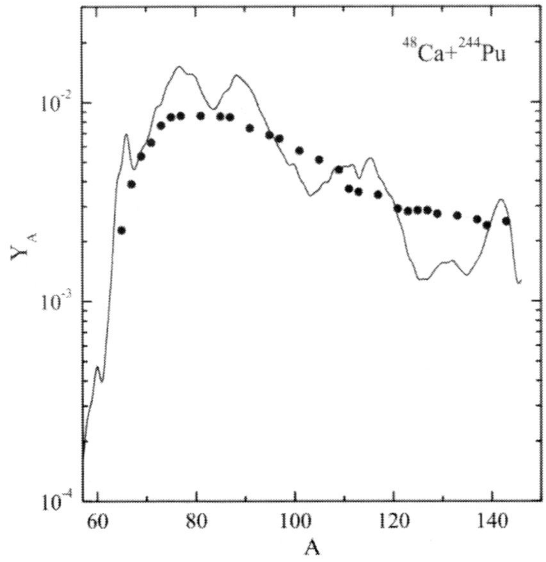

Figure 2. The same as in Fig. 1, but for ^{48}Ca+^{244}Pu reaction at bombarding energy $E_{\text{c.m.}}$=201 MeV.

The multinucleon transfer products of the quasifission reactions ^{48}Ca($E_{\text{c.m.}}$= 190.2 MeV)+^{238}U and ^{48}Ca($E_{\text{c.m.}}$= 201 MeV)+^{244}Pu at incident energies close to the Coulomb barrier are correctly described within our model. Figures 1 and 2 show the mass yield Y_A as functions of the mass number of the light fragment. The calculated data in Figs. 1 and 2 are related to the primary (before neutron emission) fragments. Therefore, the maxima and minima in the calculated functions Y_A are more pronounced. The postneutron evaporation washes out some peculiarities of these functions. Taking into consideration the experimental uncertainties in the identification of quasifission and fusion-fission products, and the measurement of mass, the agreement between the calculated and experimental data [41] is quite good. In the experiment besides the quasifission and fusion-fission, the fission of the heavy nucleus in the DNS with a subsequent fusion of one of the fission fragments with the light nucleus of the DNS and ternary processes with the emission of a light particle are identified as two-body processes of complete momentum transfer. It should be noted that the small oscillations in experimental data are comparable with accuracy of the measurements [41]. The maximum yield of the quasifission fragments occurs around the nucleus ^{208}Pb for the heavy fragment where the DNS potential energy has a deep minimum. The evolution of the DNS is hindered by this minimum to go to smaller mass asymmetry and, correspondingly, the decay probability from the configuration with ^{208}Pb is increased. In the reaction ^{48}Ca+^{238}U (^{48}Ca+^{244}Pu), the height of the peak around A=80 is 4.5 (3.5) times larger than the height of peaks in the symmetric mass region. The main contribution to the symmetric and near symmetric fragmentations comes from the multinucleon transfer process.

3. Production cross sections of new neutron-rich ^{84,86}Zn and ^{90,92}Ge isotopes

As shown above, the suggested method is suitable to predict the mass and charge yields and the production cross sections for the products of multinucleon transfer reactions. The calculated production cross sections of neutron-rich isotopes in the reactions ^{48}Ca+^{238}U,^{244}Pu at incident energies near the Coulomb barrier are presented in Figs. 3-5. We treat only the reactions leading to excitation energies of light neutron-rich nuclei equal to or smaller than their neutron separation energies ($E^*_L(Z,N,J) \leq S_n(Z,N)$). In this case W_{sur}=1 and the primary and secondary yields coincide. In Figs. 3 and 4 the values of $E_{c.m.}$ provide the condition $E^*_L(Z,N,J) = S_n(Z,N)$. The predicted values of $S_n(Z,N)$ for unknown nuclei are taken from the finite range liquid drop model [42]. If $E^*_L(Z,N,J) > S_n(Z,N)$, the primary neutron-rich nuclei are transformed into the secondary nuclei with less number of neutrons because of the de-excitation by neutron emission. The DNS evolution in the reactions treated can be schematically presented in the following way: ^{48}Ca+^{238}U→^{78,80}Zn+^{208,206}Pb→^{82,84,86}Zn+^{204,202,200}Pb and ^{48}Ca+^{244}Pu→^{84,82}Ge+^{208,210}Pb→^{86,88,90,92}Ge+^{206,204,202,200}Pb. The system initially moves to the deep minimum of the potential energy surface (energetically favorable) which is caused by the shell effects around the DNS with magic heavy ^{208}Pb and light ^{80}Zn or ^{82}Ge nuclei then from this minimum it reaches the DNS with exotic light nucleus by fluctuations in mass asymmetry. For low excitation energy, the evolution of the dinuclear system towards symmetry is hindered by this minimum.

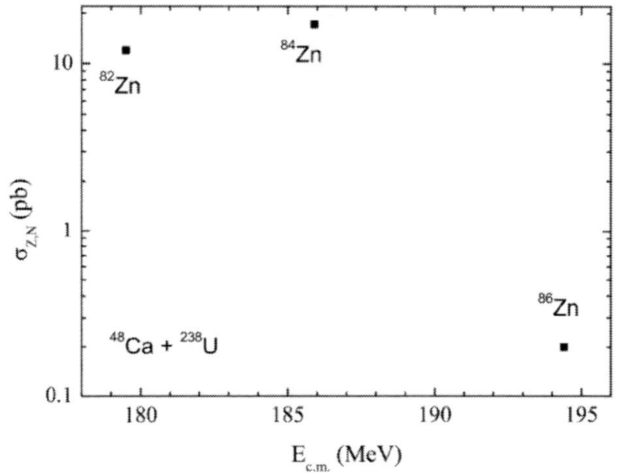

Figure 3. The expected cross sections for the indicated neutron-rich isotopes of Zn produced in the ^{48}Ca+^{238}U reaction at values of $E_{c.m.}$ providing the excitations of these isotopes to be equal to the corresponding thresholds for the neutron emission.

The production cross section for ^{82}Zn (^{86}Ge) is about 2 (3) orders of magnitude larger than the production cross section for ^{86}Zn (^{92}Ge) (Figs. 3 and 4). Although $P_{30,52} \gg P_{30,54}$ ($P_{32,54} \gg P_{32,56}$), the cross sections of production of ^{82}Zn (^{86}Ge) and ^{84}Zn (^{88}Ge) are comparable because in the case of ^{82}Zn (^{86}Ge) the optimal bombarding energy is considerably below the Coulomb barrier for the spherical nuclei and, correspondingly, σ_c(^{82}Zn (^{86}Ge))$\ll$ σ_c(^{84}Zn (^{88}Ge)). At $E_{c.m.}$ smaller than the value of this barrier, only the collisions at certain mutual orientations leads to the capture. Since the yields of ^{82}Zn and ^{86}Ge isotopes as primary products is suppressed by the capture process, the production of these isotopes as secondary products at higher bombarding energies seems be possible. However, the formation of the DNS with heavier isotopes of Zn or Ge occurs with smaller probability. Therefore, the productions of ^{82}Zn and ^{86}Ge as secondary products and as primary products seem to be with comparable cross sections.

The dependence of the production cross section of the neutron-rich ^{84}Zn isotope versus $E_{c.m.}$ is presented in Fig. 5. The overall trend of dependence of cross section on the neutron separation energy is good visible. The solid arrow indicates the value of $E_{c.m.}$ at which value of $E^*_L(Z,N,J)$ reaches $S_n(Z,N)$ = 3.99 MeV. Since the predictions of $S_n(Z,N)$ have some uncertainties, we indicate by dashed arrow the value of $E_{c.m.}$ at which the value of $E^*_L(Z,N,J)$ reaches $0.5S_n(Z,N)$. One can see that the decrease of neutron binding energy by 2 MeV leads

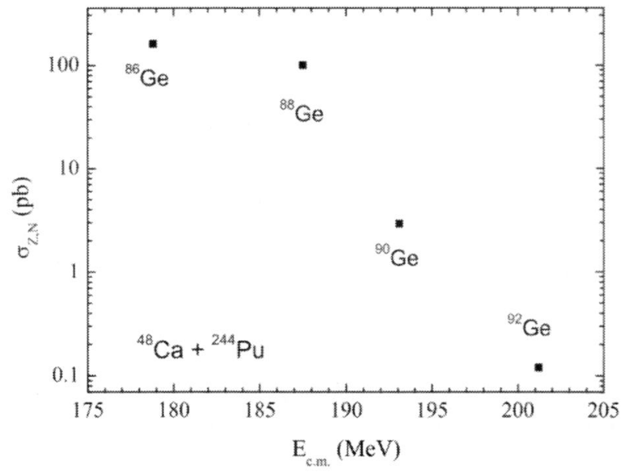

Figure 4. The same as in Fig. 3, but for the indicated neutron-rich isotopes of Ge produced in the ^{48}Ca+^{244}Pu reaction.

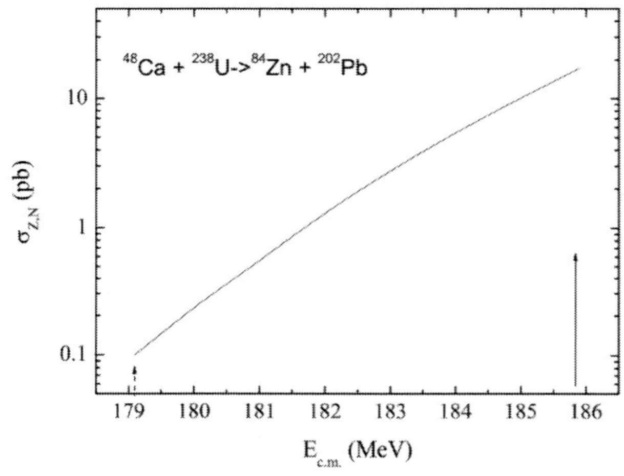

Figure 5. The excitation function for producing ^{84}Zn in the multinucleon transfer reaction ^{48}Ca+^{238}U. The solid (dashed) arrow indicates the expected cross section at the value of $E_{\rm c.m.}$ providing the excitation of ^{84}Zn to be equal to the threshold (half of the threshold) for the neutron emission.

to the shift of $E_{\rm c.m.}$ by about 7 MeV and decrease of the production cross section by about 2 orders of magnitude. This decrease of $\sigma_{Z,N}$ is mostly due to the strong decrease of the capture cross section with decreasing $E_{\rm c.m.}$ below the spherical Coulomb barrier. The measurement of the excitation function would be useful to estimate $S_n(Z,N)$ in the neutron-rich nuclei since at $E^*_L(Z,N,J) > S_n(Z,N)$ the excitation function strongly drops down.

In our calculations the theoretically predicted [42] mass excesses and neutron separation energies are used for neutron-rich nuclei. The uncertainties of these predictions mainly contribute to the uncertainty of our results. We assume that the excitation energy of DNS is shared between the DNS nuclei proportionally to their mass numbers. Since the isotopes of interest are formed from the projectile by the acceptance of nucleons, they can be slightly more excited than in the limit of thermal equilibrium. Taking into consideration these facts, we estimate the uncertainty within a factor of 3-5 in the calculated cross sections. If the neutron-rich isotope is close to the region of known nuclei, then the predictions for it have less uncertainties and the cross section is estimated with higher accuracy.

4. Predicted yields of new neutron-rich isotopes of nuclei with Z=64-80

The calculated production cross sections of the primary isotopes in the reactions ^{48}Ca+^{238}U at $E_{\mathrm{c.m.}}$=189 MeV are presented in Figs. 6-13. The primary neutron-rich nuclei of interest are excited and transformed into the secondary nuclei with less number of neutrons without a loss of the cross section because of the neutron emission is dominant over other deexcitation channels. The neutron emission channels are indicated in Figs. 6-13 for primary neutron-rich isotopes. The predictions are done by assuming the excitation energy of the DNS is divided proportionally to the mass numbers of the fragments.

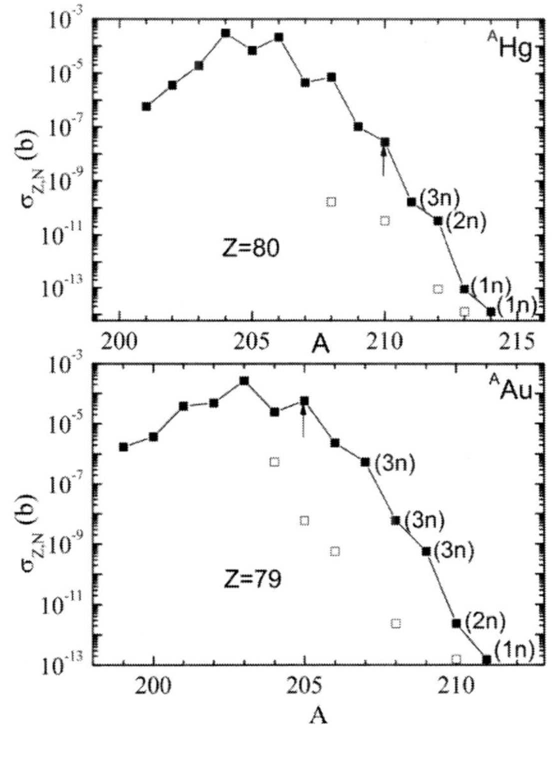

Figure 6. Calculated production cross sections of the primary Hg and Au isotopes versus the mass number. Neutron evaporation channels for neutron-rich primary isotopes are indicated. The heaviest known isotopes are marked by arrows.

Since the predicted production cross sections for new exotic isotopes ^{193}W, ^{195,196}Re, ^{198}Os, and ^{200}Ir are at the microbarn level, they can be easily identified. For these nuclei, the known heaviest isotopes are in the vicinities of maxima of the primary isotopic distributions (Figs. 7-9). Because the calculated production cross sections for new exotic isotopes ^{178}Er, ^{180,181}Tm, $^{182-184}$Yb, $^{185-187}$Lu, ^{190}Hf, $^{191-193}$Ta, ^{194,196}W, ^{197,199}Re, ^{199,200}Os, ^{201,202}Ir, ^{203}Pt (Figs. 7-12) are between microbarn and nanobarn levels, they can be also detected with the present experimental setups. The group of new isotopes ^{173}Tb, ^{174,176}Dy, ^{176,177,179}Ho, $^{179-182}$Er, ^{182,183}Tm, $^{185-187}$Yb, ^{188,189,191}Lu, ^{191,192}Hf, ^{195,196}Ta, ^{197}W, ^{200}Re, ^{202}Os, ^{204}Ir, ^{204,206}Pt, ^{206,208}Au (Figs. 6-13) is produced with the cross sections in the interval between nanobarn and picobarn. The unknown neutron-rich isotopes of Hg can be produced in the reaction treated

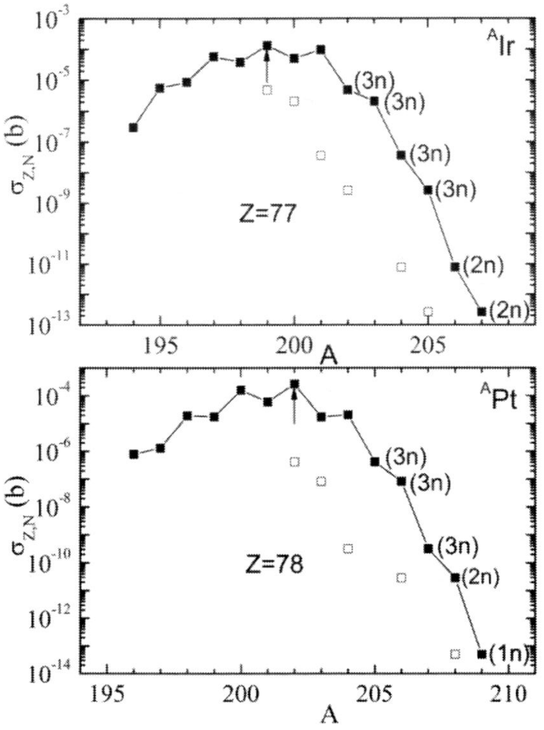

Figure 7. The same as in Fig. 6, but for the primary Pt and Ir isotopes.

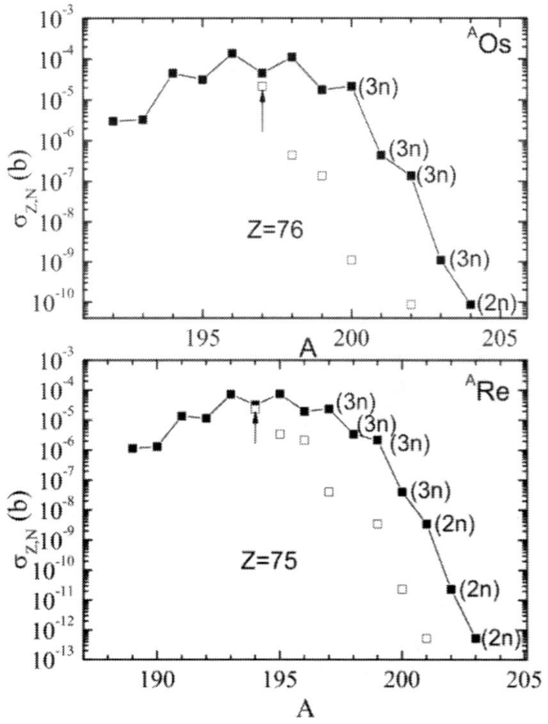

Figure 8. The same as in Fig. 6, but for the primary Os and Re isotopes.

with the cross sections less than 1 pb. In this case the known heaviest isotope ^{210}Hg is far to the right side from the maximum of the calculated isotopic distribution (see Fig. 6).

5. Transfer-induced fission of superheavy nuclei

In Fig. 14 the cross sections σ_f of transfer-induced fission of superheavy nuclei are predicted in the reactions ^{48}Ca+^{244,246,248}Cm at beam energies near the corresponding Coulomb barriers. The value of σ_f drops down by approximately three orders of magnitude as Z increases from 101 to 108. As the value of $E_{\rm c.m.}$ increases by 20 MeV the values of σ_f grow by about two orders of magnitude. The experimental data [19] as well as our analysis show the advantages of reactions with a smaller number of neutrons in the target for the synthesis of unknown isotopes of superheavy nuclei. The probability of a large number of nucleons transferring from the projectile to the target nucleus is correlated with the dependence of the DNS potential energy on Z. For 102< Z <110, the DNS potential energy decreases as the total number of DNS neutrons decreases, therefore one can expect larger yields of superheavy nuclei in transfer type reactions with the ^{244,246}Cm nuclei than when the ^{248}Cm target is used. In the reaction ^{48}Ca + ^{248}Cm, the calculated production cross section for the ^{254}Fm is close to the experimental results presented in Refs. [19] where the yields of transfermium nuclei were not measured.

Note that isotopes of nuclei with Z=103–108 listed in Fig. 14 were not observed yet in the experiments. The complementary to the fission fragments of these nuclei are the O, F, Ne, Na, Mg, Al, Si, and P nuclei. Figure 14 shows that fission of unknown isotopes with Z=103–108 can be measured with acceptable cross sections ($\sigma_f \approx$ 100 nb – 100 μb) in the reactions with ^{244,246,248}Cm targets and ^{48}Ca beam at energies which are about 28 MeV above the corresponding Coulomb barriers. If the excited superheavy nucleus undergoes fission in

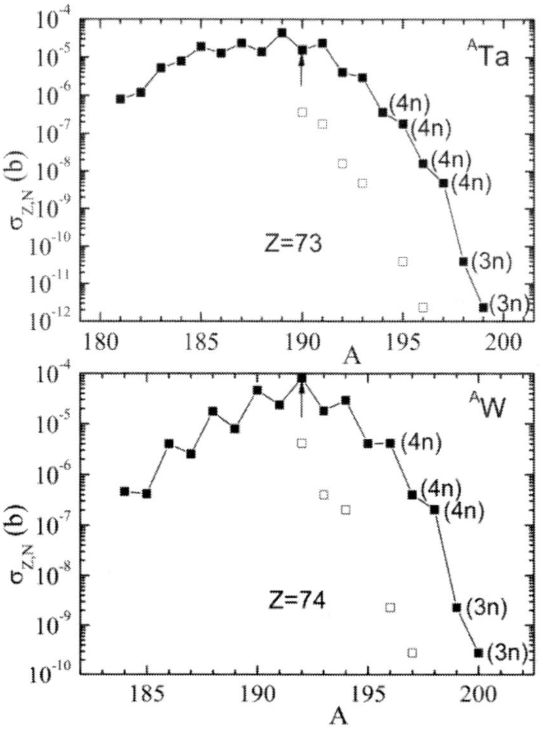

Figure 9. The same as in Fig. 6, but for the primary W and Ta isotopes.

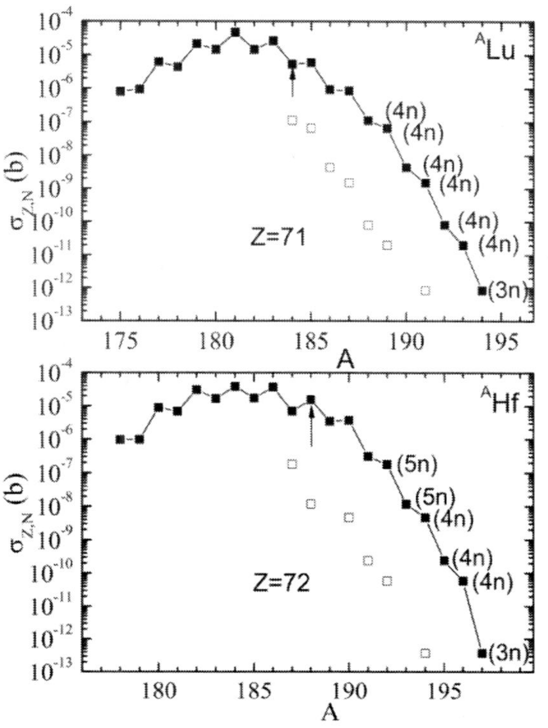

Figure 10. The same as in Fig. 6, but for the primary Hf and Lu isotopes.

10^{-20} s, its distance from the light complementary fragment is about 300 fm, i.e. the real quasi-ternary-process occurs. In Ref. [43] the transfer-induced fission was called as the quasi-ternary-fission process by analogy to the binary quasifission process. Note that isotopes of nuclei with Z=103–108 produced undergo fission with probability of more than 99% at the excitation energies treated [40]. In the experiment one can observe the light nucleus from O to P in coincidence with two fission fragments resulting from superheavy nucleus.

The average excitation energies E^*_H of the fissioning superheavy nuclei are presented in Fig. 15. It was assumed that the excitation energies of the DNS nuclei are proportional to their masses. Since the DNS potential energy increases towards Z larger than Z_i, the excitation energy of the heavy fragment (the fissioning nucleus) decreases from (16–18) MeV to (8–11) MeV at the smaller bombarding energies treated in Fig. 14. Note that the peak/valley ratio of the mass distribution of the fission fragments strongly depends on the value of the excitation energy. In the transfer-induced fission one can study the multimodal structure of the mass distribution of fission fragments for certain isotopes of superheavy nuclei with Z=101–108.

In the multinucleon transfer reactions ^{48}Ca+^{244,246,248}Cm at the beam energies $E_{\rm c.m.}$=207, 205.5, 204 MeV, respectively, one can produce the new isotopes of superheavies (the evaporation residues) with Z=104–108 which are not reachable in the hot and cold complete fusion reactions with the stable projectiles and targets [40]. While σ_f drops down by approximately three orders of magnitude as Z increases from 102 to 108, the evaporation residue cross section decreases only about of 100 times due to the increase of survival probability in the $1n$ evaporation channel for nuclei approaching to the deformed subshell Z=108 and N=162 [40]. The experimental verification of these predictions would provide an additional information about the absolute values and the excitation energy- and Z-dependencies of shell correction in this range of

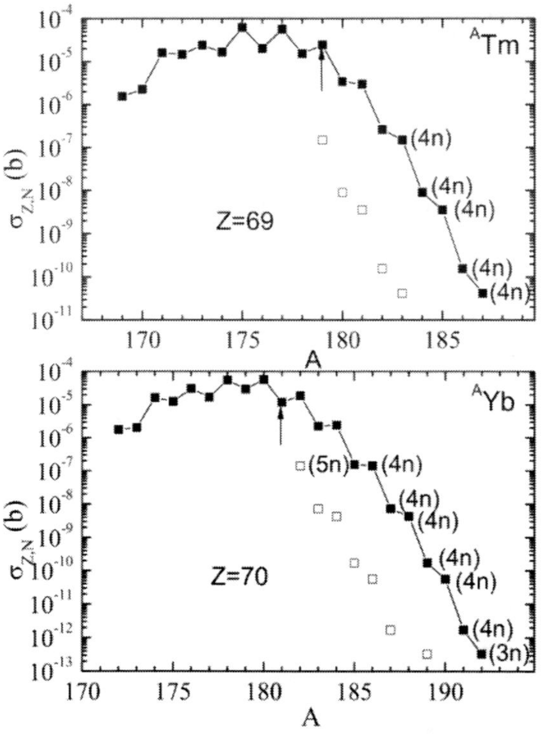

Figure 11. The same as in Fig. 6, but for the primary Yb and Tm isotopes.

superheavy nuclei.

6. Production of heaviest nuclei in quasifission reactions

The master equations also describe configurations of dinuclear systems which are more asymmetric than the DNS in the entrance channel. The processes of formation and decay are ruled by the same mechanism of diffusion in the same relevant collective coordinates: mass and charge asymmetries and relative distance.

With asymmetric-exit-channel quasifission reactions leading to nuclei with charge numbers larger than the charge number of the target, one can produce isotopes that can not be synthesized in complete fusion reactions. The direct production of transactinides in asymmetric-exit-channel quasifission reactions would give nuclei with $101 \leq Z \leq 108$ in the reactions ^{48}Ca+^{238}U, ^{243}Am, ^{244,246,248}Cm. The production of heavy actinides has been studied in the transfer-type reactions by bombarding of actinide targets with ^{16,18}O, ^{20,22}Ne, and ^{40,44,48}Ca [19]. Nuclei with $Z > 102$ have not been observed because of the small cross sections or short lifetimes in the radiochemical identification of the nuclei.

The cross section $\sigma_{Z,N}$ of the production of a primary heavy nucleus with $Z = Z_H$ and $N = N_H$ (H=heavy) in the asymmetric-exit-channel quasifission reaction is written in Eq. (1). The primary heavy nucleus is excited and evaporates neutrons in the deexcitation process. The evaporation residue cross section for the heavy nucleus with charge number Z is obtained as in Eq. (8). The actinide targets proposed for such reactions are deformed. Therefore, the minimum value of the incident energy $E^{min}_{\rm c.m.}$, at which the collisions of nuclei at all orientations become possible, is larger than the Coulomb barrier calculated for spherical nuclei. In the asymmetric-exit-channel quasifission reactions which occur slightly above the Coulomb barrier, only partial waves with $J \leq J_{cap} = 20$ contribute to the production of superheavy nuclei. For $J_{cap} = 20$, the

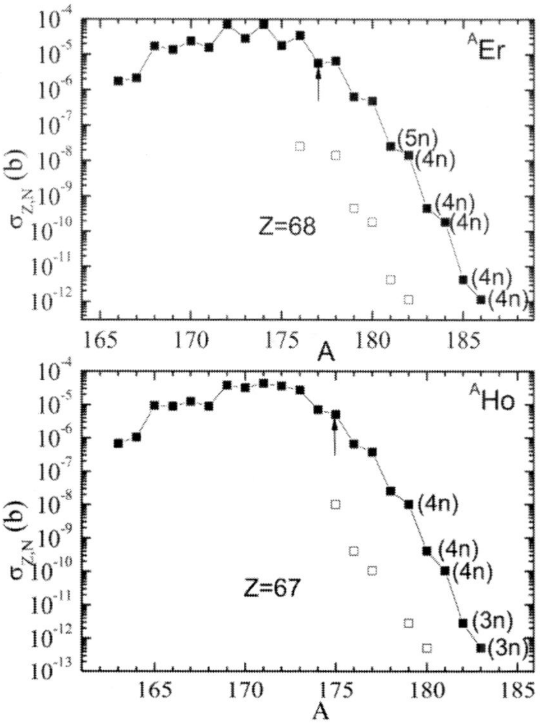

Figure 12. The same as in Fig. 6, but for the primary Er and Ho isotopes.

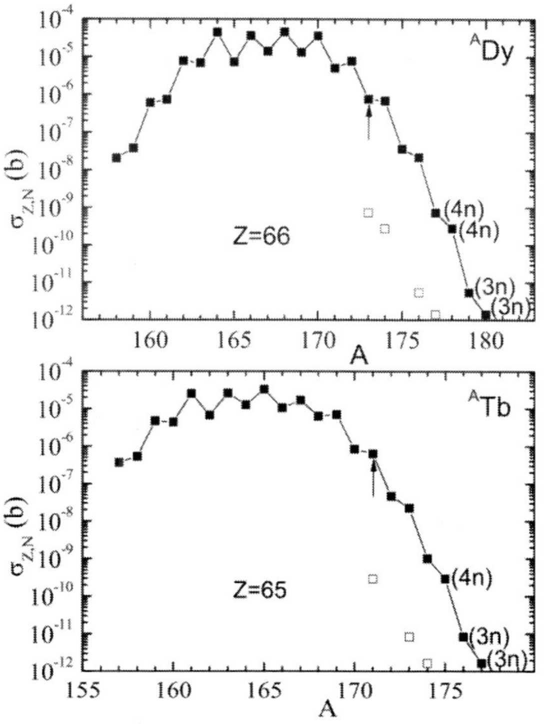

Figure 13. The same as in Fig. 6, but for the primary Dy and Tb isotopes.

primary heavy nucleus has an angular momentum of about 10.

For $102 < Z < 110$, the potential energy decreases with the total number of neutrons of the DNS and a larger primary yield of superheavy nuclei is expected in the reactions with ^{244,246}Cm rather than with ^{248}Cm. This is demonstrated in Fig. 16, where the primary yields of the most probable isotopes of heavy nuclei are calculated with the master equations and with the statistical formula

$$Y_{Z,N} \approx 0.5 \exp\left(-\frac{B_R(Z,N) - B_\eta^{sym}(Z_i,N_i)}{T(Z_i,N_i)}\right), \tag{9}$$

where B_R is given by $B_R(Z,N) = U(R_b,Z,N,J) - U(R_m,Z_i,N_i,J)$ and T is temperature calculated by using the Fermi-gas expression $T = \sqrt{E^*/a}$ with the excitation energy E^* of the initial DNS and the level-density parameter $a = A_{tot}/12$ MeV^{-1}. Similar results are obtained with both methods.

In Fig. 16 the excitation energies of primary heavy nuclei correspond to $E_{\rm c.m.} = 204 - 207$ MeV. The excitation energy of the primary heavy nucleus is defined proportionally to its mass A_H: $E_H^*(Z,N) = (E^*(Z_i,N_i) - B_R(Z,N))A_H/(A_L + A_H)$. In this case $E_H^*(Z,N)$ is related to the maxima or to the right hand sides of excitation functions for one-neutron emission. For example, for ^{262}No we find $E_H^* = 16$ MeV and $W_{sur}(1n) = 2.4 \times 10^{-4}$.

Although $Y_{Z,N}$ decreases by about 3 orders of magnitude with increasing Z from 102 to 108, the evaporation residue cross section decreases only by about 30 times because of the increase of W_{sur} with approaching the deformed subshell closure Z=108, N=162. The experimental data as well as our treatment indicate the preference of a smaller number of evaporated neutrons to produce superheavy nuclei. So, one can see that with the asymmetric-exit-channel quasifission

Journal of Physics: Conference Series **282** (2011) 012002 doi:10.1088/1742-6596/282/1/012002

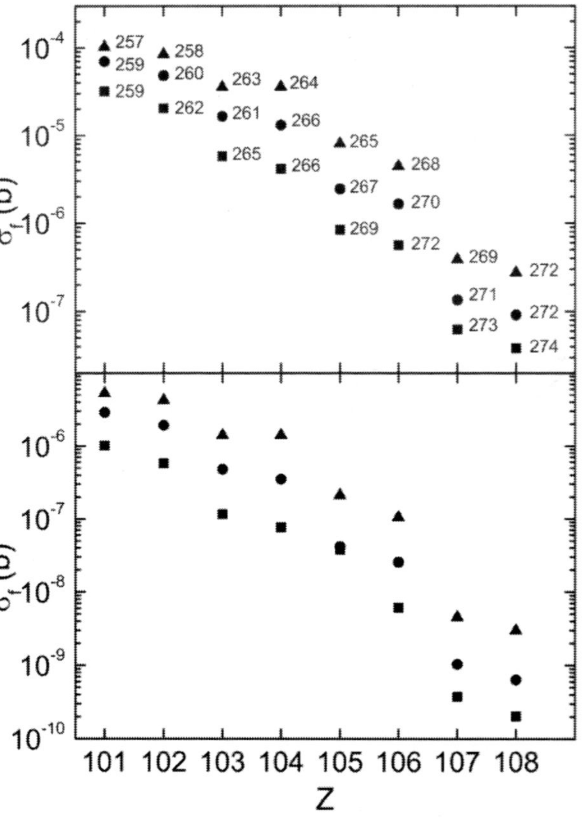

Figure 14. The cross sections σ_f of transfer-induced fission of indicated isotopes of superheavy nuclei (the mass number near the symbol) in the reactions ^{48}Ca($E_{\rm c.m.}$=207, 227 MeV)+^{244}Cm (triangles), ^{48}Ca($E_{\rm c.m.}$=205.5, 225.5 MeV)+^{246}Cm (circles), and ^{48}Ca($E_{\rm c.m.}$=204, 224 MeV)+^{248}Cm (squares). The results at larger bombarding energies are shown in the upper part.

reactions on actinide targets unknown isotopes of superheavy nuclei can be produced with suitable cross sections. The quasifission reactions ^{48}Ca+^{244,246,248}Cm at beam energies close to the corresponding Coulomb barriers, one can produce the new isotopes of superheavies with $Z = 103-108$, which undergo fission (the fission width is much larger than the neutron emission width). The calculated results indicate that these quasifission reactions provide a very efficient tool for the study of new isotopes of superheavy nuclei that fill the gap between the isotopes produced in the cold and hot complete fusion reactions. The predicted cross sections of the fission, which follow multinucleon transfer are on the level (100 nb-100 μb). One can propose the experiments on the quasiternary fission in which the fission fragment mass and the angular distributions in coincidence with the complementary transfer products, which range from O to P ions can be measured. Since the fission barrier of the superheavy nuclei is mainly determined by the shell correction in the ground state, in these experiments, one can study the dependence of the value of shell correction on the average excitation energy, which is easily calculated, and (Z, N) of the fissioning nucleus.

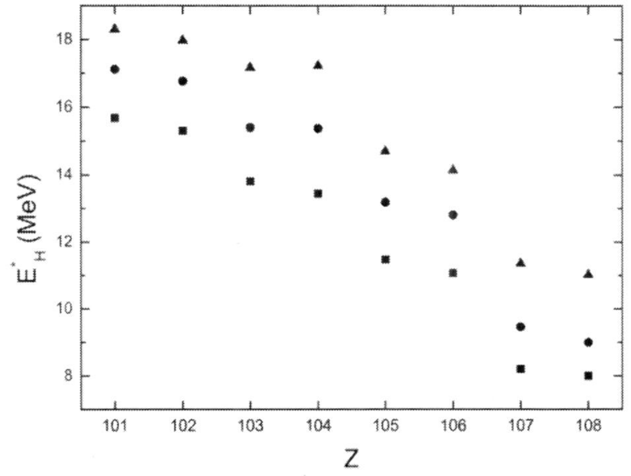

Figure 15. The average excitation energies of the fissioning superheavy nuclei produced in the reactions ^{48}Ca($E_{\rm c.m.}$=207 MeV) + ^{244}Cm (triangles), ^{48}Ca($E_{\rm c.m.}$=205.5 MeV)+^{246}Cm (circles), and ^{48}Ca($E_{\rm c.m.}$=204 MeV)+^{248}Cm (squares).

7. Summary

We demonstrated the possibilities for producing neutron-rich isotopes ^{82,84,86}Zn and ^{86,88,90,92}Ge in the reactions ^{48}Ca+^{238}U,^{244}Pu at incident energies near the Coulomb barrier. Note that ^{84,86}Zn and ^{90,92}Ge isotopes were not observed yet in the experiments. One can also produce new heavy neutron-rich isotopes of nuclei with Z=66–82 as complementary fragments in the actinide-based multinucleon transfer reactions with ^{48}Ca beam. The predicted cross sections are on the level (0.1–160) pb. Therefore, the multinucleon transfer reaction at low energies provides an efficient tool for producing nuclei far from stability and may be the fruitful method to reach the neutron drip line. The production of isotopes treated can be a supplementary information in the experiments on production of superheavy nuclei which run a long time with the same reactions. The multinucleon transfer reactions can provide detailed information about the dynamics of dinuclear system in the mass and charge asymmetry degrees of freedom.

Due to the large neutron excess and smaller losses because of the quasifission near the entrance channel, the use of ^{48}Ca projectile is more preferable than the use of heavier projectiles to reach the neutron-rich region of nuclide in the actinide-based reactions. It is apparent that the use of the heavier actinide target, for example, ^{244}Pu or ^{248}Cm, with ^{48}Ca beam at energies near the Coulomb barrier one can reach more neutron-rich region of nuclide. Irradiating the heavier actinide targets by ^{48}Ca beam for producing neutron-rich isotopes, we gain in the Q-value and the known heaviest isotopes are closer to the maxima of isotopic distributions.

In the multinucleon transfer reactions ^{48}Ca+^{244,246,248}Cm at beam energies close to the corresponding Coulomb barriers one can produce new isotopes of superheavies with Z=103–108 which undergo fission (the fission width is much larger than the neutron emission width). The calculated results indicate that the multinucleon transfer reactions provide a very efficient tool for the study of new isotopes of superheavy nuclei which fill the gap between the isotopes produced in the cold and hot complete fusion reactions. The predicted cross sections of the fission following multinucleon transfer are on the level (100 nb – 100 μb). One can propose the experiments on the quasi-ternary-fission in which the fission fragment mass and angular distributions in coincidence with the complementary transfer products ranging from O to P ions can be measured. Since the fission barrier of the superheavy nuclei is mainly determined by the shell correction in the ground state, in these experiments one can study the dependence of the

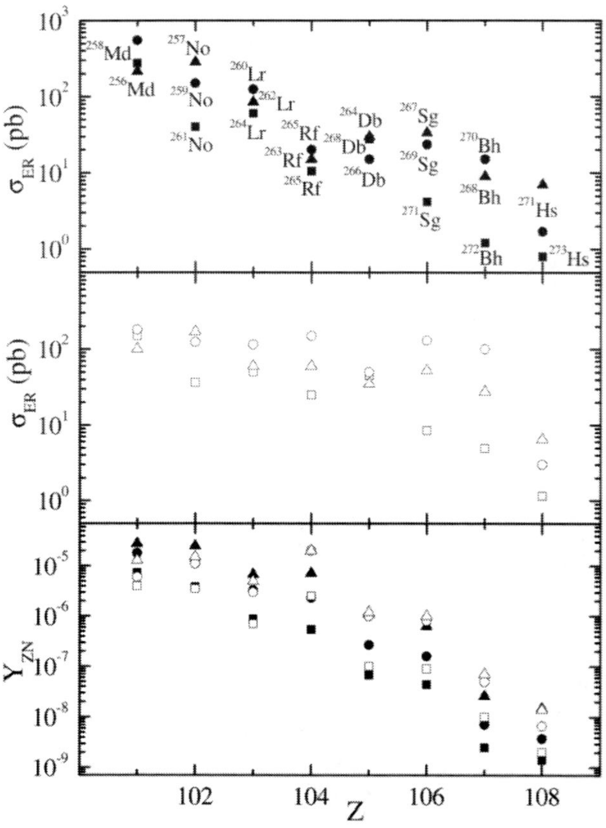

Figure 16. The calculated primary yields $Y_{Z,N}$ (lower part) and evaporation residue cross sections σ_{ER} (middle and upper parts) are shown by triangles, circles and squares for the reactions ^{48}Ca + ^{244,246,248}Cm ($E_{\rm c.m.}$=207, 205.5 and 204 MeV, respectively). The heavy fragments after $1n$ evaporation are indicated in the upper part of the figure. The results obtained with (9) and (3) are indicated by closed and open symbols, respectively.

value of shell correction on the average excitation energy, which is easily calculated, and (Z, N) of the fissioning nucleus.

In the complete fusion-fission reactions it is difficult to separate the true fission fragments from the quasifission fragments. This problem is absent in the transfer-induced fission reactions. Therefore, comparing mass distributions of fragments of the same fissioning nucleus produced in the complete fusion-fission and transfer-induced fission reactions, one can distinguish the quasifission products.

Acknowledgments

We thank S.Heinz, S.Szilner, and L.Corradi for fruitful discussions. This work was supported in part by DFG and RFBR. The IN2P3 (France)-JINR (Dubna) and Polish - JINR (Dubna) Cooperation Programmes are gratefully acknowledged as well.

References

[1] Guillemaud-Mueller D *et al.* (1989) *Z. Phys.* A **332** 189; Guillemaud-Mueller D *et al.* (1990) *Phys. Rev.* C **41** 937.
[2] Mocko M *et al.* (2006) *Phys. Rev.* C **74** 054612
[3] Tarasov O B *et al.* (2007) *Phys. Rev.* C **75** 064613; Tarasov O B *et al.* (2009) *Phys. Rev. Lett.* **102** 142501

[4] Tsang M B *et al. Phys. Rev.* C **76** 041302(R)
[5] Baumann T *et al.* (2007) *Nature* **449** 1022
[6] Adamian G G, Antonenko N V, Lukyanov S M and Penionzhkevich Yu E (2008) *Phys. Rev.* C **78** 024613
[7] Lukyanov S M *et al.* (2009) *Phys. Rev.* C **80** 014609.
[8] Antonenko N V, Nasirov A K, Shneidman T M and Toneev V D (1998) *Phys. Rev.* C **57** 1832.
[9] Corradi L *et al.* (1999) *Phys. Rev.* C **59** 261
[10] Corradi L *et al.* (2002) *Phys. Rev.* C **66** 024606; Corradi L, Pollarolo G and Szilner S (2009) *J. Phys.* G **36** 113101
[11] Fornal B *et al.* (2004) *Phys. Rev.* C **70** 064304
[12] Penionzhkevich Yu E, Adamian G G and Antonenko N V (2005) *Phys. Lett.* B **621** 119; (2006) *Eur. Phys. J.* A **27** 187
[13] Rejmund M *et al.* (2007) *Phys. Rev.* C **76** 021304(R)
[14] Lunardi S (2009) *AIP Conference Proceedings* vol. **1120** 70
[15] Volkov V V (1978) *Phys. Rep.* **44** 93
[16] Schröder W -U and Huizenga J R *Treatise on Heavy-Ion Science*, edited by D.A. Bromley, Vol. 2 (Plenum Press, New York, 1984) p.115
[17] Volkov V V *Treatise on Heavy-Ion Science*, edited D.A. Bromley, Vol. 8 (Plenum Press, New York, 1989) p.255
[18] Bock R *et al.* (1982) *Nucl. Phys.* A **388** 334; Zheng Z *et al.* (1984) *Nucl. Phys.* A **422** 447; Guarino G *et al.* (1984) *Nucl. Phys.* A **424** 157; Toke J *et al.* (1985) *Nucl. Phys.* A **440** 327; Shen W Q *et al.* (1987) *Phys. Rev.* C **36** 115
[19] Gäggeler H *et al.* (1986) *Phys. Rev.* C **33** 1983; Hoffman D C *et al.* (1985) *Phys. Rev.* C **31** 1763; Türler A *et al.* (1992) *Phys. Rev.* C **46** 1364
[20] von Oertzen W *et al.* (1987) *Z. Phys.* A **326** 463; Künkel R *et al.* (1990) *Z. Phys.* A **336** 71; Speer J *et al.* (1991) *Phys. Lett.* B **259** 422
[21] de Souza R T, Huizenga J R and Schröder W -U (1988) *Phys. Rev.* C **37** 1901; de Souza R T, Schröder W -U, Huizenga J R, Töke J, Datta S S and Wile J L (1989) *Phys. Rev.* C **39** 114
[22] Broda R *et al.* (1990) *Phys. Lett.* B **251** 245
[23] Scherer U W *et al.* (1990) *Z. Phys.* A **335** 421
[24] Gippner P *et al.* (1990) *Phys. Lett.* B **252** 198
[25] Oganessian Yu Ts (2007) *J. Phys.* G **34** R165
[26] Loveland W *et al.* (2002) *Phys. Rev.* C **66** 044617; Gregorich K E *et al.* (2005) *Phys. Rev.* C **72** 014605
[27] Yakushev A B *et al.* (2002) *Radiochim. Acta* **91** 443; Eichler R *et al.* (2007) *Nature* **447** 72
[28] Hofmann S *et al.* (2007) *Eur. Phys. J.* A **32** 251; Hofmann S (2009) *Lect. Notes Phys.* **764** 203
[29] Adamian G G *et al.* (1996) *Int. J. Mod. Phys.* E **5** 191
[30] Adamian G G, Nasirov A K, Antonenko N V and Jolos R V (1994) *Phys. Part. Nucl.* **25** 583
[31] Volkov V V (1986) *Izv. AN SSSR ser. fiz.* **50** 1879
[32] Adamian G G, Antonenko N V and Scheid W (1997) *Nucl. Phys.* A **618** 176; Adamian G G, Antonenko N V, Scheid W and Volkov V V (1997) *Nucl. Phys.* A **627** 361; Adamian G G, Antonenko N V, Scheid W and Volkov V V (1998) *Nucl. Phys.* A **633** 409
[33] Adamian G G, Antonenko N V and Scheid W (2000) *Nucl. Phys.* A **678** 24
[34] Adamian G G, Antonenko N V and Scheid W (2003) *Phys. Rev.* C **68**, 034601
[35] Heinz S *et al.* (2010) *Eur. Phys. J.* A **43** 181
[36] Nasirov A K *et al.* (2009) *Phys. Rev.* C **79** 024606
[37] Fazio G *et al.* (2005) *Phys. Rev.* C **72** 064614
[38] Adamian G G, Antonenko N V and Sargsyan V V (2009) *Phys. Rev.* C **79** 054608
[39] Raman S, Nestor C W and Tikkanen P (2001) *At. Data Nucl. Data Tables* **78** 1
[40] Adamian G G, Antonenko N V and Zubov A S (2005) *Phys. Rev.* C **71** 034603
[41] Itkis M G *et al.* (1999) *JINR, Report* No. E15-99-248; *Proceedings of the Seventh International Conference on Clustering Aspects of Nuclear Structure and Dynamics* edited by Korolija M, Basrak Z and Caplar R (World Scientific, Singapore, 2000) p. 386; *Proceedings of the International Symposium on Exotic Nuclei* edited by Penionzhkevich Yu E and Cherepanov E A (World Scientific, Singapore, 2001) p.143; *Proceedings of the International Conference on Nuclear Physics at Border Lines* edited by Fazio G *et al.* (World Scientific, Singapore, 2002) p.146; Itkis I M *et al. Proceedings of the International Conference on Nuclear Physics at Border Lines* edited by Fazio G *et al.* (World Scientific, Singapore, 2002) p.142; Kozulin E M *et al.* ibid. p.157
[42] P. Möller *et al.* (1995) *At. Data Nucl. Data Tables* **59** 185
[43] Kratz J V, Norris A E and Seaborg G T (1974) *Phys. Rev. Lett.* **33** 502

Experimental Signatures of Quasifission

Birger B. Back
Argonne National Laboratory, 9700 S. Cass Avenue, Argonne, IL 60439, USA
E-mail: back@anl.gov

Abstract. Three decades ago, it was first recognized that the observation of fission fragments in heavy-ion induced reactions does not necessarily mean that they originate from the fission decay of a compound nucleus formed by the fusion of the projectile and the target nuclei. This conclusion was based on several different observations. First, it was recognized that the fission cross section exceeded the upper bound imposed by the existence of a stabilizing pocket in the ion-ion potential thus some of the fission cross section must originate from partial waves that do not proceed through a compound nucleus and, in addition, that the fission mass distribution in these cases was observed to be wider than expected on the basis of a compound nucleus model. Concurrently, it was noted that the fission fragment anisotropy in heavy-ion induced fission substantially exceed expectations based on the transition state model. Subsequent studies of the two-dimensional mass-angle distribution of fission fragments clearly demonstrated that these fragments are the result of a dynamic process, in which the system evolves toward mass symmetry on a time scale that is commensurate with the rotational period of the complex. This process is now referred to as "quasi-fission" although the terms "fission without a barrier" and fast fission were also used. Recently, much progress in the theoretical description of this process has been achieved and further precise experiments have been conducted, which provide further constraints on our understanding of these complex processes that also play a critical role in attempts to synthesize heavy and super-heavy nuclei via heavy-ion fusion processes. In this talk I will discuss some of the history and recent developments in the study of the quasi-fission process.

1. A little history

The history of fission research goes back more than seventy years with the discovery of this new, unexpected and very spectacular process, in which an atomic divides into two almost equal size parts or fragments. The discovery is attributed to Hahn and Strassmann [1]. Science historians would, however, argue that also Lise Meitner made decisive contributions to this work, but that she was not given due credit because of the societal and political situation in Germany at the time. In retrospect, earlier hints of the fission process were probably missed in work by Enrico Fermi *et al.* at Rome [2], who were disturbed by some large "noise" pulses in his Geiger-Müller tube when bombarding uranium with slow neutrons, but as experimenters can testify, aluminum foil can very often get rid of such large electronic noise, and this method also worked for Fermi when he inserted a this Al foil between the uranium and the detector. With the benefit of hindsight it is evident that the large "noise" pulses were probably caused by fission fragments reaching his detector. However, Ida Noddack [3] did suggest the revolutionary idea that the observation of reaction products with barium-like chemistry could indeed be Ba that was created by splitting the uranium nucleus into two large, almost equal fragments, but

Journal of Physics: Conference Series **282** (2011) 012003 doi:10.1088/1742-6596/282/1/012003

the idea was too outrageous to get serious consideration at the time and her suggestion was essentially ignored.

When Hahn and Strassmann again observed Ba products in neutron irradiated uranium [1], and the news of the results reached Lise Meitner and her nephew Otto Frisch, who were both in Sweden at the time, they quickly understood the implications of the experimental observation and concluded that what was seen was a process in which the uranium nucleus splits in half by neutron bombardment. Meitner and Frisch quickly published a letter in Nature [4] giving this explanation of the Berlin group's results. In addition, within a couple of weeks of learning about the Hahn and Strassmann results, Otto Frisch was able to confirm the results [5] when he repeated the experiment in Stockholm. In the meantime, Niels Bohr learned about the new process from Lise Meitner over the Christmas holidays in 1938, just in time before he departed for a longer trip to the United States in January 1939. During his stay at Princeton University he devoted his time to understand and describe the fission process in detail in collaboration with John A. Wheeler. This collaboration resulted in the famous Physical Review article published in June 1939 [6], only seven months after the discovery, that gives a detailed exposition and understanding of this newly discovered process - a remarkable feat. Figure 1 shows the potential energy surface for ^{235}U and the fission barrier that they calculated for this system.

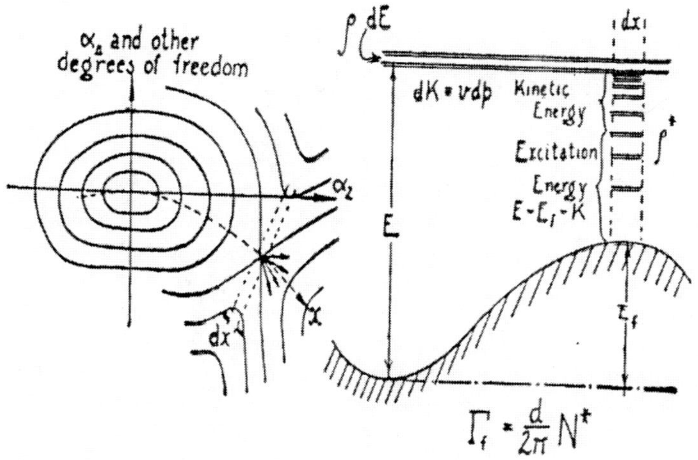

Figure 1. In this figure from Ref. [6] the potential energy surface obtained in the liquid drop model is shown along with the resulting fission barrier.

If is difficult to overestimate the consequences of the discovery of fission. Of course, it has led to a long and very fruitful research field in nuclear physics studies that we all love and which is continuing at this symposium. But it also launched nuclear physics into the public eye because of the far-reaching implications for the nuclear chain reaction in electric power production and nuclear explosions. Hardly a day goes by that nuclear fission is not the center focus of a news story in terms of either of these applications, even more than seventy years hence. But in some way, I believe that the whole field of nuclear physics, no matter how removed from studies of the fission process, has benefitted from this discovery in terms of justifying the continued study of this field, also the branches that do not have applications that are of interest for society as a whole.

After this brief diversion, I should get back on track and discuss the topic of "quasi-fission" and the experimental signatures of this process as the title of my talk promises. In the following, I will therefore touch on some of the early indications that the fission fragments seen in heavy-ion bombardments originate from a slightly different process, although several properties do justify the "fission" part of the name.

2. Large fission cross sections and wide mass distributions

One of the first indications that fission fragments can come from non-equilibrium fission were seen in the work of Heusch *et al.* in the reaction of ^{132}Xe+^{56}Fe at 5.73 MeV/u [7]. They found a fission-like cross section of 1040 mb, which corresponds to a maximum compound nuclear spin of 120 $\hbar$ (in the sharp cutoff model). This agrees well with fusion models [8], [9] assuming that fission exhausts the total fusion cross section. But they also realized that not all of these reactions could come from an equilibrated compound nucleus since the fission barrier disappears for spins larger than about 72 $\hbar$ according to the Rotating Liquid Drop Model [10]. Therefore, the fission-like processes originating from total angular momenta in the range $72\hbar < l < 120\hbar$ must be characterized as "fission without a barrier" In the paper they state that:

"The strongly damped component which constitutes the major part of the reaction cross section exhibits characteristics of a fusion-fission reaction with typical fission fragment kinetic energies and 1/sinθ angular distributions... We tentatively conclude that an essential part of the fully damped cross section originates from partial waves for which the compound nucleus has no fission barrier".

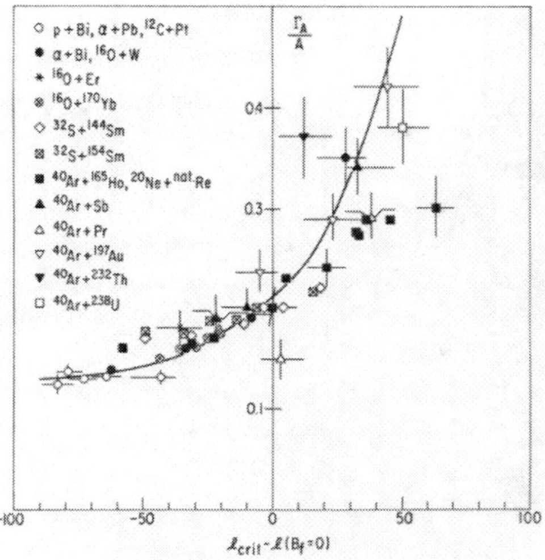

Figure 2. Systematics of the width of fission fragment mass distribution obtained in different systems. The data are shown as a function of the difference between the critical angular momentum for fusion, l_{crit} and the angular momentum for which the fission barrier vanishes $l_{Bf=0}$, such that the "fission without a barrier" contributes for positive values of the abscissa.

In addition it was noticed that the width of the mass distribution of fission fragments increased strongly for the partial waves where the fission barrier vanishes. Lebrun *et al.* [11] studied fission mass distributions of ^{20}Ne+^{nat}Re and ^{40}Ar+^{165}Ho reactions, each at two bombarding energies, and they found that the width of the fragment mass distributions increased noticeably with angular momentum and that it appeared to widen considerably above the partial wave for which the fission barrier vanishes. Later, Borderie *et al.* [12] reviewed a larger data set and found this conclusion fully supported and concluded that "fast fission" occurs for l-values, where the fission barrier vanishes, see Fig. 2.

Further evidence was found by Shen *et al.* [13] who analyzed the mass width of fission-like reaction products seen in U induced reactions and showed that, while the expected width was seen for reactions with ^{16}O and ^{24}Mg, heavier targets lead to wider mass distributions. The width of the mass distribution may be estimated in a thermal model assuming a quadratic mass asymmetry potential

$$U(A) = \frac{1}{2}k(A - A_s)^2, \tag{1}$$

where A_s is the symmetric mass, and k is the stiffness coefficient, which based on theoretical liquid drop model estimates [14] is about k =0.0035 MeV/u^2. At a temperature T one obtains a variance of the mass distribution of

$$\sigma_A^2 = \frac{T}{k} = \frac{1}{k}\sqrt{\frac{8.5E^+}{A}}, \tag{2}$$

where E^+ is the nuclear excitation energy at scission given by $E^+ = E_{exc} + Q_{sym} - E_K - E_{def} - E_{rot}$. Here E_{exc} is the excitation energy of the compound system, Q_{sym} is the Q-value for symmetric fission, E_K is the total kinetic energy estimated from the Viola systematics [15], E_{def} accounts for the fragment deformation energy taken to be 12 MeV, and E_{rot} is the rotational energy at scission. One thus expects only a weak (fourth root) increase in the mass width as a function of excitation energy. This dependence is found to agree well with data, as demonstrated for the reactions $^{238}U+^{16}O$, ^{26}Mg [13] and for $^{32}S+^{182}W$ and $^{48}Ti+^{166}Er$ [16], see Fig. 3 left panel, but a somewhat larger stiffness coefficient of k=0.0048 MeV/u is needed to reproduce the data.

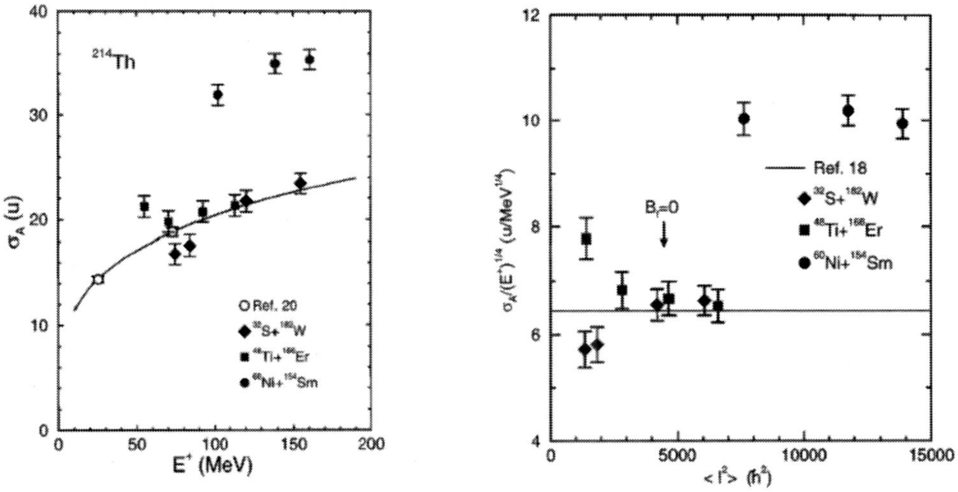

Figure 3. Left panel: The standard deviation of fission fragment mass distributions are shown as a function of the available excitation energy at the scission point, E^+ for reactions of $^{32}S+^{182}W$, $^{48}Ti+^{166}Er$, and $^{60}Ni+^{154}Sm$, all leading to the compound system ^{214}Th, from [16] (solid symbols) and ^{16}O and $^{12}C+^{208}Pb$,[17](open circles). The solid curve represents Eq.2 using a value of k=0.0048 MeV/u^2 obtained by fitting the ^{16}O and $^{12}C+^{208}Pb$ data. Right panel: The excitation energy dependence has been removed and the data are shown as a function of the variance of the spin distribution.

One observes that the ^{60}Ni reaction, also leading to the ^{214}Th compound system, has substantially wider fission mass distributions. However, since a heavier projectile is used in this case, one may argue that this increase is caused by probing larger angular momenta in the fission process. To test this hypothesis one may remove the excitation energy dependence by dividing with $(E^+)^{1/4}$ and display the results as a function of the mean squared angular momentum, $\langle I^2 \rangle$, as illustrated in the right panel of Fig. 3. Using this analysis, it becomes evident that the main factor determining the large mass widths seen for ^{60}Ni+^{154}Sm is most likely associated with the heavier projectile mass used in this system.

The conclusion from this analysis appears to be that there is good agreement with expectations based on thermal excitations in the mass asymmetry mode at scission for reactions induced by relatively light ions. For heavier projectiles, this thermal model fails and the observed large mass widths point to a dynamical origin of this effect, one that we now associate with the quasi-fission process.

3. Fission angular distributions

The angular distributions of fission fragments from excited nuclei is expected to be determined at the transition state in the process, namely the saddle point, where the orientation of the nuclear symmetry axis is given by statistical arguments following Halpern and Strutinsky [18]. In this formulation, the angular distribution may be written

$$W(\theta) = \sum_I (2I+1) T_I \sum_K \rho_I(K) \left| d^I_{0K}(\theta) \right|^2, \tag{3}$$

where I is the spin of the nucleus, K its projection onto the symmetry axis, T_I is the transmission coefficient for fusion with spin I, $\rho_I(K)$ is the probability of spin projection K for total spin I, and d^I_{0K} is the d-function. The level density at the saddle point may be written $\rho(E, I, K) \propto e^{(E-B_f-E_{rot})/T}$ where the rotational energy, E_{rot} is

$$E_{rot} = \frac{\hbar^2}{2J_\parallel}K^2 + \frac{\hbar^2}{2J_\perp}(I(I+1) - K^2) = \frac{\hbar^2}{2J_\perp}I(I+1) + \frac{\hbar^2}{2}K^2\left(\frac{1}{J_\parallel} - \frac{1}{J_\perp}\right), \tag{4}$$

such that

$$\rho(K) \propto e^{-K^2/2K_0^2}, \tag{5}$$

where

$$K_0^2 = \frac{T}{\hbar^2}J_{eff}; \frac{1}{J_{eff}} = \frac{1}{J_\parallel} - \frac{1}{J_\perp}. \tag{6}$$

Measurements of the fission anisotropy therefore gives information about the moments of inertia *i.e.* the deformation of the saddle point configuration. To a good approximation, the fragment anisotropy, $W(0°)/W(90°)$, is related to the variance of the K-distribution, K_0^2, and the mean square of the spin distribution according to

$$\frac{W(0°)}{W(90°} \approx 1 + \frac{\langle I^2 \rangle}{4K_0^2}. \tag{7}$$

The Halpern-Strutinsky model is based on the assumption that the angular distribution is determined at the saddle point, but in the early days it was also proposed that the scission point could take this role. However, a systematic set of data obtained by Reising *et al.* [19] settled this question, at least for light-ion induced reactions. They used 42.3 MeV α-particles to bombard a wide range of targets from ^{197}Au to ^{249}Cf and derived transition state shapes using the Halpern-Strutinsky model. The results, represented in terms of the parameter J_0/J_{eff}, where J_0 is the moment of inertia corresponding to the spherical shape, are given in Fig. 4 (open circles) as a function of the fissility parameter, x. The solid curve represents the expectation on the basis of the liquid drop model, namely that the deformation of the saddle point increases with x up to the Businaro-Gallone point at about x=0.7 whereafter it assumes less elongated shapes until it

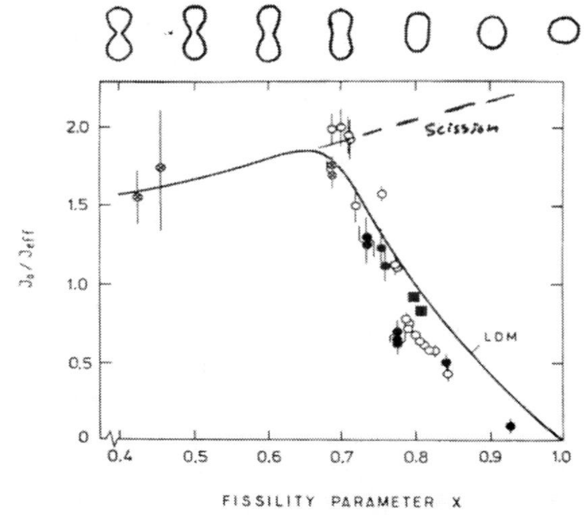

Figure 4. The saddle point shape, given by J_0/J_{eff} as predicted by the liquid drop model, is shown as the solid curve as a function of the fissility, x and compared with experimental data.

reaches sphericity at x=1. Here, at the spherical shape, the repulsive Coulomb forces are exactly balanced by the nuclear attractive forces represented by the surface tension in the liquid drop model and the fission barrier vanishes. It is clear that the equilibrium shape, that is reflected in the fission angular distributions, coincide with the saddle point as predicted by the liquid drop model and not at the scission shape represented by a dashed line for x >0.7.

It is, however, interesting to ask what the fission anisotropy would be for systems with even higher fissility, x=1 or above. Taken at face value, the K-distribution should become flat such that no fission direction relative to the spin axis would be preferred resulting in a flat fission

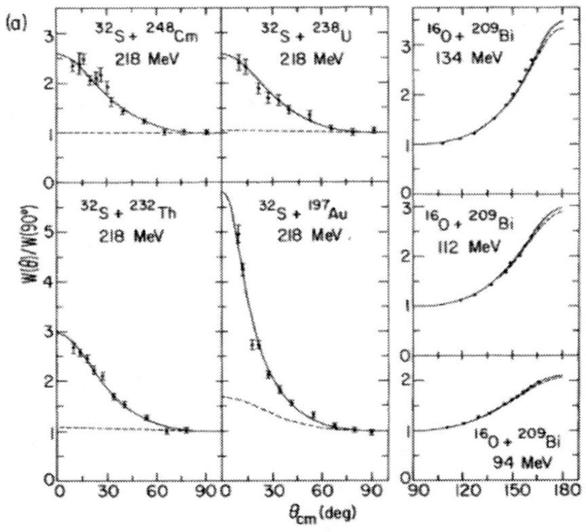

Figure 5. The angular distributions for 218 MeV ^{32}S-induced fission on targets of ^{197}Au, ^{232}Th, ^{238}U, and ^{248}Cm [20] are shown in the left part of the figure. Solid curves are the optimal fits to the data using Eq. 3 whereas the dashed curves are expected from the liquid drop model saddle shapes. Data for the ^{16}O+^{209}Bi [21] system are shown in the right panels.

angular distribution. On the other hand, it seems unlikely that the system would not be able to re-adjust its orientation to minimize its rotational energy during the descent from saddle to scission, which for such heavy systems corresponds to a substantial change in the nuclear shape. In order to address this question, we studied four reactions using ^{32}S beams at 218 MeV on targets of ^{197}Au, ^{232}Th, ^{238}U, and ^{248}Cm with fissilities of x= 0.817, 0.899, 0.914, and 0.948, respectively, thus approaching the critical value of x=1 [20]. The resulting angular distributions are shown in Fig.5. Moreover, recent studies of anisotropies at near and below barrier energies in the ^{32}S+^{232}Th system by Hinde *et al.* [22] also lead to the conclusion that this system bypasses the compound nucleus as well as studies of the ^{32}S+^{184}W reaction by Zhang *et al.*[23], which find a substantial quasifission contribution.

In these ^{32}S-induced reaction we clearly see a strong deviation from the naive expectation based on the saddle point model (dashed curves), whereas ^{16}O-induced fission on a ^{209}Bi target [21], a system with a slightly lower fissility of x=0.774, appears to agree well with the theory.

It should be mentioned here that at the time these data were originally published, alternate explanations for the enhanced anisotropy in some heavy-ion induced reactions were put forward. These explanations were focused on a proposed failure of the saddle point model and suggested that the angular distributions were determined closer to or at the scission point [24, 25]. Although models of this type are able to explain a small subset of the data, it was shown [26] that scission-type models are not able to reproduce the overall systematic trend of a larger set of data.

To investigate the role of angular momentum, we carried out experiments that compare the ^{16}O+^{238}U and ^{32}S+^{208}Pb reactions [27], which lead to compound systems with almost identical fissility of x=0.84 and they should therefore have the same saddle point properties. The results are given in Fig. 6 where the angular distributions are shown on the left at two energies for each system, such that a similar range of angular momenta are populated. One observes that while the ^{16}O+^{238}U data are quite well represented by the liquid drop model predictions for the saddle point shape, this is not the case for ^{32}S+^{208}Pb. Here, the observed fission anisotropies are substantially larger than expected, even at beam energies that populate the same range of angular moments. This is also illustrated in the right side of Fig. 6 where the values of J_0/J_{eff}, derived from the fission angular distributions, are shown as a function of the mean square of the spin distribution of the fissioning system, $\langle I^2 \rangle$ in comparison with the rotating liquid drop model expectation. One again observes that the ^{32}S induced reaction lead to values of J_0/J_{eff} that are substantially larger than expected from theory [10].

From these examples, and numerous others, the following conclusion can be drawn: 1) For fission induced by light ions and heavy ions up to perhaps mass 20-24, the saddle point model gives a good account of the observed fission angular distributions, but for heavier projectiles one finds that the experimental anisotropies are too large. 2) Comparisons of systems populating compound nuclei with essentially identical fission properties (fissility) and angular momenta show that the deviation from the saddle point model prediction appears to be associated exclusively with the smaller mass asymmetry in the entrance channel. This last observation excludes any explanation that involves the formation of an equilibrated compound system as an intermediate step in the process, and one is forced to conclude that the systems with a heavy projectile proceed through a dynamical process that fails to go through the compound nucleus formation. This conclusion is therefore in full agreement with those obtained form the analysis of fission cross sections and the widths of mass distributions mentioned above. Although these conclusions are quite solid, one must keep in mind that they are, to some extent, based on comparisons with theoretical models. In the following, I will therefore discuss other experimental evidence that is model independent and show that the quasifission process plays an large role in the processes that occur in heavy-ion bombardments.

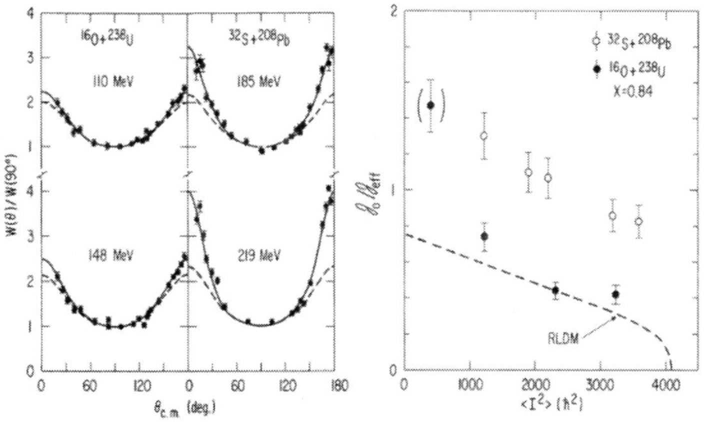

Figure 6. Fission angular distributions (left panel) and the saddle point shapes J_0/J_{eff} (right panel derived from the data for the reactions $^{16}O+^{238}U$ and $^{32}S+^{208}Pb$ are shown and compared to rotating liquid drop model predictions (dashed curves).

4. Total kinetic energy

Much of the early work on quasifission was derived from experiments performed at GSI in the early and middle 1980's [28, 29, 30, 13, 31]. The availability of Pb and U beams at energies extending up to almost twice above Coulomb barrier allowed for measurements in inverse kinematics where both fragments were detected in forward angles and fragment mass,

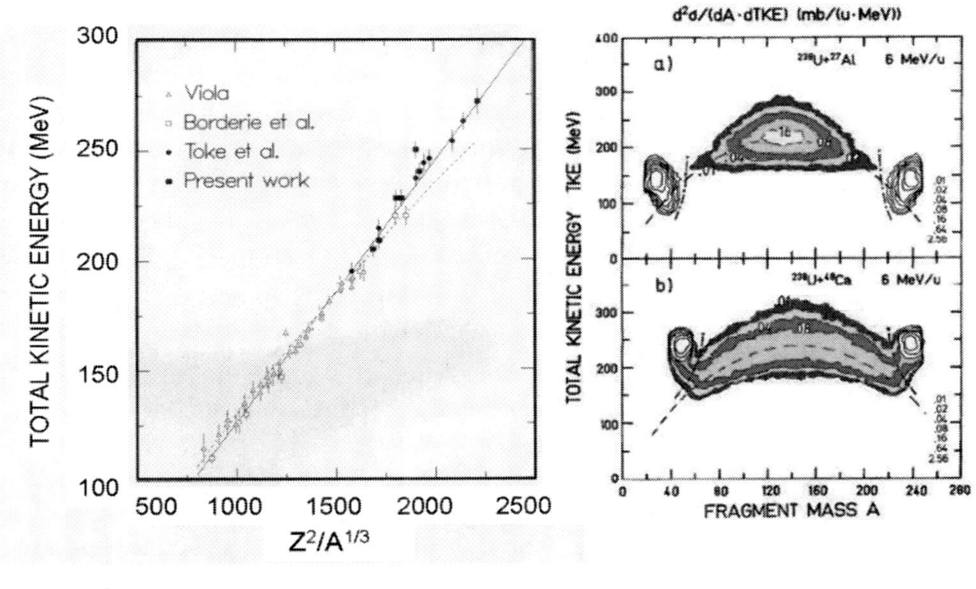

Figure 7. Left panel: The total kinetic energy of fission fragments from a wide range of nuclei are shown as a function of $Z^2/A^{1/3}$ of the fissioning system. The solid symbols are from systems studied in Refs. [29, 13]. Right panel: Cross section contours of the total kinetic energy are shown as a function of fragment mass for $^{238}U+^{27}Al$ (upper panel) and $^{238}U+^{48}Ca$ (lower panel) at E_{beam} = 6 MeV/u [29]. The dashed curves represent the Coulomb repulsion between the fragments at scission consistent with the Viola systematics at symmetry.

total kinetic energy and scattering angle could be derived from the recorded position and time of arrival of the fragments in large-area fast gas-filled detectors. These measurements provided an extended data set which I would like to discuss here.

First, let's examine the total kinetic energy of the fission-like fragments seen in these studies. Viola et al. [15] have performed a comprehensive study of fission fragment kinetic energy for a wide range of systems and found that data for symmetric mass splits may be described by a simple analytic expression that essentially reflect the Coulomb repulsion between the deformed fragments at scission. In the GSI studies, this data set was extended substantially toward heavier systems and found to agree well with the Viola systematics. For the heavier targets used in these experiments, the main contribution to the fission-like fragments clearly come from quasifission processes. This is illustrated in the left panel of Fig. 7 where the open symbols were from Viola's previous systematics and the solid points were obtained in the GSI work, mostly representing quasifission. Also the fragment mass dependence of the total kinetic energy is well described on the basis of the Coulomb repulsion, both for mostly fusion-fission reactions in $^{238}U+^{27}Al$ (right top panel in Fig. 7 as well as for the wide fragment mass distribution seen in $^{238}U+^{48}Ca$ (right, lower panel). It is therefore important to realize that the total kinetic energy does not separate compound fission and quasifission.

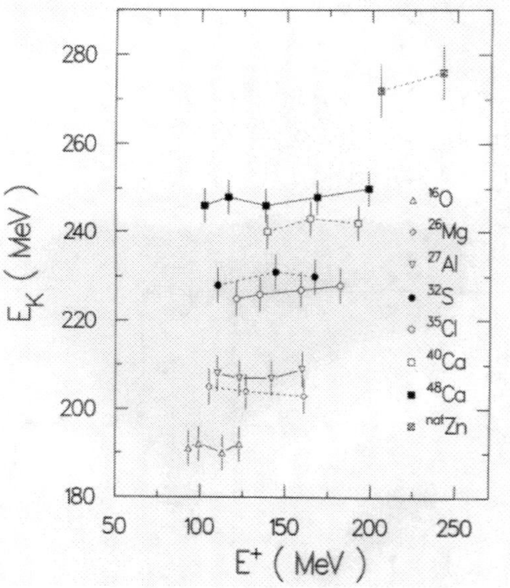

Figure 8. The average kinetic energy for symmetric mass division observed in ^{238}U bombardment of various targets is given as a function of the excitation energy at the scission point, E^+ [13].

It is also interesting to examine the total kinetic energy as a function of the excitation energy of the system. This is done in Fig. 8 where the total kinetic energy, E_K, is shown as a function of the excitation energy, E^+, at the scission point (see Eq. 2 for definition) for the reactions studied with ^{238}U beams in Ref. [13]. It is interesting to note that the increase in total available energy associated with increased beam energy is not reflected in the total kinetic of the final fragments indicating that the motion toward scission is almost completely damped, or in other words, that the pre-scission kinetic energy appears to be small. A further analysis of the pre-scission kinetic energy may be found in Ref. [13].

5. Mass-angle correlations

Niels Bohr's [32] compound nucleus model is based on the assumption that formation and decay of the intermediate system are decoupled and that only certain quantities, such as total angular momentum, linear momentum and energy are conserved during the process. In fission decays of the compound nucleus it is thus required that there is no preference for forward or backward emission of fragments, because any information about the initial direction and mass asymmetry is lost during the intermediate compound nucleus step. Experimentally this means that, in a two-body exit channel, such as fission, the mass distribution must be symmetric (about A_s) for any center of mass angle and that the angular distributions must be forward-backward symmetric (about θ_{cm}=90°) for any fragment mass. Note that for two-body exit channels, one of these requirements implies the other. These conditions may also be expressed as,

$$\sigma_f(A, \theta_{cm}) = \sigma_f(A, \pi - \theta_{cm}) = \sigma_f(A_{cn} - A, \theta_{cm}) = \sigma_f(A_{cn} - A, \pi - \theta_{cm}), \tag{8}$$

where A_f is the fragment mass, A_{cn} is the total mass, and θ_{cm} is the emission angle in the center-of-mass system. In contrast, for any two-body exit channel one has the less stringent requirement

$$\sigma_f(A, \theta_{cm}) = \sigma_f(A_{cn} - A, \pi - \theta_{cm}). \tag{9}$$

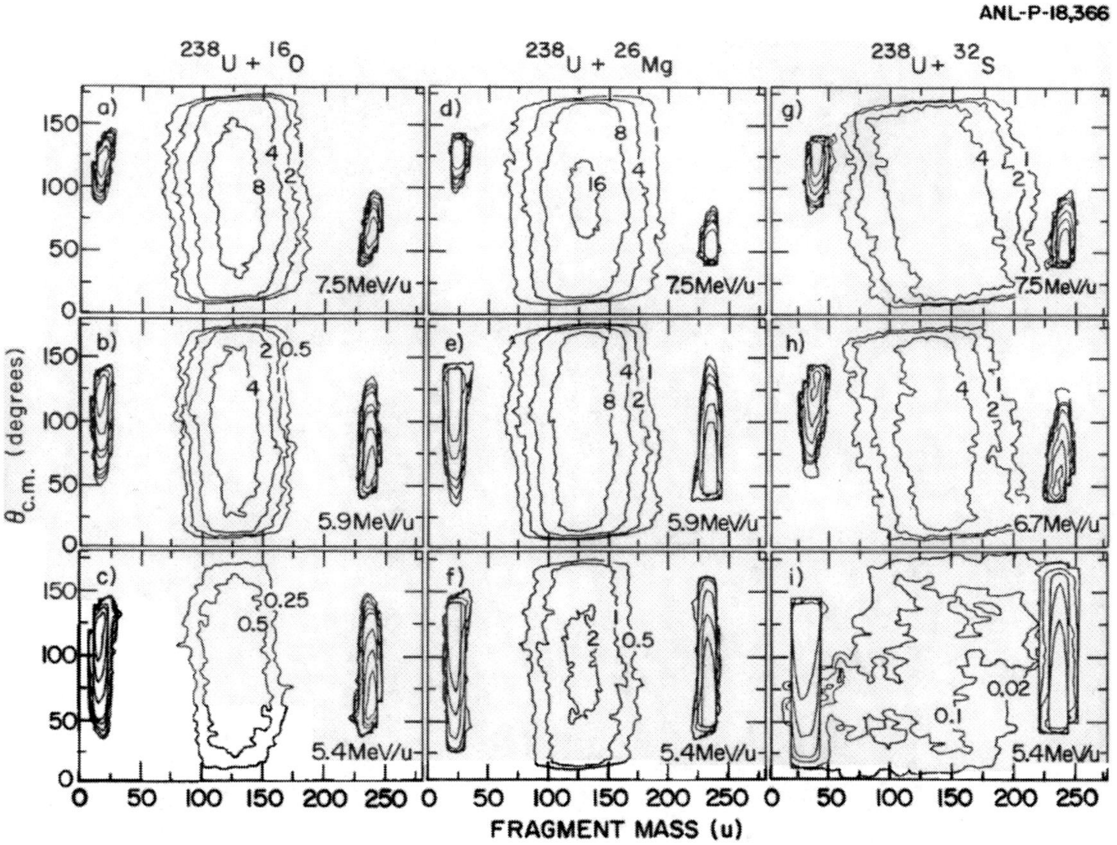

Figure 9. Double differential cross section contours are shown for ^{238}+^{16}O, ^{26}Mg, and ^{32}S in the left, middle and right columns, respectively. The top, middle, and bottom rows correspond to ^{238}U-beam energies of, 7.5, 6.7, and 5.4 MeV/u, respectively [13].

In Fig. 9 the complete mass-angle distributions for reactions of ^{238}U+^{16}O,^{26}Mg, and ^{32}S are shown [13]. Although very symmetric cross section distribution are seen at all three energies for the ^{238}U+^{16}O and probably also for ^{238}U+^{26}Mg it is clear that the ^{238}U+^{32}S mass-angle distribution is skewed. At the lowest energy, two different fission components may even be visible. One may also note that the center-of mass energy that we used in the fission anisotropy study of ^{32}S+^{238}U [20] (E_{cm}=192 MeV) is only about 10% lower than the one corresponding to 7.5 MeV/u ^{238}U beams. Mass-angle distributions for ^{32}S+^{232}Th has also recently been measured at beam energies between E_{beam} = 143-168 MeV at Australian National University [22]. In this work is was concluded that two different quasifission components contribute to the cross section, possibly associated with multi-turn rotation of the intermediate system.

For even heavier targets, it becomes even more clear that a fast, dynamical process is responsible for the fission-like products as shown in Fig. 10. Although these distributions clearly show that the main part of the fission-life cross section comes from quasi-fission, one cannot rule out that a fraction of the cross section leads to the formation of a compound nucleus which may decay by fission or particle-gamma emission.

Let me emphasize that many studies of fission fragment mass measurements fail to expose the asymmetry in the mass-angle correlation by concentrating on the angular region close to θ_{cm} = 90°. Likewise, many fission angular distribution studies do not measure the fragment masses. Because of the two-body nature of quasi-fission, such measurements will always result in mass-symmetric distributions and thus miss this important clue that a short time-scale process

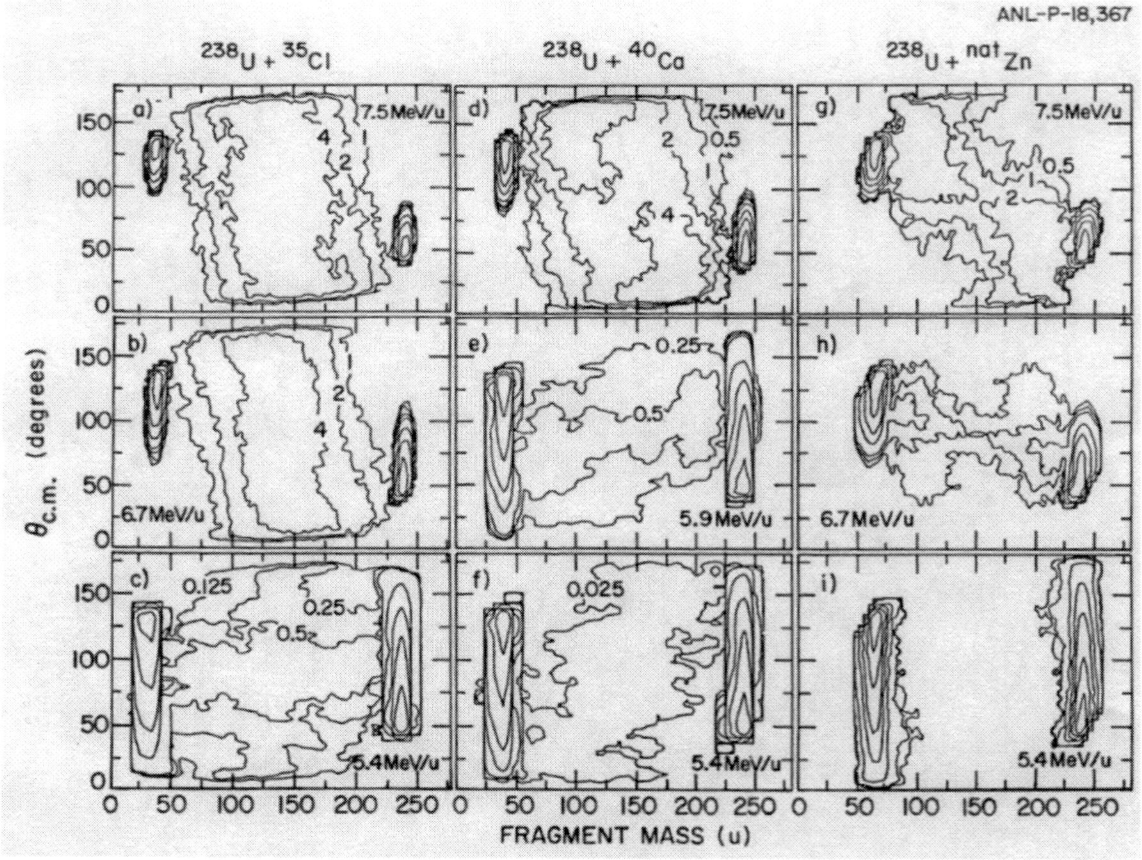

Figure 10. Same as Fig. 9 but for ^{35}Cl, ^{40}Ca, and ^{nat}Zn targets [13].

is observed.

6. Quasi-fission and synthesis of super-heavy elements

In recent years, incontrovertible evidence for the synthesis of heavy elements formed via the so-called "hot fusion" reaction has been found at Dubna and elsewhere [33]. The formation of these elements occurs by several successive small branches of neutron evaporation in competition with fission. So it is clear that some fusion-fission cross section must be present underneath the often much larger quasifission branch. From the fission measurements, one cannot obtain accurate estimates of this fusion-fission branch, but it is possible to set upper limits under the assumption that it follows the expected behavior in terms of width of the mass distribution and angular anisotropy. One such analysis was carried out by us for reactions leading to the ^{214}Th system [34] using three different entrance channels. For the most mass symmetric system ^{60}Ni+^{154}Sm we found that only of the order of 10-15% of the fission-like cross section could come from fusion-fission - however, it could also be substantially smaller. Data analysis of this type has not often been employed, but it may help provide some limits on fission barrier heights in systems populated in the "hot fusion" reactions that are successfully being used for the synthesis of the heaviest elements.

7. Time-scale of quasifission processes

The most striking signature that distinguishes quasifission from normal fusion-fission is the asymmetry in the mass-angle distributions as discussed above. This asymmetry arises because the intermediate system rotates only a finite angle and the mass drift toward symmetry is disrupted when re-separation occurs. In principle, these distributions therefore carry information about the time scale of the process under assumptions about the angular momenta involved and the moment of inertia of the rotating complex. In our work on the GSI data [29], we developed a simple prescription for such an analysis in which the rotation angle $\Delta\theta$ of the intermediate system is derived from the mass-angle distributions using the relation

$$\Delta\theta = \pi - \theta_i - \theta_f - \theta, \tag{10}$$

where θ is the observed scattering angle and θ_i and θ_f are half of the calculated Coulomb deflection angle in the entrance and exit channels. The reaction time may be written as

$$t_{reac} = \Delta\theta/\omega, \tag{11}$$

where $\omega = l/J$ is the angular velocity, l is the angular momentum of the complex and J its moment of inertia. Of course, although l is conserved, one expects that both ω and J varies during the rotation because of shape changes caused mainly by the mass transfer from the light to the heavy reaction partner. In the spirit of simplicity, we take two different estimates for J, either an average of the entrance and exit channel rigid moment of inertia assuming touching rigid spheres, $J = 0.5(J_{in} + J_{out})$ or a value of $J = 1.4J_{sph}$ *i.e.* 40% larger than the rigid sphere value for the total system. In order to perform this analysis, the observed quasifission cross section was divided into three equal size bins and for each bin the rotation angle $\Delta\theta$ and the associated average mass and angular momentum of the system was evaluated as illustrated in Fig. 11 (see Ref. [29] for details).

For each bin, one may compute the average mass transfer ΔA toward symmetry and normalize relative to the maximum drift reaching symmetry, $\Delta A_{max} = |A_p - A_s|$, where A_p is the projectile mass, see Fig 11. Using the data from 6.0 MeV ^{238}U induced reactions of Ref. [29] to extract the quantities $\Delta A/\Delta A_{max}$ and $\langle l \rangle$ and correlating them with the angle of rotation, $\Delta\theta$, and

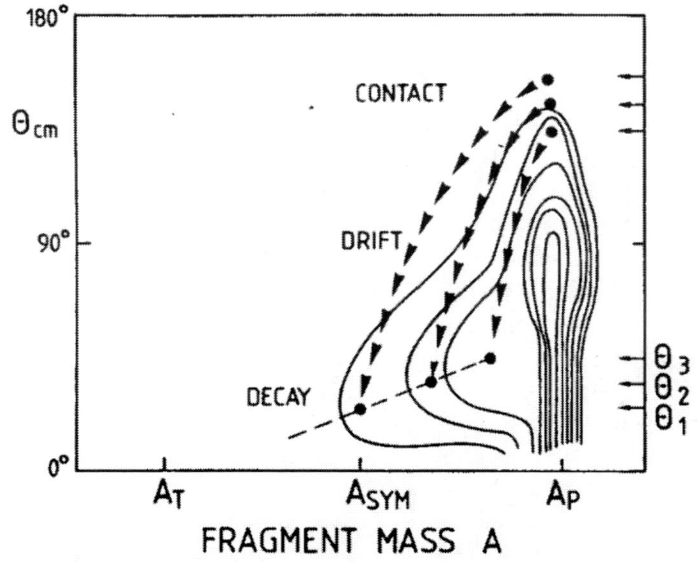

Figure 11. This figure illustrates the correspondance between mass drift toward symmetry and the angle of rotation of the intermediate complex in quasifission. The final emission angles, $\theta_1, \theta_2, \theta_3$ for the three different mass regions are shown. [29].

reaction time, t_{reac}, we see that a tight correlation is obtained for $\Delta A/\Delta A_{max}$ plotted *vs.* t_{reac}. This dependence is well described as an exponential mass drift toward symmetry given by

$$\Delta A/\Delta A_{max} = 1 - \exp[-(t - t_0)/\tau], \quad (12)$$

where t_0 is a short time offset of about $t_0 \sim (1 - 2) \times 10^{-21}$ s before the mass drift starts - maybe to allow the connecting window between the two reaction partners to open, and τ is a characteristic mass drift time of $\tau = 5.3 \times 10^{-21}$ s, which appears to fit the data.

This analysis method was also applied to the larger data set from GSI [13] where it was shown that the mass relaxation time of $\tau \approx 5 \times 10^{-21}$ sec also holds for a wide range of beam energies, E_{beam}=4.6-7.5 MeV/u and therefore appears to be temperature independent.

Theoretically, it is of interest to ask whether this mass drift rate is consistent with theoretical estimates. Such estimates may be based on the assumption of an over-damped motion by balancing the driving force $F_c = dU/d\langle A\rangle = -k(\langle A\rangle - A_s)$ in the mass-asymmetry potential, Eq. 1 with dissipative force, $F_d = m^{-1}d\langle A\rangle/dt$, that acts in the opposite direction. Solving for the time dependence one finds the observed dependence given in Eq. 12, with

$$\tau = \frac{1}{mk}, \quad (13)$$

where m is the mobility and k is the restoring force constant. A theoretical estimate of the mobility may be obtained on the basis of the one-body dissipation mechanism, which arises from particle collisions with the moving nuclear surface [35]. In the application to quasi-fission, the moving surfaces are represented by the increase of the lighter fragment and the shrinking the heavier partner that occurs during the mass transfer phase. Applying this mechanism to the simple geometry of two touching spheres of un-equal size one arrives at the following expression for the inverse mobility [36]

$$\frac{1}{m} = \frac{3}{8}\left(\frac{\pi}{3}\right)^{1/3}\left[\frac{1}{A_1^{2/3}} + \frac{1}{A_2^{2/3}}\right]\hbar \quad (14)$$

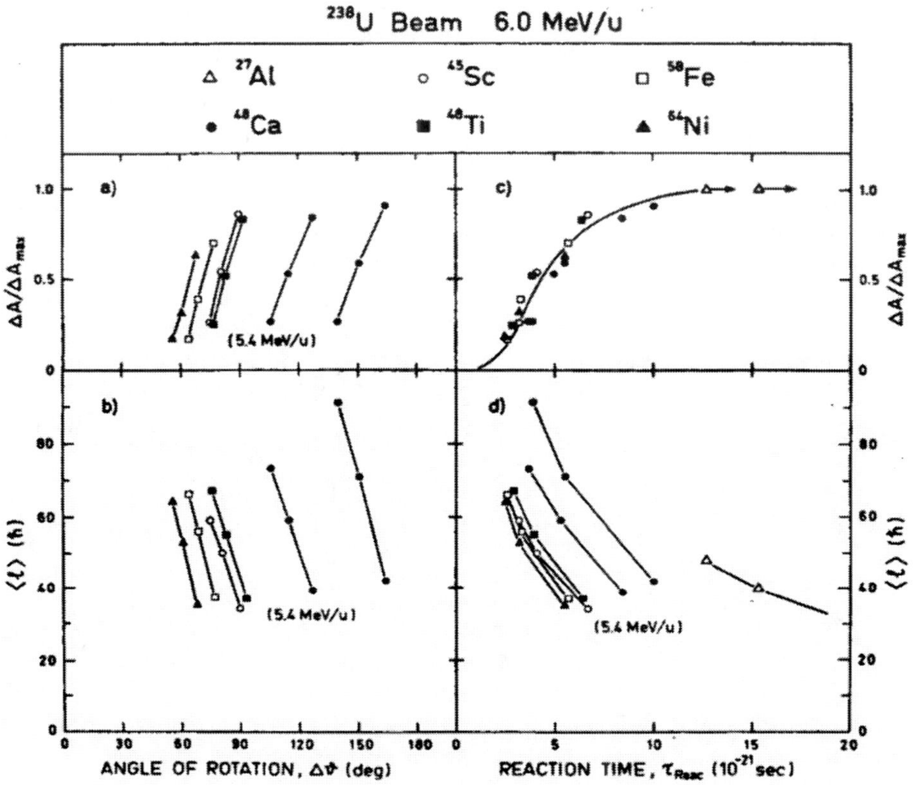

Figure 12. The correlations between the normalized mass drift, angle of rotation, angular momentum, and reaction times for quasifission products formed by 6 MeV/u and 5.4 MeV/u indicated in figure) ^{238}U-beams on various targets are shown [29].

or numerically

$$\frac{1}{m} = 2.5 \times 10^{-22} \left[\frac{1}{A_2^{1/3}} + \frac{1}{A_2^{2/3}} \right] [\text{MeVsec}]. \tag{15}$$

Using a mass asymmetry restoring force coefficient of k=0.04 [MeV/u^2] this results in a characteristic mass relaxation time of τ=4.6$\times$10^{-21} sec for the ^{48}Ca+^{238}U system in the entrance channel and τ=6.4$\times$10^{-21} sec at mass symmetry, which is a very good agreement with the experimental value of τ_{exp}=5.3$\times$10^{-21} sec. Here it should be noted that the dissipation associated with particles crossing the neck that is formed between the two interacting partners has been ignored; see Ref. [13] for further discussion of the validity of this assumption.

From the study of the mass drift and rotation of the intermediate complex in quasifission, we thus find a surprisingly good agreement with the expected energy dissipation strength and lack of temperature dependence predicted by the one-body wall dissipation mechanism. It is, however, clear that this analysis of the data is rather crude and prone to some uncertainty in the derived reaction times. Maybe a direct comparison between the observed mass-angle distribution with those predicted by modern reaction mechanism calculations [37] could provide more stringent tests of the assumed dissipation mechanism.

8. Summary

Even after seventy years of experimental study and theoretical developments, the fission process continues to reveal new and intriguing properties. Both normal compound fission and the more dynamical quasi-fission reaction probe many aspects of nuclear physics from the nuclear structure properties, that gives rise to the shell structure that modify the potential energy landscape, as well as many aspects of the dynamics associated with the drastic re-arrangement of the nuclear shape. In this talk, I have attempted to review the early studies of quasi-fission with emphasis on the experimental signatures for the process. Although much of the early thinking were guided by theories developed by W. Swiatecki and Bjørnholm [38], W. Nörenberg [39], H. Feldmeier [40] and others, I have not attempted to review the theoretical work in this talk but concentrated on the experimental signatures.

We have seen that there are a number of ways, in which quasi-fission distinguishes itself from the standard fusion-fission process. These are all associated with the shorter time scales of the quasifission process and the fact that the intermediate shapes, that the system assumes during the process, do not include the equilibrated compound nucleus captured behind a fission barrier. As we have seen, this has consequences for the fission anisotropies, the width of the mass distributions and most importantly, the mass-angle distributions that directly reveal the short time-scale, dynamical nature of the quasifission process. Indeed, for the heavier systems this feature allows us to study the time evolution of the mass transfer and compare the time scales with simple theoretical estimates, but much work remains to be done experimentally to study the process in greater detail and to develop even more sophisticated theoretical models describing the process.

Acknowledgement

In most of my work on the quasi-fission progress I have benefitted greatly from the mentorship, guidance and close collaboration with Sven Bjørnholm, Niels Bohr Institute, Copenhagen. This work was supported by the U.S. Department of Energy, Office of Nuclear Physics under DE-AC02-06CH1135

References

[1] O. Hahn and F. Strassmann, Naturwissenschaften, **27**, 11 (1939).
[2] E. Fermi, E. Amaldi, O. D'Agostino, F. Rasetti, and E. Segré, La Ricerca Scientifica, **5**, 452, (1934)
[3] Ida Noddack, Zeitschrift fr Angewandte Chemie, **47**, 653(1934)
[4] L. Meitner and O.R. Frisch, Nature, **143**, 239 (1939)
[5] O. R. Frisch, Nature, **143**, 276 (1939)
[6] N. Bohr and J. A. Wheeler, Phys. Rev. **56**, 426 (1939)
[7] B.Heusch *et al.*, Z. Phys. **A288**, 391 (1978)
[8] R. Bass, Phys. Lett. **B47**, 139 (1973); Nucl. Phys. **A231**, 45 (1974)
[9] J .Blocki, J. Randrup, W.J.Sweiatecki, and C.F.Tsang, Ann. Phys. (NY) **105**, 427 (1977)
[10] S.Cohen, F.Plasil and W.J. Swiatecki, Ann. Phys. **82**, 557 (1974)
[11] C. Lebrun *et al.*, Nucl. Phys. **A321**, 207 (1979)
[12] Borderie *et al.*, Z. Phys. **A299**, 263 (1981)
[13] W.Q.Shen *et al.*, Phys. Rev. C **36**, 115 (1987)
[14] P. Möller and J.R. Nix, Nucl. Phys. **A281**, 354 (1977)
[15] V. E. Viola, Jr., Nucl. Data Sect. **A 1**, 391 (1966); V. E. Viola, K. Kwiatkowski, and M. Walker, Phys. Rev. C**31**, 1550 (1985).
[16] B.B.Back *et al.*, Phys. Rev. C**53**, 1734 (1996)
[17] M. G. Itkis *et al.* Yad. Fiz. **52**, 23 (1990) [Sov. J. Nucl. Phys. **52**, 15 (1990)]
[18] I.Halpern and V.M. Strutinski, in Proc. 2nd UN Conf. on Peaceful uses of atomic energy (UN Geneva, Switzerland, 1958), p. 408.
[19] Reising *et al.*, Phys. Rev. **141**, 1161 (1966)
[20] B.B.Back *et al.*, Phys. Rev. Lett. **46**, 1068 (1981)
[21] S.A. Karamyan *et al.*, Yad. Fiz. **6**, 494 (1967) [Sov. J. Nucl. Phys. **6**, 360 (1968)]
[22] D. Hinde *et al.*, Phys. Rev. Lett. **101**, 092701 (2008)

[23] H.Q. Zhang, Phys. Rev. C **81**, 034611 (2010)
[24] P. Bond, Phys. Rev. Lett. **52**, 414 (1984)
[25] H.H.Rosner *et al.* Phys. Rev. Lett. **53**, 38 (1984)
[26] B.B.Back, Phys. Rev. C **31**, 2104 (1985)
[27] B.B.Back *et al.*, Phys. Rev. Lett. **50**, 818 (1983)
[28] R. Bock *et al.* Nucl. Phys. **A388**, 334 (1982); H.Sann *et al.*, Phys. Rev. Lett. **47**, 1248 (1981)
[29] J. Toke *et al.*, Nucl. Phys. **A440**, 327 (1985)
[30] J. Toke *et al.*, Phys. Lett. **B142**, 258 (1984)
[31] B.B.Back *et al.*, Phys. Rev. C**41**, 1495 (1990)
[32] N. Bohr, Nature, **136**, 344 (1936)
[33] Yu. Ts. Oganessian *et al.*, Phys. Rev. Lett. **104**,142502 (2010)
[34] B.B.Back *et al.*, Phys. Rev. C **53**, 1734 (1996)
[35] J. Błocki *et al.*, Ann. Phys. (NY) **113**, 330 (1978)
[36] This derivation is due to R.Donangelo, private communication (1985)
[37] A.K. Nasirov *et al.*, Phys. Rev. **79**, 024606 (2009) and references therein.
[38] W.J. Swiatecki, Proc. Nobel Symp. on nuuclei at very high spin, Phys. Scripta, **24**, 113 (1981); W.J. Swiatecki, Nucl. Phys. **A376**, 275 (1982); S. Bjørnholm and W.J. Swiatecki, Nucl. Phys. **391**, 471 (1982)
[39] W. Cassing and Nörenberg, Nucl. Phys. **A401**, 467 (1983)
[40] H. Feldmeier and H. Spangenberger, Nucl. Phys. **A 428**, 223c (1984)

Non-equilibrium processes in Heavy-ion induced Fission Reactions

R.K.Choudhury and R.G.Thomas
Nuclear Physics Division, Bhabha Atomic Research Centre, Trombay, Mumbai - 400085, India
E-mail: rkc@barc.gov.in

Abstract. Heavy ion induced reactions offer a wide range of possibilities in exploring the collective behavior of nuclei and nuclear processes. Recent years have seen a spurt in the experimental and theoretical studies to understand the mechanism of the fission process in heavy target projectile systems, arising from the interest to produce elements in the superheavy region. A number of new features are observed experimentally with respect to the kinetic energy, mass and angular distributions of fission fragments and their correlations, implying prevalence of non-equilibrium phenomena in the fission process in many target projectile systems. There are many theoretical attempts both in terms of static potential energy considerations and dynamics to understand the important degrees of freedom that govern the evolution of the nuclear system from initial interaction to the final fission stages. Depending on the degree and nature of equilibration, various processes such as fast-fission, quasi-fission and pre-equilibrium fission have been invoked that can compete with the fully equilibrated compound nuclear fission process. In what we call a C-T Fissility plot, we show the remarkable dependence of the onset of these non-equilibrium processes on the entrance channel parameters.

1. Introduction

Heavy ion induced reactions offer a wide range of possibilities in exploring the collective behavior of nuclei and nuclear processes [1]. The time evolution of the composite system formed after the initial interaction, and the parameters on which the dynamics depends are still not fully understood. A systematic study of these reactions can reveal information about the complex processes involved in the heavy ion collisions. Worldwide there are considerable efforts in the synthesis of heavy and superheavy elements using heavy ion collisions [2]. The success of synthesising heavier and heavier elements, therefore depends crucially on the accurate estimation of the non-equilibrium processes which take away the flux from the formation of superheavy evaporation residues.

A major challenge in superheavy element production is the presence of non-equilibrium processes such as quasifission [3, 4, 5, 6]. In terms of reaction time scales, quasifission lies between deep inelastic collisions (DIC) and Compound Nucleus (CN) formation. DIC reactions do not involve capture but represent the energy relaxation mode, exhibiting a wide spectrum of kinetic energy losses while mass distributions show little relaxation that peak around the projectile and target masses. On the other hand, CN formation and subsequent fission and/or ER formation is characterized by full equilibration in all degrees of freedom. Quasifission is a fission like process that proceeds without the formation of a compact mono-nuclear system and is characterized by total energy relaxation and mass division ranging from initial entrance channel mass asymmetry and all the way to symmetry. This process by its very own nature is critically dependent on the entrance channel conditions. The probability of this process in turn determines the probability of formation of evaporation residues in heavy and superheavy systems.

2. Characteristics of equilibrium fission

Equilibrium fission is characterised by the formation of a compound nucleus (CN) which is equilibrated in all degrees of freedom. The mass, energy and angular distribution of fragments are expected to exhibit properties which are independent of the entrance channel [8].

2.1. Mass-TKE distributions

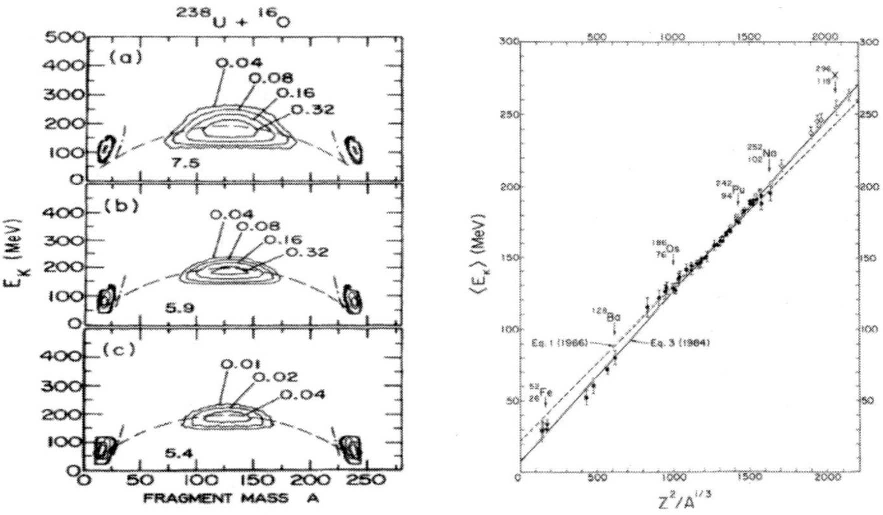

Figure 1. The Mass-TKE correlation plot for a system undergoing compound nucleus fission. The average kinetic energy shows a remarkable dependence on the charge and mass of the composite system and is independent of the entrance channel parameters. Figures taken from [16] and [15] respectively.

There have been extensive measurements on the mass and kinetic energy distribution of fission fragments in heavy ion induced reactions. Based on these systematics, the mass distributions from the fission of a compound nucleus formed at moderate excitation energies ($E^* = 30 - 60MeV$) are well described by Gaussian peaking around symmetric mass division, and with the variance depending linearly on the temperature and the square of the angular momentum [9, 10, 11, 12, 13, 14].

$$\sigma_A^2 = \mu T + \nu < l^2 > \tag{1}$$

where T is the transition state temperature and l the angular momentum.

The average kinetic energy of the fragments is found to depend on the charge and mass of the compound nucleus and independent of the entrance channel properties. The Viola systematics gives remarkably close estimate of the experimentally obtained $< TKE >$ over the entire range of fissility [15]. As a typical example, Figure 1 shows the mass-$< TKE >$ correlations the fission fragments in the $^{238}U + ^{16}O$ reaction [16] (left). The $< TKE >$ of a number of systems along with the Viola systematics [15] is plotted in the right part of Figure 1. Figure 2 shows the variance of the mass distribution as a function of excitation energy for different systems, along with the fit to the above expression. The mass distribution and the mass-angle correlation for $^{16}O +^{186} Os$ are shown in Figure 2(right), which demonstrate the general features observed for the compound nuclear fission reactions.

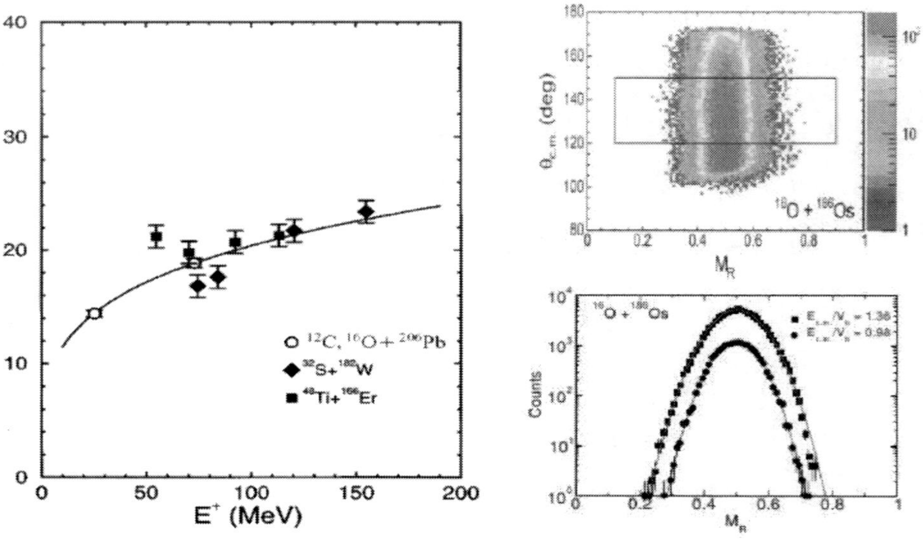

Figure 2. The dependence of the variance of the mass distribution on the temperature of the fissioning sytem. From [9]. The mass angle distribution showing no dependence of angle of emission and mass split in the case of compound nucleus fission. Figure from [17]

2.2. Angular distributions

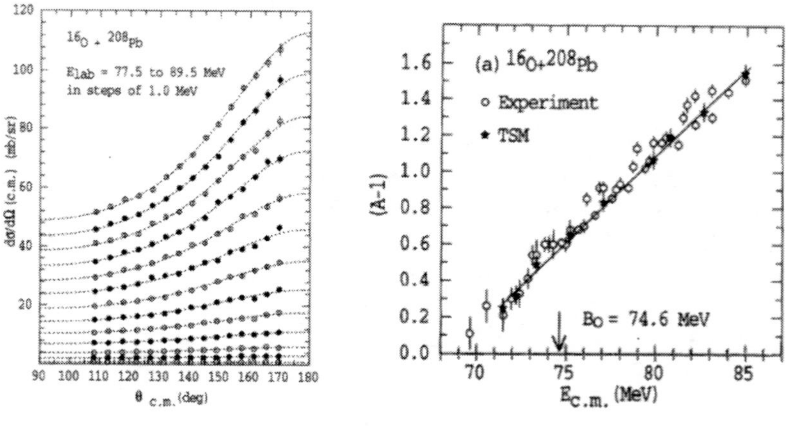

Figure 3. Fission fragment angular distributions well described by the statistical model [8] in case of compound nucleus fission. Figure taken from [23].

For compound nucleus fission, the transition state model (TSM) assumes that statistical equilibrium is achieved at the saddle point transition state and the K degree of freedom (projection of total angular momentum onto the symmetry axis) of the system is a good quantum number during the fission process. The angular distribution of fragments is described by the rotor wave function ($|D^J_{M,K}(\theta)|^2$) and the angular anisotropy reduces to

$$A \simeq 1 + \frac{< l^2 >}{4K_0^2} \tag{2}$$

where, K_0^2 is the variance of the Gaussian K distribution peaked at K=0. It can also be shown that K_0^2 is proportional to the effective moment of inertia at the saddle point and the temperature at the saddle point [8]. The model, often known as statistical saddle point model (SSPM) successfully explains the angular distribution of a large number of systems at moderate excitation energies [8].

3. Characteristics of non-equilibrium fission (fission-like processes)

3.1. Strong mass angle correlations

The dynamical models proposed in early eighties predicted the onset of quasifission process for heavier systems, when the product $Z_P Z_T > 1600$ (where Z_P and Z_T are the atomic charges of the projectile and target, respectively) [4, 5, 6]. This has been experimentally verified in the mid eighties by a number of experimental groups [16, 18, 19] where strong mass angle correlations were observed for systems satisfying the above mentioned criterion. However, there are very recent measurements where strong mass angle correlations are also observed for systems with $Z_P Z_T < 1600$.

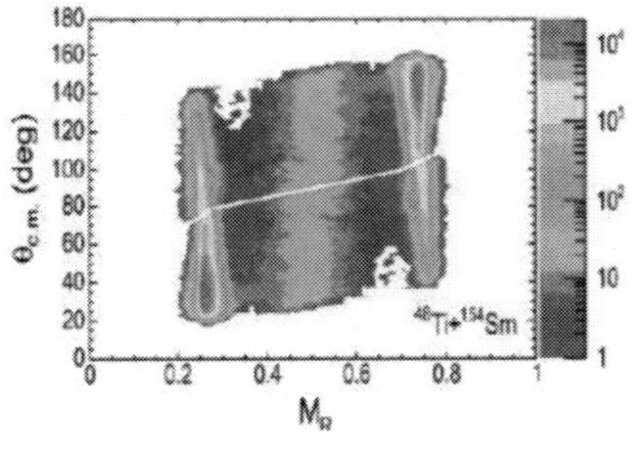

Figure 4. Mass angle correlation observed in the $^{48}Ti +^{154} Sm$ reaction. Figure taken from [17].

3.2. Broader mass distributions

It has been observed in a number of cases that even when there is no apparent mass angle correlation, the variance of the mass distributions for a near symmetric entrance channel reaction is higher than that for more asymmetric entrance channel for the same composite system at same excitation energy and angular momentum [17, 20, 21]. The apparent broadening of mass distributions may be caused by the presence of a small fraction of non-equilibrated mass asymmetry degree of freedom.

3.3. Highly anisotropic angular distributions

In the mid-eighties Back *et al.* [9, 25, 26, 27] have reported the observation of anomalously large anisotropies in systems where the charge product is much less than 1600. These results indicated the occurrence of fission-like events which are occurring in a time scale which is comparable to or less than that of the characteristic K equilibration time τ_K. This non-equilibrium feature is ascribed to the presence of quasifission. However, for highly fissile systems when the temperature becomes comparable to the temperature of the CN, a new process called pre-equilibrium fission (PEF) has been also invoked to describe the angular distribution results [28].

4. Macroscopic variables defining dynamics

Extensive experimental investigations carried out by Toke *et al.* [16] Bock *et al.* [18] and Shen *et al.* [19] using U and Pb projectiles on a number of targets ranging from O to Cr in the mid eighties had brought out the dependence for the onset of quasifission on various entrance channel parameters. For completeness, in the following we describe them briefly. According to the macroscopic model of Swaitecki and others, [4, 5, 6], the dynamic evolution of the di-nuclear sytem after contact is principally governed by the multi dimensional potential energy surface (PES) of the combined system described by the three variables viz, the radial separation, the neck radius and the mass asymmetry of the fragments. While the choice of the variables may vary with models [7], the dynamics is depicted by the three milestone configurations viz, the contact configuration, the conditional saddle point and the un-conditional saddle point. The location of these points on the PES, and the injection point onto this multi dimensional surface are the governing factors for the subsequent evolution of the composite system. From the experimental observations in the eighties, Blocki *et al.* have parameterized the onset of quasifission using a macroscopic variable called mean fissility given by $\chi_m = \frac{1}{3}\chi_{eff} + \frac{2}{3}\chi_{CN}$, which incorporates the target-projectile and compound nucleus fissilities:

$$\chi_{eff} = (Z^2/A)_{eff}/(Z^2/A)_{crit} \tag{3}$$

with

$$(Z^2/A)eff = 4Z_1Z_2/(A_1^{1/3}A_2^{1/3}(A_1^{1/3} + A_2^{1/3})) \tag{4}$$

while the compound nucleus fissility is

$$\chi_{CN} = (Z^2/A)/(Z^2/A)_{crit} \tag{5}$$

where

$$(Z^2/A)_{crit} = 50.883(1 - 1.7826I^2), I = (A - 2Z)/A \tag{6}$$

From the analysis of Blocki *et al.*, it was found that $\chi_m = 0.72$ marks the onset of the need for extra push energy (excess energy over the Coulomb barrier) to achieve fusion. It was also noticed that the manifestation of the non-equilibrium nature of the re-separation process on the mass and spatial distributions of the fragments depends critically on the sticking time τ_s of the di-nuclear system compared to the characteristic equilibration time of various degrees of freedom. As an example, Figure 5 shows the evolution of the mass asymmetry degree with time for various systems. The characteristic equilibration time for mass asymmetry is seen to be $\tau_M = (5.3 \pm 1)10^{-21}s$.

5. Non-equilibrium fission at different fissility regions

In the following we try to classify the observation of non-equilibrium fission based on two quantities, one describing the entrance channel (χ_{eff}) and the other corresponding to the compound system (χ_{CN}).

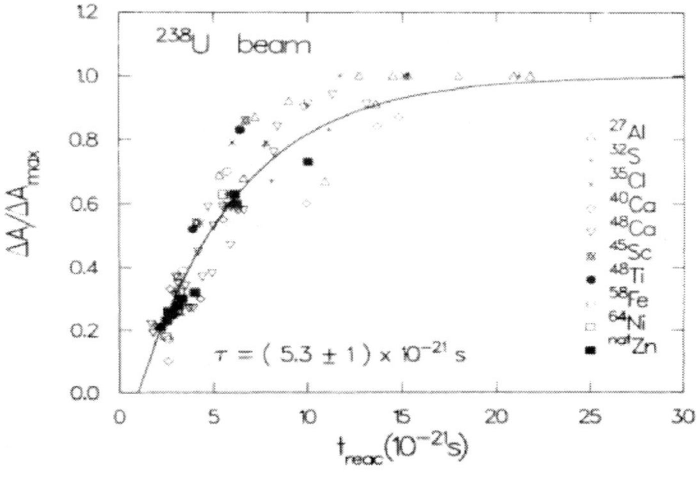

Figure 5. Characteristic equilibration time (τ_M) of mass equilibration obtained by Shen *et al.* [19]. Figure taken from [19].

5.1. The C-T Fissility Plot

Figure 6 shows the plot of entrance channel fissility (χ_{eff}) vs the compound nucleus fissility (χ_{CN}) (referred to as C-T fissility plot). Different targets are characterized by arcs whereas projectiles are described by lines cutting the arcs for various target projectile combinations. $Z_1Z_2 \geq 1600$ (thick orange line) and $\chi_m = 0.72$ (solid blue line) are also shown in the plot. The onset of non equilibrium fission can be very well classified into different regions in this plot. The locus of $\alpha = \alpha_{BG}$ for different composite systems are shown by the thick black curve. α_{BG} is calculated by the following equation [24],

$$\alpha_{BG} = p\sqrt{\frac{x - x_{BG}}{(x - x_{BG}) + q}} \tag{7}$$

where $x_{BG} = 0.396, p = 1.12$ and $q = 0.24$.

For this calculation, following Blocki *et al.* [6], we have taken the charge to mass ratio of target and projectile to lie along the valley of beta stability as approximated by Green's formula,

$$Z/A = \frac{1}{2}[1 - 0.4A/(200 + A)] \tag{8}$$

For a given pair of nuclei with mass numbers A_1, and A_2, we first calculate the corresponding atomic numbers according to the above equation and the sum of these numbers is the total atomic number Z, which is then divided between the two fragments in the ratio of their masses, viz. $Z_i = A_i(Z/A)$.

We also show on the right hand side of Figure 6, the Coulomb barrier obtained using the parametrization following Swaitecki [29].

$$V_b = 0.85247z + 0.001361z^2 - 0.00000223z^3 MeV. \tag{9}$$

where the Coulomb parameter, z is given by $z = Z_1Z_2/(A_1^{1/3} + A_2^{1/3})$. On the top along the x axis, the liquid drop fission barrier is also shown for the corresponding fissility [30].

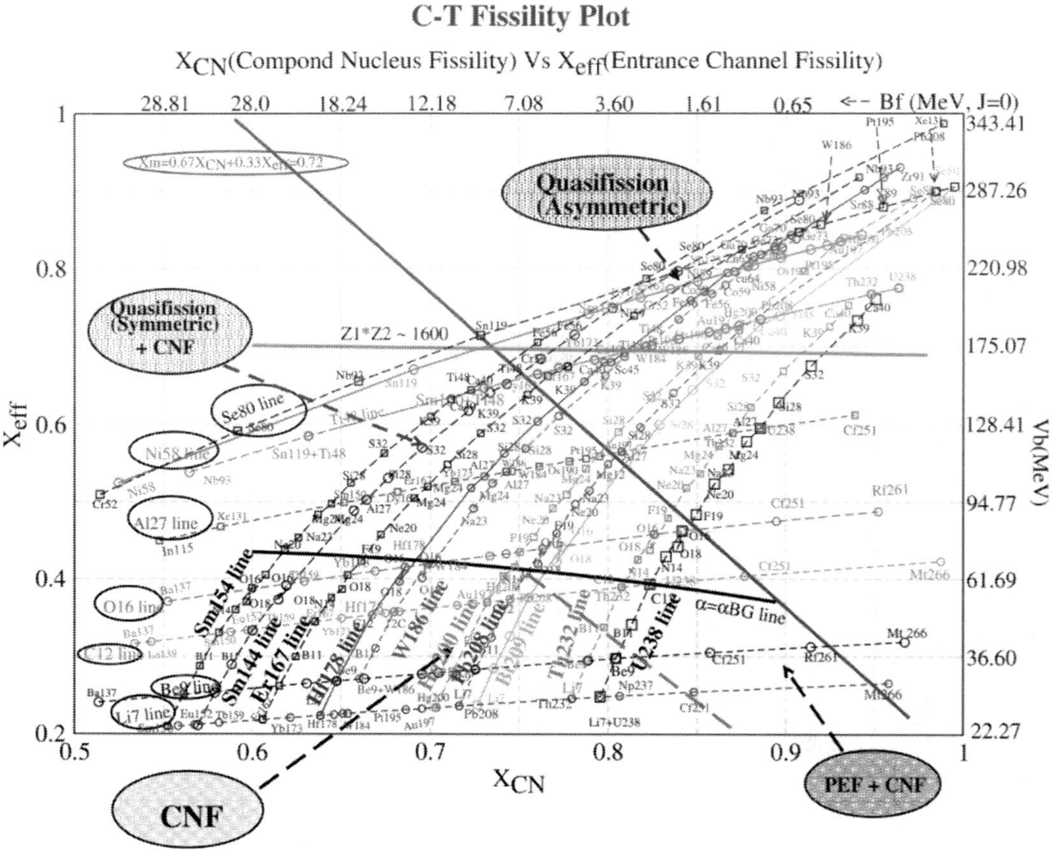

Figure 6. Entrance channel fissility χ_{eff} vs compound system fissility χ_{CN} for various target projectile systems.

5.2. Less fissile systems ($\chi \sim 0.7-0.8$): Evidence of fission before mass and K equilibration for $l > l_{B_f=0}$

It was predicted in the early eighties that when the angular momentum l of the system exceeds the angular momentum at which fission barrier vanishes ($l_{B_f=0}$), the di-nuclear system re-separates prematurely giving rise to fission like fragments [31]. This non-equilibrium process known as fast fission (fission occurring at a much shorter time scale) is characterized by a sudden broadening of mass widths (Figure 7) and highly anisotropic angular distributions [9]. This phenomenon, occurring due to the vanishing fission barrier is different from quasifission process, predicted by Swaitecki, where the dynamical trajectories end up outside the unconditional saddle point even for zero angular momentum.

While fast fission is predominant at very high energies, quasifission is predicted to occur when the mean fissility, χ_m, exceeds 0.72. As can be seen in the C-T plot, in this fissility range, the systems above the $\alpha = \alpha_{BG}$ line are expected to evolve to a di-nuclear complex and a re-separation before reaching the compound nucleus may be probable. However, it must be noted that the fission barrier is considerably larger than the temperature (T $\sim 1MeV$) in this region, thereby reducing the probability of PEF considerably. Nevertheless, evidence of non-equilibration is reported in a number of systems, e.g. $Mg, Si, S + W$ [20, 21, 22] in this region though they lie way below the $\chi_m = 0.72$ line (blue solid line). The two factors that

Journal of Physics: Conference Series **282** (2011) 012004 doi:10.1088/1742-6596/282/1/012004

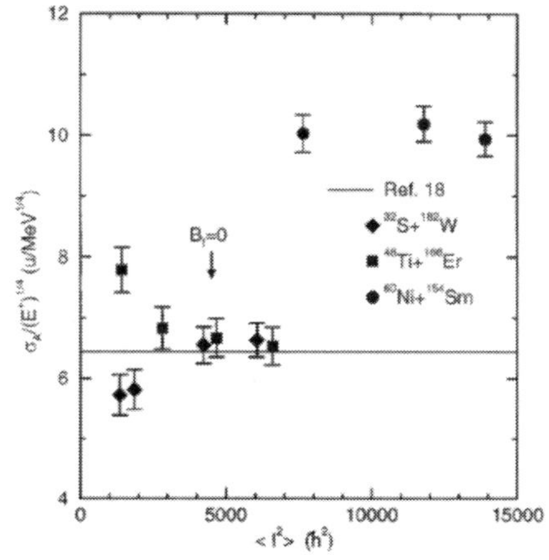

Figure 7. Sudden rise in the widths of the mass distributions when l exceeds $l_{B_f=0}$ clearly indicating the presence of fast fission. Figure taken from [9].

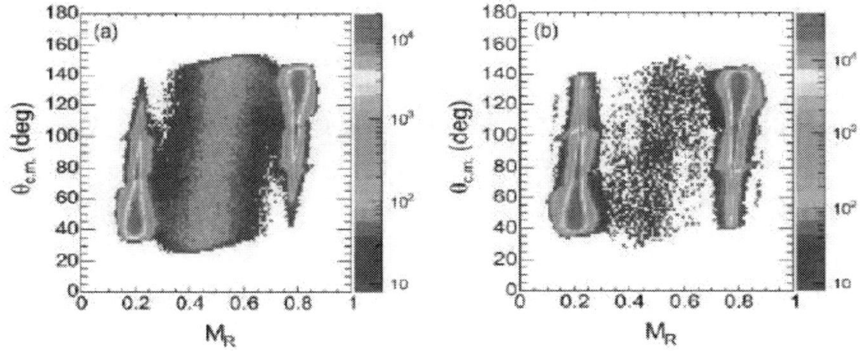

Figure 8. Strong mass angle correlations observed in less fissile systems above barrier (a) and below barrier (b) for the ^{48}Ti+^{170}Er reaction.

are facilitating early re-separation may be the elongated contact configuration [17, 21] and the lowering of fission barrier at higher angular momentum. The mass distribution is symmetric for both CN and quasifission reactions

5.3. Fissile systems (trans-actinide region): Fission before K equilibration

As we move to the fissile systems where the fissility $\chi \sim 0.8-0.9$, anomalously large anisotropies were reported in several cases were actinide targets were involved [32, 33, 34, 35, 36] even for systems that lie way below the red and blue lines (right-bottom region of the C-T plot). However, no mass angle correlation or anomalously large variances were observed in these systems [16]. This points to the fact that the sticking time of the di-nuclear complex in this region is considerably larger that the characteristic rotation time and mass equilibration time, τ_M. Anomalously large anisotropies observed in very asymmetric systems like $B, C+Th, B, C+Th$ [36], can only be understood if we assume a fraction of events re-separating before the characteristic K equilibration time $\tau_K \sim 10^{-20}s$, a non-equilibrium process known as pre-

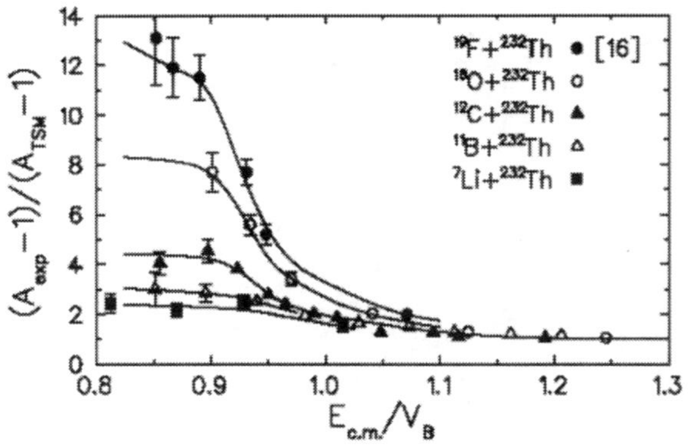

Figure 9. Anomalously large anisotropies irrespective of the entrance channel mass asymmetry in the trans-actinide region showing evidence of fission before K equilibration. Figure taken from [36].

equilibrium fission (PEF) . This also implies the evolution towards a di-nuclear system, due to an apparent shift of the Businaro-Gallone point towards larger mass asymmetries due to the elongated touching configuration (large static deformation of actinide targets in this region indicated by the green thick dashed line in Figure 6.). It should also be noted that in this region the temperature ($\sim 1MeV$) becomes comparable to the fission barrier and a substantial portion of the flux can diffuse over the barrier before reaching the compound nucleus phase. This is schematically illustrated in Figure 10 [37].

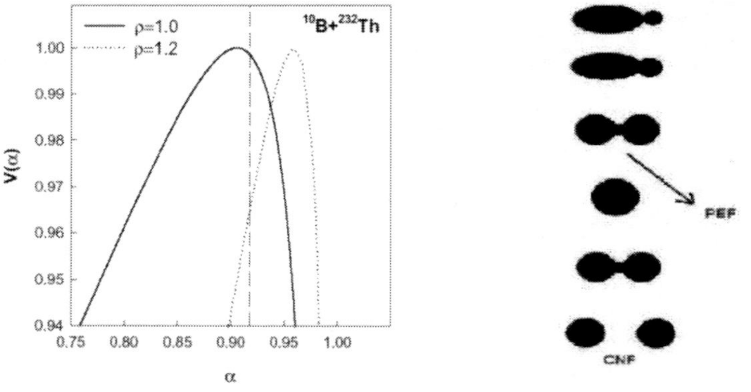

Figure 10. Figure illustrating the shift of the Businaro-Gallone mass barrier towards higher mass asymmetries for elongated contact configurations and the consequent occurrence of PEF even in very asymmetric systems at energies close to the Coulomb barrier.

Once the entrance channel is more symmetric beyond the red line, evidences are reported for strong mass angle correlation as well as the emergence of an asymmetric component particularly at sub-barrier energies [16, 38, 39] indicating a much shorter sticking time for complexes in this region. Figure 11 shows the mass angle correlation for $^{32}S+^{232}Th$ system where the asymmetric component is clearly visible at sub-barrier energies [38].

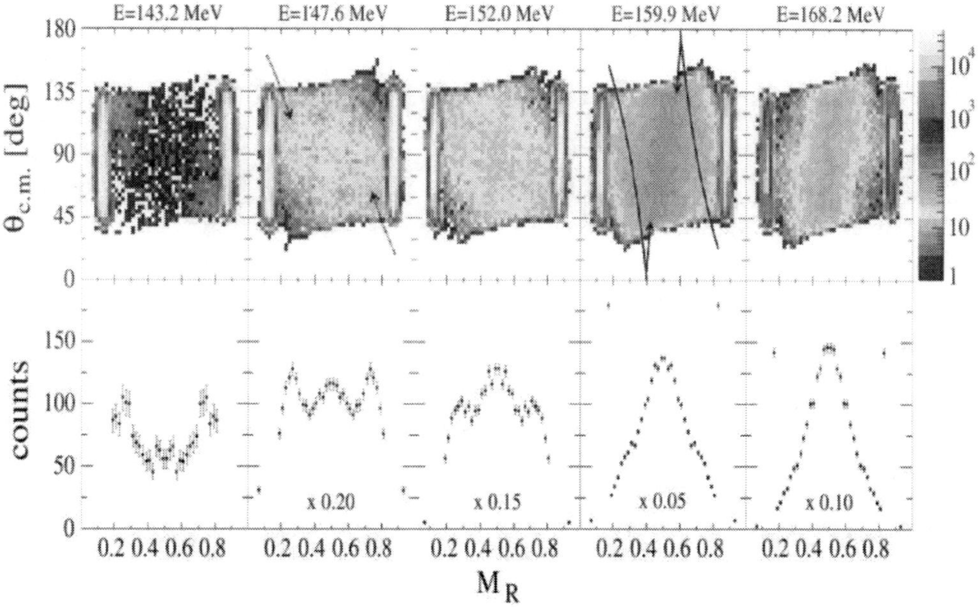

Figure 11. Strong mass angle correlation observed in case of highly fissile systems ($^{32}S+^{232}Th$) [38]. Notice the appearance of the asymmetric component at sub-barrier energies. Figure taken from [38].

5.4. Highly fissile systems (superheavy region)

The top-right hand side region of the C-T Fissility plot indicates the region of high fissility (also the superheavy region). It is quite evident that the region is heavily dominated by quasifission (above the thick red line). Several experimental studies have revealed the strong mass angle correlations and also the asymmetric nature of mass distribution in this region [38, 39, 40]. It is also interesting to note that the average kinetic energy of fission-like fragments produced in the $Se + Th$ reaction shows consistently less value compared to the Viola systematics value indicating a much larger elongation at re-separation, than what is expected from the standard scission configuration for this fissility.

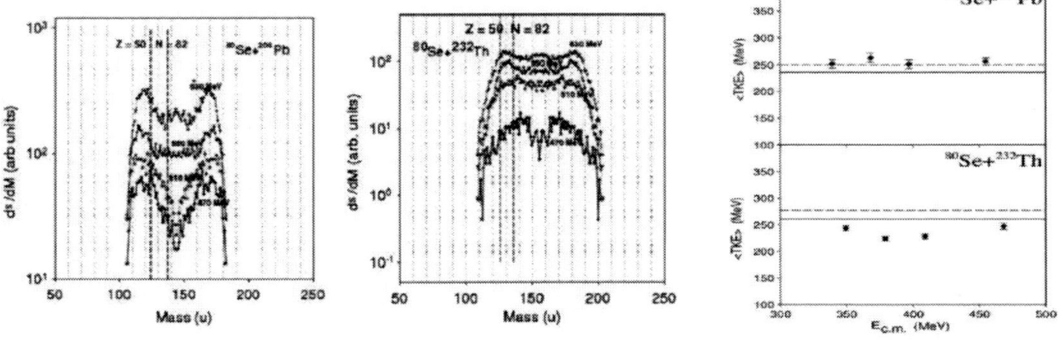

Figure 12. Highly asymmetric mass distributions observed in the superheavy region. Figure taken from [40].

6. Summary

It is quite apparent that the experimental observables in quasission reactions are found to crucially depend on the sticking time of the dinuclear complex with respect to the characteristic relaxation time of various degrees of freedom, before it decays into binary fission-like fragments. This interaction time in turn depends on the entrance channel parameters of the colliding nuclei. Thus a system that remains together for a time comparable or shorter than τ_M exhibits strong mass-angle correlations, large mass widths, and large anisotropies (the top-right region in C-T plot). The central region in the plot is dominated by systems whose sticking time is between τ_M and τ_K that may not show any signicant correlation of mass with angle but can show broadened mass distributions and large angular anisotropies. The bottom-right of the plot indicates systems that stick together for time scales comparable to τ_K but re-separate before reaching a compound nucleus. These systems may not exhibit any signs of incomplete relaxation of mass but can still exhibit angular anisotropies larger than predicted by the TSM for compound nucleus fission. It is evident that non-equilibrium fission is present in many more systems than that predicted by earlier dynamical models. True compound nucleus fission may be confined only to a small region bounded by the thick black line ($\alpha = \alpha_{BG}$) and the thick dashed green line (speculative) as shown in Figure 6.

Acknowledgments

The authors acknowledge Mr. Chandrabhan Yadav for his contribution in preparing the C-T Fissility plot.

References

[1] Yuri. Ts. Oganessian and Yuri. A. Lazarev, Heavy-ions and Nuclear Fission, Treatise in heavy-ion science, edited by D. A. Bromly, Vol. 4.
[2] Peter Armbruster, C. R. Physique **4** (2003) 571594.
[3] Peter Moller and Arnold J. Sierk, Nature **422**, 485 (2003).
[4] W. J. Swiatecki, Phys. Scr. **24**, 113 (1981).
[5] S. Bjornholm and W. J. Swiatecki, Nucl. Phys. **A 391**, 471 (1982).
[6] J. P. Blocki et al., Nucl. Phys. **A 459**, 145 (1986).
[7] P. Moller *et al.* , Nature **409**, 785 (2001).
[8] R. Vandenbosch and J. R. Huizenga, Nuclear Fission (Acadamic, New york, 1973).
[9] B. B. Back *et al.*, Phys. Rev. **C 53**, 1734 (1996).
[10] R. K. Choudhury et al., Phys. Rev. C**60**, 054609 (1999).
[11] M. G. Itkis and A. Ya. Rusanov, Phys. Part. Nuclei **29**, 160 (1998).
[12] G. D. Adeev, I. I. Gontchar, V. V. Pashkevich, N. I. Pischasov, O. I. Serdyuk, Phys. Part. Nuclei **19**, 1229 (1988).
[13] E. G. Ryabov, A. V. Karpov and G. D. Adeev, Nucl. Phys. **A765**, 39 (2006).
[14] G. N. Knyazheva, E. M. Kozulin, R. N. Sagaidak, A. Yu. Chizhov, M. G. Itkis, N. A. Kondratiev, V. M. Voskressensky, A. M. Stefanini, B. R. Behera, L. Corradi, E. Fieretto, A. Gadea, A. Latina, S. Szilner, M. Trotta, S. Bhegini, G. Montagnoli, F. Scarlassara, F. Hass, N. Rowley, P. R. S. Gomes, and A. Szanto de Toledo Phys. Rev. **C75**, 064602 (2007).
[15] V. E. Viola, K. Kwiatkowski, and M. Walker, Phys Rev. C **31**, 1550 (1985).
[16] J. Toke et al., Nucl. Phys. **A440**, 327 (1985).
[17] R. Rafiei, R. G. Thomas, D. J. Hinde, M. Dasgupta, C. R. Morton, L. R. Gasques, M. L. Brown, and M. D. Rodriguez, Phys. Rev C **77**, 024606 (2008).
[18] R. Bock et al., Nucl. Phys. **A 388**, 334 (1982).
[19] W. Q. Shen *et al.*, Phys Rev. C **36**, 115 (1987).
[20] A. C. Berriman, D. J. Hinde, M. Dasgupta, C. R. Morton, R. D. Butt, and J. O. Newton, Nature (London) **413**, 144 (2001).
[21] R. G. Thomas, D. J. Hinde, D. Duniec, F. Zenke, M. Dasgupta, M. L. Brown, M. Evers, L. R. Gasques, M. D. Rodriguez, and A. Diaz-Torres, Phys. Rev C **77**, 034610 (2008).
[22] E. Prasad et al., Phys. Rev. C **81**, 054608 (2010)

[23] C. R. Morton *et al.*, Phys. Rev. **C 52**, 243 (1995).
[24] M. Abe, KEK Report No. 86-26, KEK TH-28 (1986).
[25] B. B. Back, R. R. Betts, K. Cassidy, B. G. Glagola, J. E. Gindler, L. E. Glendenin, and B. D. Wilikins, Phys. Rev. Lett **50**, 818,(1983).
[26] M. B. Tsang et al., Phys. Lett. **B 129**, 18 (1983).
[27] B. B. Back, R. R. Betts, J. E. Gindler, B. D. Wilkins, S. Saini, M. B. Tsang, C. K. Gelbke, W. G. Lynch, M. A. McMahan, and P. A. Baisden, Phys. Rev. C **32**, 195 (1985).
[28] V. S. Ramamurthy and S. S. Kapoor, Phys. Rev. Lett. **54**, 178 (1985).
[29] W.J. Swaitecki *et al.* Phys. Rev. C **71**, 014602 (2005)
[30] A. J. Sierk, Phys. Rev. C**33**, 2039 (1986).
[31] C. Ngo *et al. Nucl. Phys.* **A 400**, *259c (1983).*
[32] D. J. Hinde *et al.*., Phys. Rev. Lett **74**, 1295 (1995).
[33] D. J. Hinde, M. Dasgupta, J. R. Leigh, J. C. Mein, C. R. Morton, J. O. Newton, and H. Timmers, Phys. Rev. **C 53**, 1290 (1996).
[34] Z. Liu *et al.*., Phys. Lett **B 353**, 173 (1995).
[35] J. C. Mein, D. J. Hinde, M. Dasgupta, J. R. Leigh, J. O. Newton, and H. Timmers, Phys. Rev. **C 55**, R995 (1997).
[36] J. P. Lestone *et al.*, J. Phys. G. **23**, 1349 (1997)
[37] R. G. Thomas, R. K. Choudhury, A. K. Mohanty, A. Saxena, and S. S. Kapoor, Phys. Rev. C **67**, 041601(R) (2003).
[38] D. J. Hinde *et al.*., Phys. Rev. Lett **101**, 092701 (2008).
[39] K. Nisho *et al.*, Phys. Rev. **C 82**, 44604 (2010).
[40] R. G. Thomas *et al.*, Phys. Rev. **C 75**, 24604 (2007).

The evolution from quasi-elastic to deep inelastic processes and its connection to fusion

L. Corradi

INFN - Laboratori Nazionali di Legnaro, Viale dell'Universita' 2 - 35020, Legnaro (Padova) Italy

E-mail: `corradi@lnl.infn.it`

Abstract. Substantial experimental and theoretical efforts have been made in the last decade to study into detail the transition from quasi-elastic to deep inelastic processes in heavy ion collisions. The understanding of this transition is relevant not only to probe which degrees of freedom play a role in the multinucleon transfer process but also for its connection with near barrier fusion reactions. A boost in the field has been given with the implementation of large solid angle magnetic spectrometers coupled to γ arrays with which extensive investigations have been carried out for nuclear structure and reaction dynamics. In the present paper aspects of these studies will be presented, with emphasis on processes undergoing significant energy loss in binary collisions.

1. Introduction

Multinucleon transfer reactions at Coulomb barrier energies is an important field of research in low energy heavy-ion physics [1]. Through this mechanism one can in fact investigate nucleon-nucleon correlation in nuclei [2, 3], the transition from the quasi-elastic to the deep-inelastic regime [4] and channel coupling effects in sub-barrier fusion reactions. Different aspects of the correlation between reaction channels have been extensively discussed at the recent Fusion06 [5] and Fusion08 [6] conferences. A yet poorly investigated question is what are the relevant degrees of freedom acting in the transfer process, i.e. single nucleon, pair or even cluster transfer modes.

From the theoretical point of view different regimes have been historically treated on completely different bases (see Fig.1). In the quasi-elastic regime (few transferred nucleons and small energy losses) one or, at most, two-nucleon transfer has been treated using Distorted Wave Born Approximation or Coupled Channel codes, while in the deep inelastic regime (many transferred nucleons and large energy losses) stochastic, friction or diffusion models have been used [7]. From the experimental point of view, nuclei produced in transfer reactions have been identified with good A, Z and Q-value resolutions for medium-light systems, but channel detection was limited to the transfer of few nucleons. For medium-heavy systems, mainly studied in deep-inelastic processes, the transfer of several nucleons becomes available in the reaction, but, at best, nuclear charge identification was possible without mass or energy resolution.

The "transition" regime, where many nucleons are transferred and where nuclear structure effects still play a significant role in the dynamics, represents a window not well studied in its detail and for which both experimental and theoretical advances have been recently made. Thanks to the development of high resolution and high efficiency experimental set-up's one

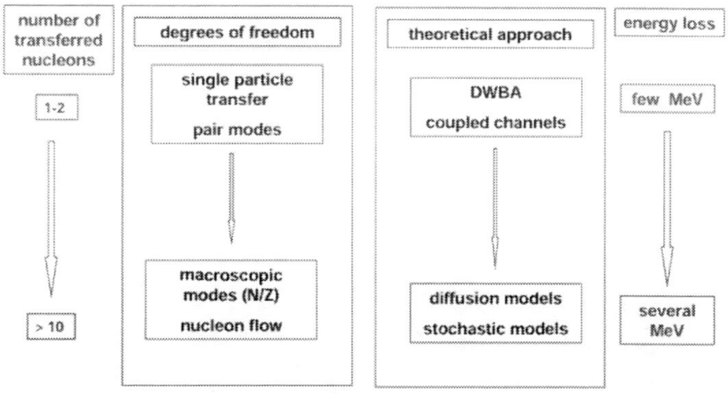

Figure 1. A naive conceptual frame identifying the two regimes of quasi-elastic and deep inelastic processes, which have been treated on different footing from the theoretical and experimental point of view, in dependence on the number of transferred nucleons and energy loss involved in the reaction.

could unambiguously detect in mass and charge the nuclei produced in transfer reactions up to the pick-up of six neutrons and the stripping of six protons (see e.g. [8, 9, 10] and references therein). The advent of the last generation large solid angle magnetic spectrometer PRISMA [11, 12] allowed to increase the detection limit by more than an order of magnitude, with a significant gain in mass resolution for very heavy ions. Further, the coupling of this spectrometer to the large gamma array CLARA [13] allowed to perform gamma-particle coincidences, thus detecting the transfer strength to the lowest excited levels of binary products and performing gamma spectroscopy for nuclei moderately far from stability produced via nucleon transfer or deep-inelastic reactions, especially in the neutron-rich region [14, 15, 16, 17]. These studies are of primary importance for reactions to be done with radioactive ion beams, where multinucleon transfer has been shown to be a competitive tool for the study of neutron-rich nuclei, at least for certain mass regions (see e.g. Ref. [18] and references therein).

2. Quasi elastic and deep inelastic processes : some concepts

Before discussing some important features present in quasi-elastic processes I here remind some basics of deep inelastic collisions, which lead to interpretations based on diffusion or stochastic models [7, 19]. In Fig. 2 it is shown, for the reaction ^{86}Kr+^{166}Er [20], how the Total Kinetic Energy (TKE) of the reaction products depends on nuclear charge Z (top left) and angle (bottom left). One clearly sees how the (mass integrated) Z distributions widen with energy loss, promptly suggesting some kind of nucleon diffusion process. Events show up at energies lower than the Coulomb barrier for two touching spheres configuration (V_c) indicating that nuclei may undergo strong deformation before reseparation. The Wilczynski plot shown on the bottom left of the figure displays a characteristic behaviour of the binary process, where quasi-elastic and deep inelastic events are associated with high TKE and broader distributions, respectively. This behaviour has a definite dependence on the size of the Coulomb field, ranging from an orbiting to a focussing condition up to a change of structure of the lower ridge for very heavy systems. This suggested that in a reaction plane defined by the beam axis and detection angle two components may show up as depicted in part a) of Fig. 2. The two branches are associated with impact parameters on the right or left hand side of the beam axis, represented by full lines and dashed lines, respectively, in part b) of Fig. 2. A dinuclear configuration can be formed, explained macroscopically as the effect of the rotation of the binary system in turn correlated with large energy losses (long interaction times). This simple picture was at the basis of various phenomenological interpretations of experimental data [19].

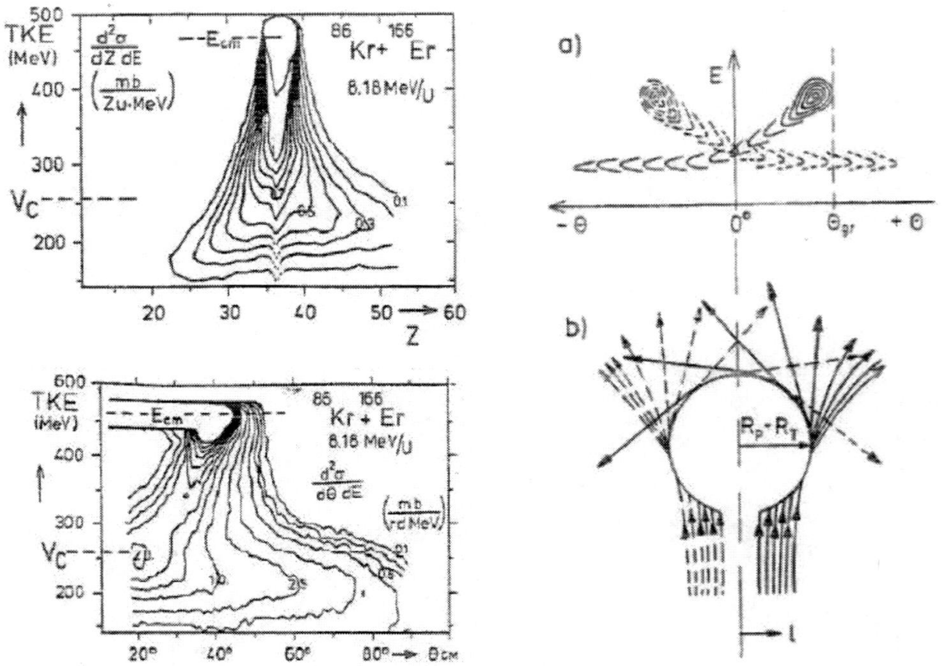

Figure 2. Left : matrixes of Total Kinetic Energy (TKE) vs. nuclear charge Z (top) and vs c.m. angle θcm (bottom) for the reaction ^{86}Kr+^{166}Er at the indicated energy. Right : Wilczynski plot indicating the distributions of events in deep inelastic collisions (panel a)) and its interpretation (panel b), adapted from Refs. [19, 20]).

To introduce the discussion on quasi elastic processes I show in Fig. 3 a representative example of the mass and charge distribution of transfer products in the ^{40}Ca+^{208}Pb system [9]. The bombarding energy is close to the Coulomb barrier and the yield, measured at the grazing angle, reflects some of the main characteristics of quasi-elastic processes. These may be appreciated by plotting on the same figure three lines, the two dash-dotted lines correspond to pure neutron pick-up and pure proton stripping channels, while the full line represents the charge equilibration, namely the location of the N/Z ratio of the compound nucleus. One sees that most nuclei are located on the left side of the charge equilibration line, indicating the dominance of a direct mechanism in the population of different fragments. Notice also that for the massive proton transfer channels the isotopic distributions drift towards lower masses, a clear indication that these distributions are affected by evaporation processes (see later).

An interpretation of these distributions can be done on the basis of semiclassical arguments, where nuclear structure and kinematic properties are taken into account via the introduction of form factors $f_{\beta\alpha}$ [21, 22] and optimum Q-value window, respectively (see Ref. [22] for details). To estimate the magnitude of a given transfer process it is not necessary to solve explicitly the

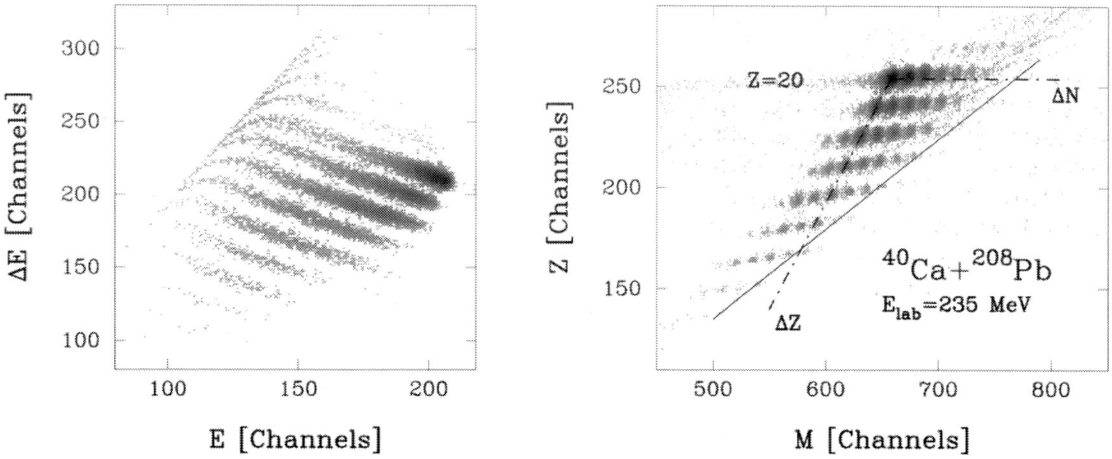

Figure 3. $\Delta E - E$ plot (left) and mass-charge distributions (right) of transfer products in the ^{40}Ca+^{208}Pb reaction at $E_{\rm lab}$ =235 MeV obtained at the grazing angle, $\theta_{\rm lab}$ = 84°. The dash-dotted lines correspond to the pure proton stripping (ΔZ) and to the pure neutron pick-up (ΔN) channels, crossing at Z=20 and A=40. The full line shows the charge equilibration, namely, the N/Z ratio of the compound nucleus (from Ref. [9]).

full system of coupled equations but it suffices to write down its first order Born approximation (here I follow the discussion more widely presented in Ref. [1]). For a given impact parameter (incoming partial wave ℓ) the probability for the transition from the entrance channel α to the channel β may be written in the form

$$P_{\beta\alpha}(\ell) = \left| \frac{i}{\hbar} \int_{-\infty}^{+\infty} dt\ e^{i\sigma_{\beta\alpha}t} f_{\beta\alpha}(0,\vec{r}) e^{i[(E_\beta - E_\alpha)+(\delta_\beta-\delta_\alpha)]t/\hbar} \right|^2 \tag{1}$$

where the time integral has to be performed along the classical trajectory for the given partial wave ℓ. In direct processes the two nuclei barely overlap, so that only the tail of the formfactor is relevant. By approximating the true trajectory with a parabolic parametrisation around the turning point the above transition probability may be written in the form

$$P_{\beta\alpha} = \sqrt{\frac{1}{16\pi\hbar^2|\ddot{r}_0|\kappa_{a_1'}}} |f_{\beta\alpha}(0,r_0)|^2\ g(Q_{\beta\alpha}) \tag{2}$$

where $\ddot{r}_0$ is the radial acceleration at the distance of closest approach r_0 for the grazing partial wave. The adiabatic cut-off function $g(Q)$ is defined as

$$g(Q) = \exp\left(-\frac{(Q - Q_{opt})^2}{\hbar^2 \ddot{r}_0 \kappa_{a_1'}} \right) \tag{3}$$

where the optimum Q-value is

$$\begin{aligned} Q_{opt} &= \left(\frac{Z_d}{Z_A} - \frac{Z_d}{Z_b}\right)E_B + \left(\frac{m_d}{m_b} - \frac{m_d}{m_A}\right)(E - E_B) \\ &+ \frac{m_d \ddot{r}_0}{m_a + m_A}(R_A m_b - R_a M_B)\ , \end{aligned} \tag{4}$$

Journal of Physics: Conference Series **282** (2011) 012005 doi:10.1088/1742-6596/282/1/012005

E_B is the Coulomb barrier and m_d and Z_d are the mass and charge of the transferred particle. The adiabatic cut-off function $g(Q)$ defines the actual value of the transition probability, the maximum being at the optimum Q-value. This derives from the requirement that the trajectory of entrance and exit channels matches smoothly close to the turning point where the contribution of the formfactor peaks. Notice that the bombarding energy dependence of the cut-off function is contained in the $\ddot{r}_0$ term that defines its width (inversely proportional to the collision time).

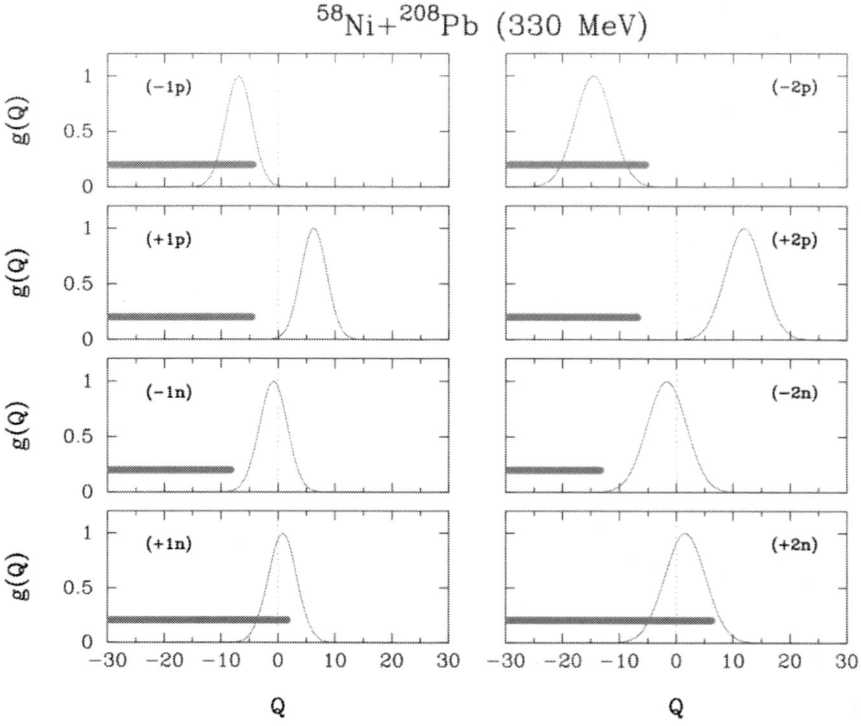

Figure 4. Adiabatic cut-off functions for one and two neutron and proton transfer channels for the reaction ^{58}Ni+^{208}Pb at the indicated energy (Q-value in MeV). The horizontal lines represent the location of all possible transitions.

A rough estimation of the total cross section for a particular transfer channel can be obtained by simply performing a Q-value integration of equation (2). In Fig. 4 the adiabatic cut-off function $g(Q)$ for all one and two particle transfer channels is shown for the ^{58}Ni+^{208}Pb reaction. In the same figure with horizontal lines are represented, for all channels, the location of all possible transitions. Since only the channels whose Q values lie below the bell shaped curve can actually occur, it is clear that the only allowed transfers are neutron pick-up and proton stripping. All the other channels are hindered by optimum Q-value consideration. From the same figure one notices that for some channels, in particular the two proton stripping and two neutron pick-up, the reaction mechanism favours transitions leading to high excitation energies.

These simple findings have been demonstrated in almost all measured systems. An example is shown in Fig. 5 where the isotopic distribution of the different charges populated in the indicated reactions [8, 23] are plotted. One observes that the strongest channels are those corresponding to the neutron pick-up and proton stripping processes as the optimum Q-value rule suggests. It is only for charges far from the entrance channel that one observes sizeable contributions that seems to derive from the stripping of neutrons, this is particularly evident for

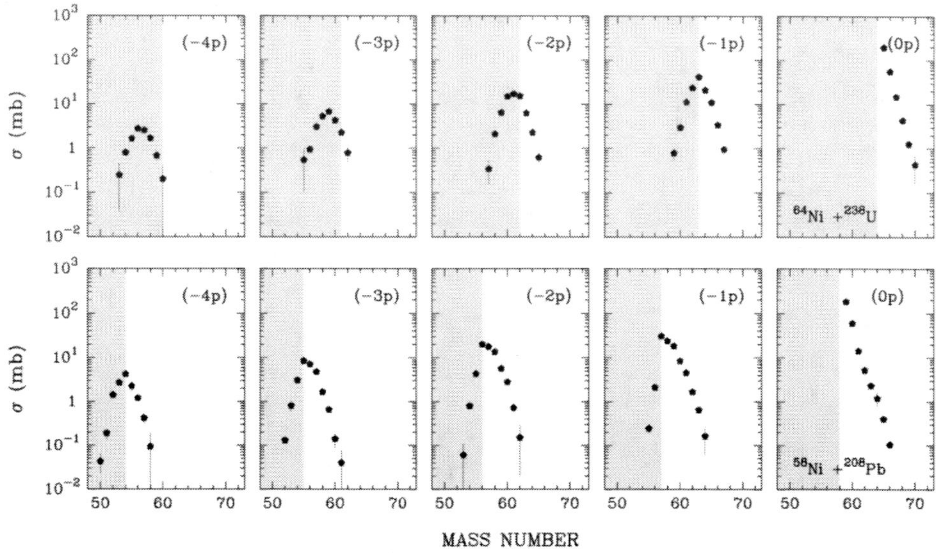

Figure 5. Isotopic distributions for the transfer channels up to the stripping of four protons are shown for the reaction ^{64}Ni + ^{238}U at 390 MeV bombarding energy (top) and for the reaction ^{58}Ni + ^{208}Pb at 328 MeV bombarding energy (bottom). The shadowed regions mark the transition from neutron stripping to neutron pick-up. Data are from Refs. [8, 23].

the collision with uranium where the isotopic distributions peak at masses lighter than the pure proton stripping channels. This should not be a surprise since these channels can be populated via other complicated processes like evaporation, deep inelastic or fission. The situation depicted in Fig. 5 holds for most of the projectile-target combinations available with stable beams, it is only by employing unstable beams that all the four kind of basic transitions shown in Fig. 4 become avaliable [24].

Concerning the angular momentum transfer one here recalls that at low bombarding energies, where the relative velocity of the two ions is much smaller than the intrinsic velocity of the transferred nucleon, the transfer reaction tries to maximize the transferred angular momentum [25], while at high bombarding energies, where the velocity of relative motion is comparable to the intrinsic velocity of the transferred nucleon, the reaction minimizes the transferred angular momentum [26]. This finding may be understood by requiring that the orbitals of the transferred nucleon match smoothly. The semi-classical theory is able to reconcile these two extreme behaviours with a correct treatment of the so called recoil [27].

3. Multinucleon transfer channels

In the near past large efforts, both experimentally and theoretically, have been devoted to the understanding of the transition from a regime of direct reactions (quasi-elastic) to the more complicated regime of deep inelastic. These last processes are characterized by a massive transfer of nucleons (toward the charge equilibration) and very large energy losses. The exit fragments emerge with energies lower than the Coulomb barrier with an angular distribution that is proper of direct reactions (bell shaped, with a maximum close to the grazing angle). While this last characteristic indicates that the process is very fast the large energy losses indicates that the two ions, before the emerging point, acquire large deformations. These two conflicting findings, short collision time and large energy losses, suggested that in the evolution of the reaction the excitation of surface modes is playing an important role being the low lying modes the main source for the formation of the large deformations.

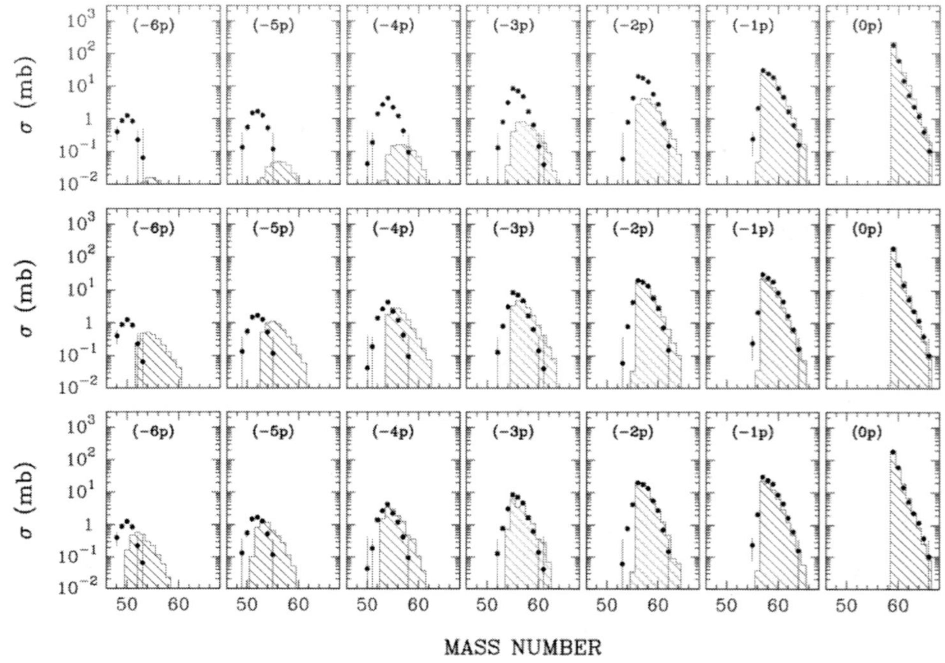

Figure 6. Total cross sections for multinucleon transfer channels populated in the $^{58}Ni+^{208}Pb$ reaction. Points are the experimental data, histograms are calculations performed with semiclassical theory taking into account independent particle transfers (top), an additional pair mode (middle) and including nucleon evaporation effects from the primary fragments (bottom) (from Ref. [8]).

It was the necessity to have a description of the reaction able to treat at the same time quasi-elastic and deep inelastic events that lead to the development of the reaction code GRAZING [28, 29, 30]. Fig. 6 shows as representative example the comparison between experimental and theoretical total transfer cross sections for the multinucleon transfer channels populated in the $^{58}Ni+^{208}Pb$ reaction [8]. In the top row the data are compared with calculations performed with the semiclassical theory. In GRAZING, the treatment of the transfer degrees of freedom is based on the assumption that in a heavy-ion collision the exchange of a nucleon proceeds via many open channels that are all quite weak, so that they may be treated independently. Surface degrees of freedom and particle transfer are treated on the same footing, and the exchange of many nucleons proceeds via a multi-step mechanism of single nucleons (both, protons and neutrons, via stripping, and pick-up processes). The trajectory is calculated by solving the system of classical equations for the variables of relative motion and the deformation parameters for the surface modes. The model includes the low-lying 2^+ and 3^- states of both projectile and target and the corresponding giant resonances. This model has been successfully applied in the description of multinucleon transfer reactions [1] and can reproduce the near-barrier fusion excitation functions [31] and extracted barrier distributions [32]. The theory describes pretty well the pure neutron transfer and the (-1p) isotopes but underestimates considerably all the other charges. However, we stress that GRAZING accounts correctly for almost 80% of the direct transfer yield. For the explanation of the discrepancies at large charge transfers, it is very difficult at this stage to say what is missing in the model, as degrees of freedom or as reaction mechanism.

To understand the charge distribution it was added in the calculation the contribution of a direct pair mode both for neutrons and protons, by using the macroscopic formulation. Since the code GRAZING does not allow to add these pair degrees of freedom, one used another code, based on the CWKB approximation. Its main characteristics is that for the single particle transfer it uses the actual distribution of single particle levels that are coupled with the true formfactors (it has been developed to check the different approximations that are included in the GRAZING program). The multinucleon transfer in obtained via a successive mechanism that mimic the approximation of GRAZING. For the relative motion the program uses the WKB approximation and it is constructed to reproduce the ordinary DWBA calculations for the one-particle transfer channels (for more details confer Ref. [9]). To keep at minimum the number of parameters one choose to add a single pair mode placed at the optimum Q-value and one used the strength of the form factor (kept the same for neutrons and protons) to fit the pure -2p channel. The results of such a calculation are shown in the middle row of the figure. Notice that the pair mode alters very little the cross section for neutron transfer but is essential for the proton transfer. Here one stresses that the pure one neutron pick-up is much larger (almost a factor 10) with respect to the pure one proton stripping transfer. This may mask the effect of the pair mode since the successive mechanism gives a much larger contribution for neutrons than for protons.

These semiclassical models provide the excitation energy of the single fragments thus one is in condition to correct the primary yield for the evaporation process. The final yields distributions are shown in the bottom row of the figure where now a reasonable overall description is obtained.

Similar kind of results have been achieved in more recent data obtained with large solid angle spectrometers, in particular in the ^{90}Zr+^{208}Pb and ^{40}Ca+^{96}Zr systems. The effects attributed to neutron evaporation, indirectly visible from the analysis of Refs.[8, 9, 23], have been directly observed by means of particle-γ coincidences [10] (see next Section).

4. Energy loss effects

The Z and A identification capability and the large detection efficiency of the magnetic spectrometer PRISMA allows to follow the evolution of the reaction from the quasi elastic to the deep inelastic regime. Here, one challenging question is to what extent the fundamental degrees of freedom (single particle, surface and pair modes) used to describe few nucleon transfer processes, holds in the presence of large energy losses and/or large number of nucleons.

The power of these new spectrometers can be appreciated from Fig. 7 which shows the total angle and Q-value integrated yields for multineutron and multiproton transfer channels populated in the reaction ^{90}Zr+^{208}Pb. Isotope identification in the proton transfer direction is visible down to $(-8p)$ stripping, but sensitivity was sufficient to observe even more proton stripping channels.

The TKEL spectra of the different transfer channels for the same reaction are shown in Fig. 8. One can here follow the evolution pattern as function of the number of transferred neutrons and protons. In the case of pure neutron transfer one sees a quasi-elastic peak and an increasing strength of large energy loss components when adding neutrons. When protons are involved one observes a faster growth of large TKEL components, up to the point that beyond three

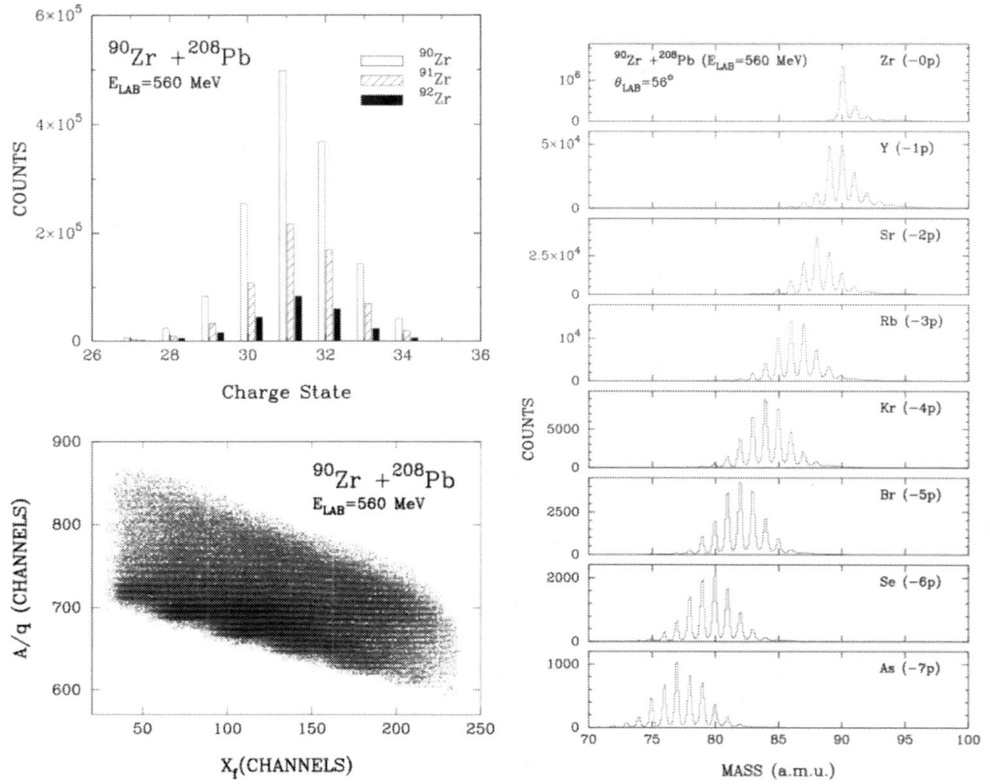

Figure 7. Left top panel: atomic charge state distribution for ^{90,91,92}Zr ions selected with PRISMA in the reaction ^{90}Zr+^{208}Pb at E_{lab}=560 MeV. Left bottom panel: mass over atomic charge state A/q vs. horizontal focal plane position X_f. Right panel: mass distributions of transfer products for different Z (adapted from Ref. [10]).

proton stripping the TKEL distributions have almost similar shapes. For channels which, due to optimum Q-values, are not directly populated, the corresponding TKEL shapes differ a lot from the smooth behaviour just decribed. Look for instance at the comparison between the (-1p+1n) channel (mainly directly populated) and the (-1p-1n) one. This different behaviour tends to smooth out with larger number of transferred protons. One should remind that one detects secondary fragments and that the TKEL spectra are constructed assuming binary reactions, so the shapes may be somewhat distorted with respect to the (unreconstructed) primary fragments. In the top row are shown the mass distribution associated to the different nuclear charges. One sees that the mass yields of pure proton stripping channels (circles) become less favourite as more protons are transferred, with a mass distributions shifting to lower values.

Large energy losses are associated with nucleon evaporation from the primary fragments. The importance of neutron evaporation in the modification of the final yield distribution was outlined in inclusive measurements [8, 9, 23]. These effects can be directly seen with PRISMA+CLARA. Gating with PRISMA on a specific Z and A (light partner) the velocity vector of the undetected heavy partner can be evaluated and applied for the Doppler correction of its corresponding γ rays. In those spectra not only the γ rays belonging to the primary binary partner are present but also the ones of the nuclei produced after evaporation takes place. An example is given in Fig. 9 for the for the $-2p + 2n$ channel populated in the ^{40}Ca+^{96}Zr reaction [10]. About 60%

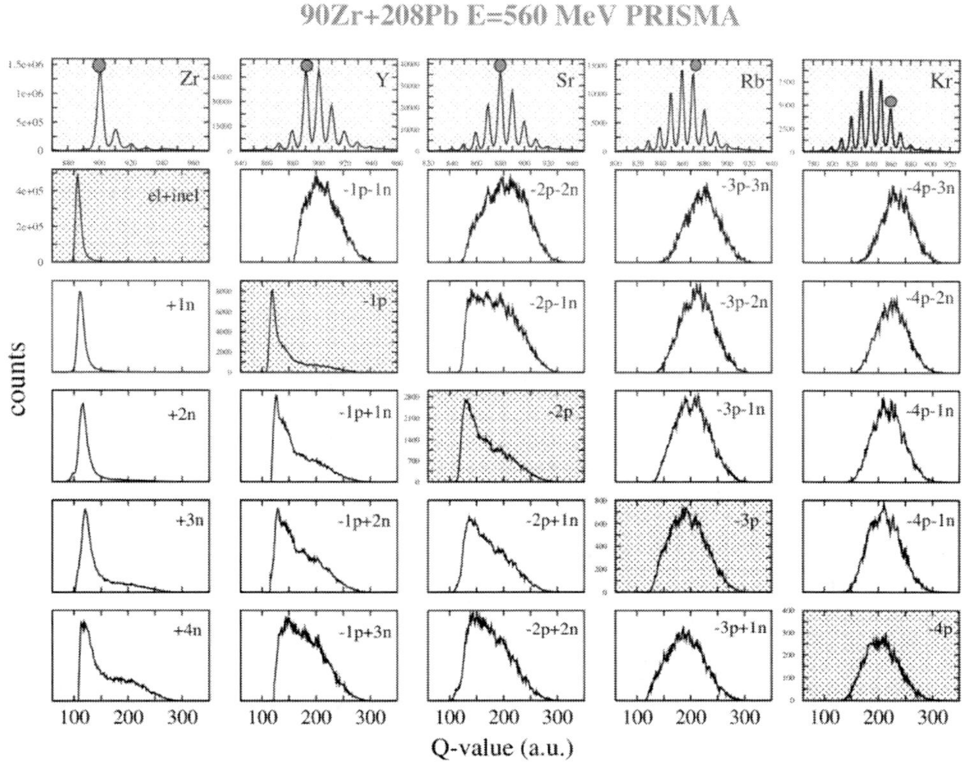

Figure 8. TKEL spectra obtained in the reaction $^{90}Zr+^{208}Pb$ for the indicated transfer channels. In the top row are shown the mass distribution associated to the different nuclear charges, while the circles indicate the specific masses corresponding to the spectra displayed along the upper-left/lower-right diagonal. The centroid of the elastic+inelastic channel corresponds to Q=0. The scale of the Q-value axis is 1 MeV/channel. The vertical scales are different for the various proton stripping channels.

of the yield corresponds to the primary ^{96}Mo, while the rest is equally shared between isotopes corresponding to the evaporation of one and two neutrons. In general, for few nucleon transfer channels most of the yield corresponds to the true binary partner. This behavior is closely connected with the observed TKEL. For the neutron pick-up channels the major contribution in the TKEL is close to the optimum Q values ($Q_{\text{opt}} \simeq 0$), while in the proton stripping channels larger TKEL are observed, thus the neutron evaporation has a stronger effect on the final mass partition.

That the exchange of nucleons may constitute the main source for the dissipation of energy can be understood with a very simple model. Let's suppose to have a system for which N independent single-particle transitions can occur all with the same probability p. In such a system the probability to have the transfer of n particles is simply given by the binomial

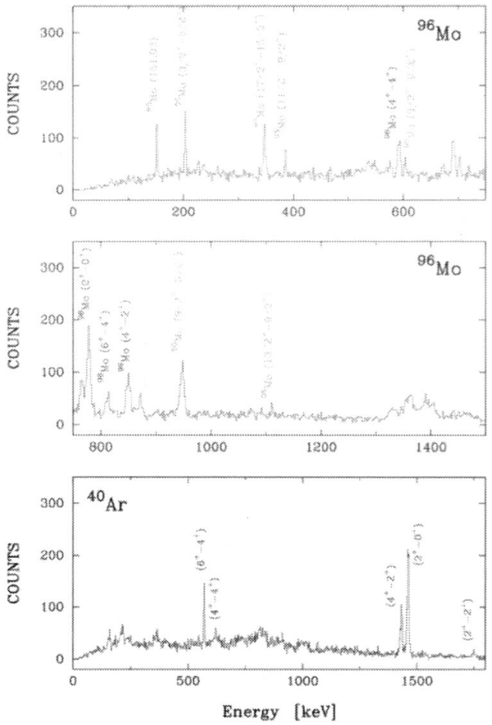

Figure 9. γ spectra for the $-2p+2n$ channel in the reaction ^{40}Ca+^{96}Zr Doppler corrected for the heavy (top two frames) and light fragments (bottom frame). To have a better identification of the different γ lines for the heavy fragment an expanded energy scale has been used (from Ref. [10]).

distribution:

$$P_n = \binom{N}{n} p^n (1-p)^{N-n} \quad . \tag{5}$$

For this distribution the average number of transferred particles is $\langle n \rangle = pN$ so that the corresponding average energy loss is $\langle E_{loss} \rangle = \langle n \rangle Q$. In a heavy-ion collision the number of open channels is very large ($N \sim 100$) so that with $p \sim 0.1$ one estimates an average number of transferred nucleons of the order of 10 and an average energy loss of 50 MeV if for each transition one looses 5 MeV. As this simple example shows, particle transfer may provide the main source for the dissipation of energy.

5. Transfer and fusion processes

I have outlined in the previous sections the importance of large energy loss components in multinucleon transfer processes. The understanding of these processes and its microscopic interpretation is important not only for the transfer in itself but also for its connection with near barrier fusion. To set the frame, for medium mass systems one can roughly distinguish three energy regions, as pictorially depicted in Fig. 10.

At energies well above the Coulomb barrier the transmitted flux T (capture) is larger than the reflected flux R (transfer), i.e. fusion dominates over transfer. However fusion does not necessarily happen when the barrier is overcome, for instance the reaction may develop large TKEL (deep inelastic, which overlap to some extent with quasi-fission processes).

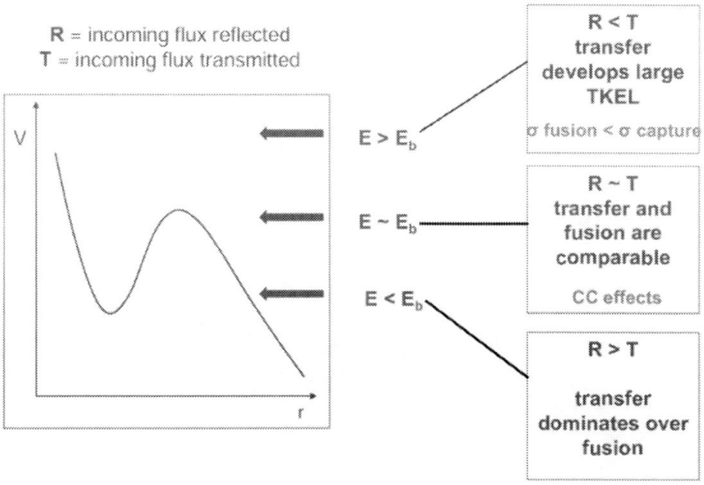

Figure 10. How transfer and fusion compare in different energy ranges. Above the Coulomb barrier (E_b) fusion can be less than capture, at the barrier fusion and transfer are roughly comparable and this is the region where coupled channel effects play a role, at sub barrier energies transfer dominates over fusion.

At energies close to the barrier transfer and fusion have comparable cross sections. This is the energy region where coupled channel effects set-in. Due to couplings of the elastic channel to inelastic and possibly transfer channels, the kinetic energy of the two colliding ions is no more well defined and the scattering process can be seen as if the barrier fluctuates. This effect is well known and has given rise to a wealth of experimental and theoretical work (cfr Refs. [5, 6] and references therein).

At energies below the barrier fusion cross sections drop exponentially and transfer dominates. This is the ideal region to study nucleon-nucleon correlation effects since on one side transfer data are not affected by evaporation (small TKEL) and on the other side theoretical predictions have less ambiguities (choice of the nuclear potential).

5.1. Energies near and above the Coulomb barrier

A debate at energies near and above the Coulomb barrier emerged in connection with the fact that experimental fusion cross sections in different cases turn out to be less than capture. Such an hindrance phenomenon has been suggested to possibly originate to an inadequacy of the nuclear potential which otherwise well fits elastic scattering data [33]. Among alternative explanations is the possibility that other competing channels, like deep inelastic or quasi fission, strongly contribute to capture and have to be taken into account.

To discuss this point one can take as representative example the ^{58}Ni+^{124}Sn system where a complete set of data exist for both evaporation residue, fission, transfer and deep inelastic [34]. These data are shown in Fig. 11, in comparison with the calculations [35] performed with the code GRAZING, which has been already shown to well reproduce the transfer cross sections and the elastic scattering for many systems. The model is also able to estimate the capture cross section, and this is shown in the same figure.

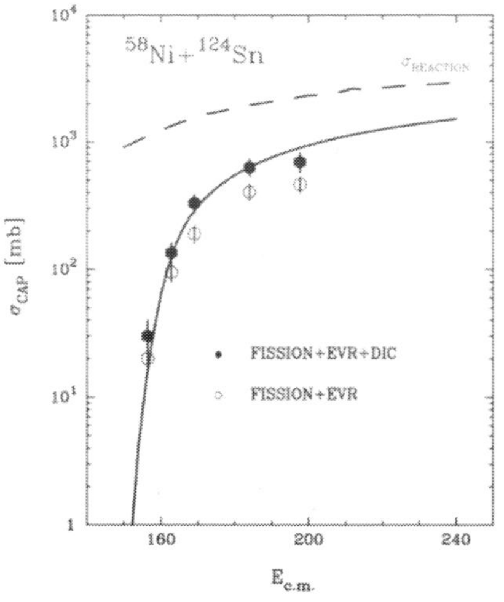

Figure 11. Experimental (points) and GRAZING calculations (lines) for the different reaction channels in the ^{58}Ni+^{124}Sn system. The dashed line represents the total reaction cross section, while the solid line is the sum of evaporation residue (EVR), fission and deep inelastic (DIC) cross sections (capture) (from Ref. [35]).

From this comparison one sees that the experimental sum of evaporation residue and fission cross sections is indeed less than the calculated capture. However, for this system deep inelastic components have been measured at different bombarding energies and shown to represent a significant fraction of the total reaction cross section [36]. By adding these cross sections to the evaporation residue and fission ones one gets a reasonable agreement with capture.

This analysis shows once more the importance of the deep inelastic components and stresses the importance to get complete set of data for other systems, since the systematics is at present too scarse. Its relevance is connected also to the concept ot barrier distributions. As said before, the effect of the couplings to the intrinsic degrees of freedom of the system can be depicted as giving rise to a smearing over several energies of the Coulomb barrier [37]. These barrier distributions can be extracted from the fusion excitation function by taking the second derivative of the energy weighted fusion cross sections [38]. It has been suggested that a similar information can be derived from the energy dependence of the quasi-elastic cross sections at backward angles [39]. The barrier distribution obtained with these two methods turn out to be in reasonable agreement with each other, although the ones extracted from quasi-elastic are somewhat broader and with less structure. A recent analysis [32] showed how the barrier distribution obtained from quasi-elastic cross sections depends on what one considers as quasi-elastic, so one may expect a difference with respect to the results obtained from the fusion cross sections. While what one extracts from fusion reactions [31] reflects how the couplings modify the transmission coefficient, what one gets from quasi-elastic reflects the modification of the reflection coefficient. If fusion and quasi-elastic scattering exhaust most of the total reaction cross section the barrier distributions are expected to be quite similar. However, for heavy systems, where the reaction is dominated by complicated processes like deep inelastic where the nuclei may overcome the Coulomb barrier but separate again with large energy losses and mass transfer, caution must be taken in the interpretation. In the analysis of different heavy ion systems where excitation functions have been measured from above to below the Coulomb barrier [40], a careful analysis had to be done in order to disentangle quasi-elastic from deep inelastic compoments, so as to get the experimental barrier distributions.

5.2. Sub-barrier transfer reactions

In recent years there has been growing interest in studying dynamic processes at energies well below the Coulomb barrier, in particular sub-barrier fusion. This same energy range is also ideal to investigate transfer processes, which are strongly connected with fusion, as they probe different but complementary ranges of nuclear overlap. To set the frame, one can write the transfer cross section as :

$$\sigma_{tr} \sim e^{-\frac{2}{\hbar}\int W(r(t))dt} \sum \left| \int F_{if}(r(t)) e^{i\omega_{if}} dt \right|^2$$

where the first exponential term gives the probability to remain in the elastic channel and the second describes the direct population of the transfer channels being $F(r)$ the transfer form factor and $e^{i\omega_{if}}$ defining the Q-value window, with the sum running over all the final channels. The integrals are performed along the Coulomb trajectory. The imaginary potential $W(r)$, that describes the depopulation of the entrance channel, at very low energies is dominated by the single-nucleon transfer channels. Since the Q-value distributions get narrower at low bombarding energies these subbarrier studies may provide important information on the nuclear correlation close to the ground state. In this energy region the multinucleon transfer channels should be dominated by a successive mechanism with negligible contribution from a cluster-like transfer [41, 42]. This fact should provide a simpler analysis of the data.

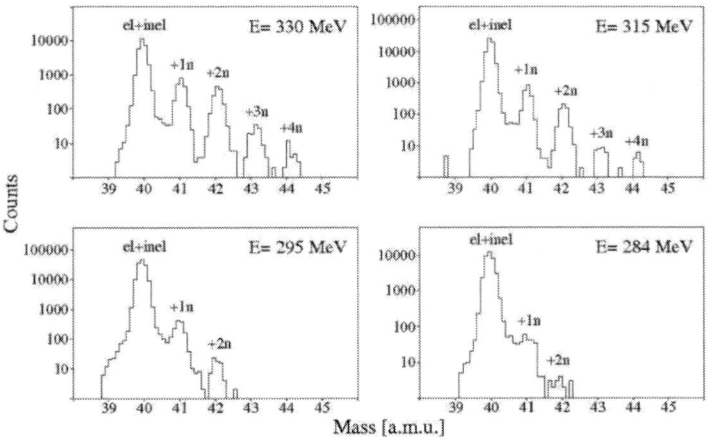

Figure 12. Mass distributions for pure neutron transfer channels obtained in the reaction $^{94}Zr+^{40}Ca$ at the indicated bombarding energies. Ca-like recoils have been detected at θ_{lab}=20° with the PRISMA spectrometer.

From the experimental point of view, measurements of heavy-ion transfer reactions at far sub-barrier energies have significant technical difficulties. At low bombarding energies angular distributions result, in the center of mass frame, in a strong backward peaking, with a maximum at $\theta_{cm} \simeq 180°$. The absolute yield gets very small, therefore high efficiency is needed. At the same time, mass and nuclear charge resolutions must be maintained at a level sufficient to distinguish the different reaction channels. For situations where the projectile has a significant fraction of the target mass, as it is in most cases, the backscattered projectile-like fragment has such a low energy that usual identification techniques become invalid. A suitable way to overcome these limitations is by means of inverse kinematics, thus we recently detected multinucleon transfer channels in the reactions $^{94,96}Zr+^{40}Ca$ at different bombarding energies below the Coulomb barrier, making use of the PRISMA+CLARA set-up. The use of inverse kinematics and the detection at very forward angles, allowed to have, at the same time, enough kinetic energy of the outgoing recoils (for energy and therefore mass resolution) and forward focused angular distribution (high efficiency). Sub barrier fusion cross sections for the same system had been previosuly measured with high precision [43] and a complete set of data for both multinucleon transfer and fusion reactions would provide an excellent basis for coupled channel calculations.

In Fig. 12 are shown the mass spectra for pure neutron transfer channels in the system $^{94}Zr+^{40}Ca$ obtained after trajectory reconstruction at four bombarding energies. While at the higher energies one observes the populations of up to four nucleon transfer, at the lower energies (below the Coulomb barrier) only one and two neutron transfer survive. The Q-value distributions for the +2n channel at the lowest energies are very narrow and at a few MeV above the ground to ground state transitions as a result of the very low excitation energy of the transfer reaction products at these sub-barrier energies. The experimental results will be compared with coupled channel calculations, in particular the comparison of two nucleon vs. one nucleon transfer should provide information on nucleon-nucleon correlation effects.

6. Summary and outlook

In this paper I discussed how the two regimes of quasi elastic and deep inelastic processes, treated historically on very different footing both experimentally and theoretically, can be connected. This transition regime has been extensively studied in the last decade.

New opportunities are offered by the implementation of large acceptance spectrometers based on trajectory reconstruction. With these devices one gained more than an order of magnitude in the total efficiency while keeping good resolution for the detection of heavy ion transfer products. In coupling them with γ arrays, one can now study the transfer yield to specific final states and their decay modes. The experimental yields have been analyzed with semiclassical theories developed to calculate at the same time quasi-elastic and deep-inelastic processes. These theories have been able to provide a consistent description of transfer and fusion reactions by using few degrees of freedom, surface modes and single particles.

It has been shown how energy loss develops in multinucleon transfer processes, in dependence on the specific mass and charge of transfer reaction products and how it can be associated to nucleon evaporation effects. The correct treatment of these processes is very important to predict yields of nuclei far from stability exploiting binary reactions. It is also important to keep in mind that, although primary yields of heavy transfer products may be large, in addition to evaporative processes the heavy fragments may be affected by fission. This fact is of relevance for the production of neutron rich nuclei, a field of growing interest, particularly in those areas connected with r-processes (for instance the region below ^{208}Pb) or with heavy elements. Therefore, experiments aiming at measuring directly the cross sections and spectroscopy of heavy partners and the survival probability of the heavy partner against fission should be pursued in the future. Interesting are the neutron rich regions of the actinides and transactinides which cannot be reached by fission or fragmentation reactions.

More in general, much wider investigations in the field are required, for instance at sub-barrier energies where nuclei enter into contact through the long tail of the nuclear densities. It is in these regions that, for very neutron-rich nuclei, information on nucleon-nucleon correlation could be obtained by studying the excitation functions of specific transfer channels.

7. Acknowledgments
Most of the material presented in this paper has been extracted from measurements performed at LNL and which is the result of a cooperative work of people from different institutions from Italy, Croatia and France. I would like to thank in particular S. Szilner, G. Pollarolo, A. M. Stefanini, E. Fioretto, G. Montagnoli and F. Scarlassara. I wish to thank as well all other people belonging to the nuclear reaction and nuclear spectroscopy group of the PRISMA+CLARA collaboration.

References

[1] Corradi L, Pollarolo G and Szilner S 2009 *J. of Phys.* G **36** 113101
[2] Wu C Y, von Oertzen W, Cline D and Guidry M 1990 *Annu. Rev. Nucl. Part. Sci.* **40** 285
[3] Von Oertzen W and Vitturi A 2001 *Rep. Prog. Phys.* **64** 1247
[4] Rehm K E 1991 *Annu. Rev. Nucl. Part. Sci.* **41** 429
[5] Fusion06: *Int. Conf. on Reaction Mechanisms and Nuclear Structure at the Coulomb barrier*, S. Servolo (Venezia), Italy, March 19-23, 2006, AIP Proceedings Series, Vol. **853**, Melville (New York), Corradi L *et al* eds
[6] Fusion08 : *Int. Conf. on New Aspects of Heavy Ion Collisions Near the Coulomb Barrier*, Chicago (USA), September 22-26, 2008, AIP Proceedings Series, Vol. **1098**, Melville (New York), Rehm K E *et al.* eds.
[7] Schröder W U and Huizenga 1984 *Treatise on Heavy-Ion Science* Vol.2 Plenum, Bromley D A ed
[8] Corradi L et al 2002 *Phys. Rev.* C **66** 024606
[9] Szilner S et al 2005 *Phys. Rev.* C **71** 044610
[10] Szilner S et al 2007 *Phys. Rev.* C **76** 024604
[11] Stefanini A M et al 2002 *Nucl. Phys.* A **701c** 217
[12] Corradi L 2010 *Nucl. Phys.* A **384c** 129
[13] Gadea A et al 2002 *Eur. Phys. J.* A **20** 193
[14] Marginean N et al 2006 *Phys. Lett.* B **633** 696
[15] Lunardi S et al 2007 *Phys. Rev.* C **76** 034303
[16] Valiente-Dobon J J et al 2009 *Phys. Rev. Lett.* **102** 242502
[17] Broda R et al 2010 *Phys. Rev.* C **82** 034319
[18] Corradi L et al 2002 *Nucl. Phys.* A **701** 109c
[19] Gobbi A and Nörenberg *Heavy Ion Collisions* Vol.2 Bock R ed, North Holland 1980
[20] Rudolf G et al 1979 *Nucl. Phys.* A **330** 243
[21] Quesada J M, Pollarolo G, Broglia R A and Winther A 1985 *Nucl. Phys.* A **442** 381
[22] Broglia R A and Winther A *Heavy Ion Reactions* (Addison-Wesley Pub. Co., Redwood City CA, 1991)
[23] Corradi L et al 1999 *Phys. Rev.* C **59** 261
[24] Dasso C H, Pollarolo G and Winther A 1994 *Phys. Rev. Lett.* **73** 1907
[25] Buttle P J A and Goldfarb L J B 1966 *Nucl. Phys.* **78** 409
[26] Brink D M 1972 *Phys. Lett.* B **40** 37
[27] Sørensen J H, Pollarolo G and Winther A 1989 *Phys. Lett.* B **225** 41
[28] Winther A program GRAZING, htpp:/www.to.infn.it/~nanni/grazing
[29] Winther A 1994 *Nucl. Phys.* A **572** 191
[30] Winther A 1995 *Nucl. Phys.* A **594** 203
[31] Pollarolo G and Winther A 2000 *Phys. Rev.* C **62** 054611
[32] Pollarolo G 2008 *Phys. Rev. Lett.* **100** 252701
[33] Newton J O et al 2004 *Phys. Rev.* C **70** 024605
[34] Esbensen H, Jiang C L and Rehm K E 1998 *Phys. Rev.* C **57** 2401
[35] Pollarolo G 2007 *Nucl. Phys.* A **787** 207c
[36] Wolfs F L H 1987 *Phys. Rev.* C **36** 1379
[37] Esbensen H 1981 *Nucl. Phys.* A **352** 147
[38] Rowley N, Satchler G R and Stelson P H 1991 *Phys. Lett.* B **254** 25
[39] Timmers H et al 1995 *Nucl. Phys.* A **584** 190
[40] Mitsuoka S et al 2007 *Phys. Rev. Lett.* **99** 182701
[41] Bayman B F and Chen J 1982 *Phys. Rev.* C **26** 1509
[42] Maglione E, Pollarolo G, Vitturi A, Broglia R A and Winther A 1985 *Phys. Lett.* B **162** 59
[43] Stefanini A M et al 2007 *Phys. Rev.* C **76** 014610

Journal of Physics: Conference Series **282** (2011) 012006 doi:10.1088/1742-6596/282/1/012006

Investigation on the quasifission process by theoretical analysis of experimental data of fissionlike reaction products

G Giardina[1,2], A K Nasirov[3,4], G Mandaglio[1,2], F Curciarello[1,2], V De Leo [1,2], G Fazio[1,2], M Manganaro[1,2], M Romaniuk[1,2], C Saccá[5]

[1] Dipartimento di Fisica, Università di Messina, I-98166 Messina, Italy
[2] Istituto Nazionale di Fisica Nucleare, Sezione di Catania, I-95123 Catania, Italy
[3] Joint Institute for Nuclear Research, 141980, Dubna, Russia
[4] Institute of Nuclear Physics, 100214, Tashkent, Uzbekistan
[5] Dipartimento di Scienze della Terra, Università di Messina, I-98166 Messina, Italy

E-mail: giardina@nucleo.unime.it

Abstract. The hindrance to complete fusion is a phenomenon presenting in the most part of the capture events in reactions with massive nuclei. This phenomenon is due to the onset of the quasifission process which competes with complete fusion during the evolution of the composed system formed at capture stage. The branching ratio between quasifission and complete fusion strongly depends from different characteristics of reacting nuclei in the entrance channel. The experimental and theoretical investigations of reaction dynamics connected with the formation of composed system is nowadays the main subject of the nuclear reactions. There is ambiguity in establishment of the reaction mechanism leading to the observed binary fissionlike fragments. The correct estimation of the fusion probability is important in planning experiments for the synthesis of superheavy elements. The experimental determination of evaporation residues only is not enough to restore the true reaction dynamics. The experimental observation of fissionlike fragments only cannot assure the correct distinguishing of products of the quasifission, fast fission, and fusion-fission processes which have overlapping in the mass (angular, kinetic energy) distributions of fragments. In this paper we consider a wide set of reactions (with different mass asymmetry and mass symmetry parameters) with the aim to explain the role played by many quantities on the reaction mechanisms. We also present the results of study of the ^{48}Ca+^{249}Bk reaction used to synthesize superheavy nuclei with Z = 117 by the determination of the evaporation residue cross sections and the effective fission barriers $< B_{\rm f} >$ of excited nuclei formed along the de-excitation cascade of the compound nucleus.

1. Introduction

In heavy ion collisions the two reacting nuclei form a rotating nuclear system which in some cases its evolution leads to complete fusion, while in other cases the nuclear system can also re-separate into two fragments different from the initial nuclei in the entrance channel [1–3]. If there is a well in the nucleus-nucleus potential at the first stage of reaction the full momentum transfer can occur (this event is defined as capture) in dependence on the values of relative kinetic energy and friction coefficients [4, 5]. During the evolution of DNS its two nuclei may change their masses A_1, A_2 and charges Z_1, Z_2 but with constant total mass $A = A_1 + A_2$ and

Journal of Physics: Conference Series **282** (2011) 012006 doi:10.1088/1742-6596/282/1/012006

charge $Z = Z_1 + Z_2$. In this context, DNS can evolve into complete fusion (depending on the intrinsic fusion barrier $B^*_{\rm fus}$ which is equal to the difference between the maximum value of the driving potential and its value corresponding to the initial charge asymmetry) through much more its mass asymmetric configurations but also during such an evolution the system can re-separate into two nuclei (quasifission fragments) [4,5]. Being in competition with the complete fusion this process is characterized by the value of the quasifission barrier $B_{\rm qf}$ (the depth of the pocket in the nucleus-nucleus potential). In this case, the quasifission fragments have a mass asymmetry parameter larger than that of nuclei in the entrance channel. On the other hand, it also exists the probability (depending on the characteristics and structures of initial nuclei) that the nuclei constituting the DNS may exchange mass and charge increasing the mass symmetry parameter. In this last case the decay of DNS into two nuclei (depending on the lifetime of DNS and the value of quasifission barrier $B_{\rm qf}$ for such reached couple of nuclei in DNS) lead to fragments with larger mass symmetry parameter than that of nuclei in the entrance channel. Of course, the rates of such different contributions of quasifission are related to the characteristics of the entrance channel (beam energy and kinds of reacting nuclei).

The capture events that survive quasifission populate the complete fusion channel. Consequently, the deformed mononucleus may reach the fully equilibrated statistical shape of the compound nucleus (CN), or if the fission barrier $B_{\rm f}$ disappears the system immediately decays into two fragments (fast fission process). The latter phenomenon occurs only at high angular momentum ℓ for which the fission barrier of the complete fusion system is zero ($B_{\rm f}(\ell > \ell_{\rm f})= 0$). Therefore, the fast fission process is present in reactions only at high angular momentum values ($\ell > \ell_{\rm f}$) while the quasifission process takes place at all values of ℓ contributing to the capture reaction.

Finally, in the last stage of nuclear reaction, the formed CN may de-excite by fission (producing fusion-fission fragments) or by emission of light particles. The reaction products that survive fission are the evaporation residues (ER) [6,7]. The registration of ER is clear evidence of the compound nucleus formation, but generally the determination of ER's only it is not enough to determine the complete fusion cross section and to understand the dynamics of the de-excitation cascade of CN if the fission fragments are not included into consideration. On the other hands, the registration of the fissionlike fragments and separation of kinds of fragments according to the mechanism of its origin on the basis of some assumptions does not allow for sure correct determination of the fusion-fission rate in the cases of overlapping of the mass fragment distributions of different processes (quasifission, fast fission and fusion-fission).

Moreover, by observing the fission products of various reactions leading to fissile nuclei formed in neutron(or very light particles) - induced reactions on fissile targets, one can conclude that the low excited compound nucleus (at about $E^*_{\rm CN} < 10$ MeV) decays into very asymmetric fission fragments (near to the shell closure), while the actinide nuclei or compound nuclei formed in heavy ion collisions at intermediate or high excitation energy ($E^*_{\rm CN} > 15$ MeV) undergo fission by symmetric fragments. Starting from these general observations some researchers put forward the idea that the complete fusion process of two colliding nuclei maybe considered as the inverse process of fission. The authors in the papers [8,9] argued that since the fission of a compound nucleus in heavy ion collisions produces just symmetric fragments, then in the collisions of two symmetric (or almost symmetric) nuclei complete fusion has to be a very probable process. But, unfortunately this is not true. For systems of colliding nuclei heavier than ^{110}Pa+^{110}Pa fusion does not occur absolutely, while for reactions like ^{100}Mo+^{100}Mo, ^{96}Zr+^{96}Zr, ^{96}Zr+^{100}Mo, ^{100}Mo+^{110}Pa or induced by projectiles higher than Zn, Ge, Kr a strong hindrance to fusion appears.

Following the previous reasonings one can affirm that the hypothetical ^{132}Sn+^{120}Cd reaction should lead to the ^{252}Cf CN since ^{120}Cd (with Z=48 near the shell closure 50) and ^{132}Sn (with double shell closure Z=50 and N=82) are produced with highest yields in spontaneous

Journal of Physics: Conference Series **282** (2011) 012006 doi:10.1088/1742-6596/282/1/012006

fission of ^{252}Cf. But calculation for this reaction does not give meaningful fusion probability ($P_{\rm CN} < 5 \times 10^{-7}$).

The simple reason resides in the peculiarities of the reaction dynamics. In the spontaneous fission of ^{252}Cf the average value of angular momentum distribution of the fragments is close to zero, but if we want to reach the ^{252}Cf compound nucleus, by the hypothetical ^{132}Sn+^{120}Cd reaction (or by the realistic ^{132}Sn+^{116}Cd reaction leading to the ^{248}Cf CN), the average value of angular momentum distribution of DNS in the entrance channel may be about $<\ell>= 50\hbar$ or higher by increasing the beam energy. The $<\ell>$ value is calculated as

$$< \ell\,(E_{\rm c.m.};\, \alpha_{\rm P}, \alpha_{\rm T}) > \;=\; \frac{\sum_{\ell=0}^{\ell=\ell_d} \ell < \sigma^{(\ell)}_{(cap)} >_{\alpha_P,\alpha_T} (E_{\rm c.m.})}{\sum_{\ell=0}^{\ell=\ell_d} < \sigma^{(\ell)}_{(cap)} >_{\alpha_P,\alpha_T} (E_{\rm c.m.})} \quad . \tag{1}$$

In this case the whole peculiarities of the reaction mechanism in the first stage of reaction should lead almost completely to quasifission (re-separation of nuclei of DNS), while the fusion probability $P_{\rm CN}$ should be lower than about 10^{-7}. But also with this low probability of complete fusion, the excited and fast rotating deformed mononucleus undergoes fast fission before the system can reach the compact shape of compound nucleus.

Also in the cases of the explored ^{22}Ne+^{250}Cf (more mass asymmetric system), ^{24}Mg+^{248}Cm, ^{28}Si+^{244}Pu, ^{34}S+^{238}U, and ^{40}Ar+^{232}Th (less mass asymmetric system) reactions, the ^{272}Hs compound nuclei formed in the different entrance channels at a defined excitation energy $E^*_{\rm CN}$ have different angular momentum distributions. Therefore, such compound nuclei decay by different rates of reaction products. The fusion-fission fragment mass distributions are peaked at around the ^{136}Xe nucleus with different dispersions and average angular momentum distributions in connection with the various entrance channels. If we calculate the formation probability of the ^{272}Hs compound nucleus in the mass symmetric ^{136}Xe+^{136}Xe reaction at the same fixed excitation energy $E^*_{\rm CN}$ as in the considered ^{22}Ne+^{250}Cf reaction (where $P_{\rm CN} \simeq 1$), we do not meaningfully reach such a compound nucleus($P_{\rm CN} < 10^{-10}$). The angular momentum distribution for the ^{136}Xe+^{136}Xe collision at the capture stage is completely different and all conditions of reaction dynamics lead to deep inelastic and quasifission products.

In this context, for the ^{136}Xe+^{132}Sn ($P_{\rm CN} < 10^{-8}$) and ^{132}Sn+^{176}Yb ($P_{\rm CN} < 5\times 10^{-11}$) reactions, one can observe the same above-described hindrance to complete fusion.

2. Capture and Deep Inelastic Collision

Our theoretical capture cross section $\sigma_{\rm cap}$ includes all damped reactions, excluding deep inelastic collisions (DIC). The partial capture cross sections are contributed by the full momentum transfer events corresponding to the trapping into a pocket in the nucleus-nucleus potential after dissipation of the relative kinetic energy. Differently, the DIC events are not characterized by the full momentum transfer and are not trapped in the pocket of the nucleus-nucleus potential. The DIC events are not connected with any pocket of a potential; for example at large values of ℓ there is no pocket but the DIC events take place. The partial capture cross section is calculated by the following formula [10, 11].

$$\sigma^{\ell}_{cap}(E_{\rm c.m.}) = \pi \lambda\!\!\!^{-2} \mathcal{P}^{\ell}_{cap}(E_{\rm c.m.}) \tag{2}$$

where

$$\mathcal{P}^{\ell}_{cap}(E_{\rm c.m.}) = \begin{cases} 1, & \text{if } \ \ell_{\rm min} \le \ell \le \ell_{\rm d} \ \text{ and } \ E_{\rm c.m.} > V_{\rm Coul} \\ 0, & \text{if } \ \ell < \ell_{\rm min} \ \text{ or } \ \ell > \ell_{\rm d} \text{ and } \ E_{\rm c.m.} > V_{\rm Coul} \\ 0, & \text{for all } \ell \quad \text{if } \ E_{\rm c.m.} \le V_{\rm Coul}\,. \end{cases}$$

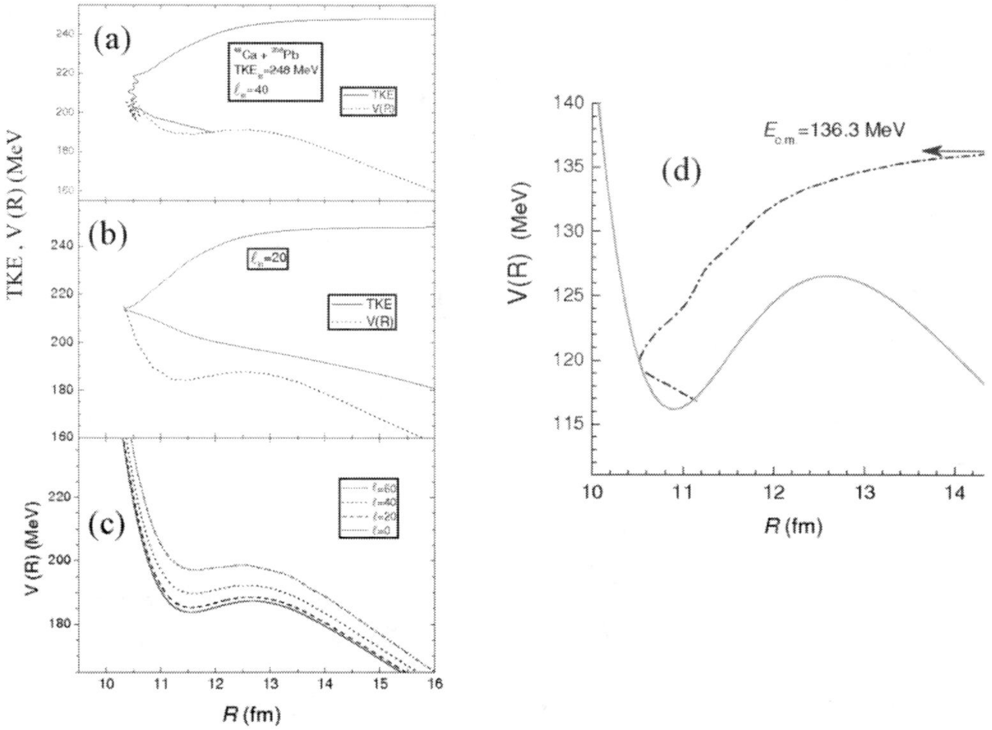

Figure 1. Capture (a) and deep inelastic collision (b) for the $^{48}Ca+^{208}Pb$ reaction. Dependence of nucleus-nucleus potential (c) on the orbital angular momentum ℓ. Illustration of capture path (d) (dot dashed line) into potential well (solid line) as obtained by the numerical solution of the equation of relative motion of colliding nuclei with the initial energy $E_{\mathrm{c.m.}}$ = 136.3 MeV and ℓ=0 for the $^{32}S+^{184}W$ reaction [5].

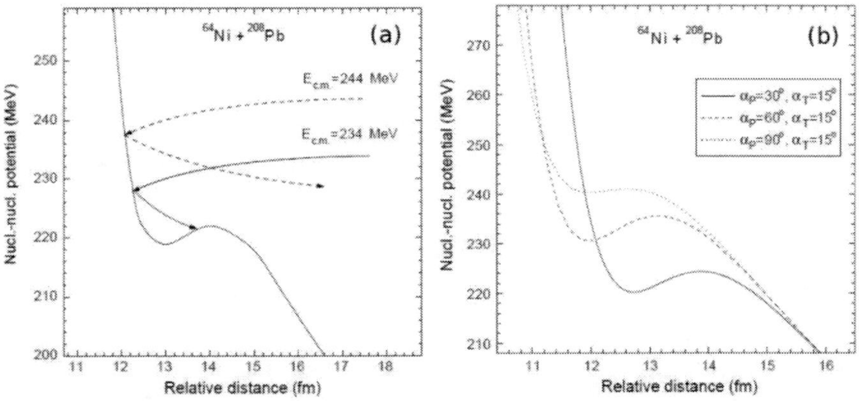

Figure 2. (a)Dependence of capture (solid curve) and quasifission (dashed curve) processes for nucleus-nucleus collisions on the internuclear potential $V(R)$ and on the initial values of the beam energy $E_{\mathrm{c.m.}}$.(b) Dependence of the nucleus-nucleus potential on the mutual orientation of colliding nuclei; solid line: $\alpha_1 = 30°$ and α_2= 15°; dashed line: $\alpha_1 = 60°$ and α_2= 15°; dotted line: α_1= 90° and α_2= 15°.

In the calculation of nucleus-nucleus potential, which includes the Coulombian $V_{\mathrm{Coul}}(Z, A, R)$, nuclear $V_{\mathrm{n}}(Z, A, R)$ and rotational $V_{\mathrm{rot}}(Z, A, R, \ell)$ parts, we take into account the static and dynamic deformations of nuclei at initial stage of interaction, shell structures of nuclei, mass (charge) asymmetry of the interacting nuclei, bombarding energy beam, orbital angular momentum, and orientation angles of the axial symmetry axes of reacting nuclei at initial stage.

In Figs. 1 and 2 we present some examples of capture events with different potential wells and its dependencies on the angular momentum and orientation angles of axes of reacting nuclei;

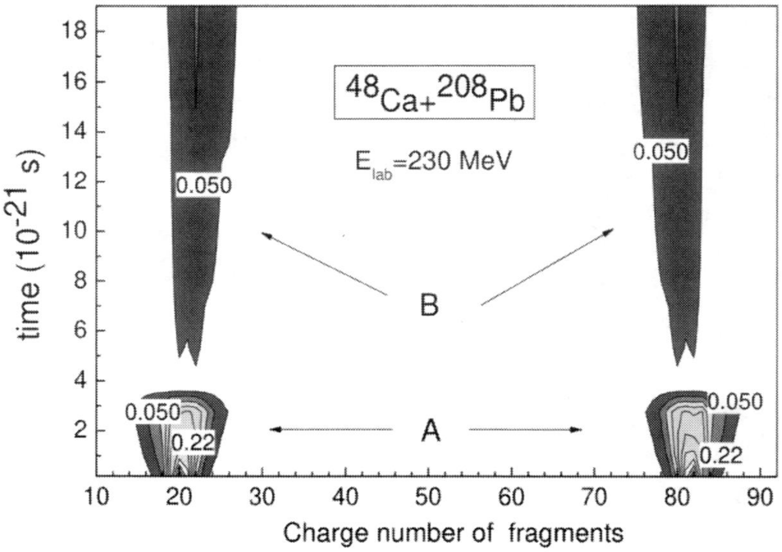

Figure 3. Time dependence of the theoretical results for the mass distribution of the deep inelastic collision (A) and quasi-fission (B) fragments in the $^{48}Ca+^{208}Pb$ reaction.

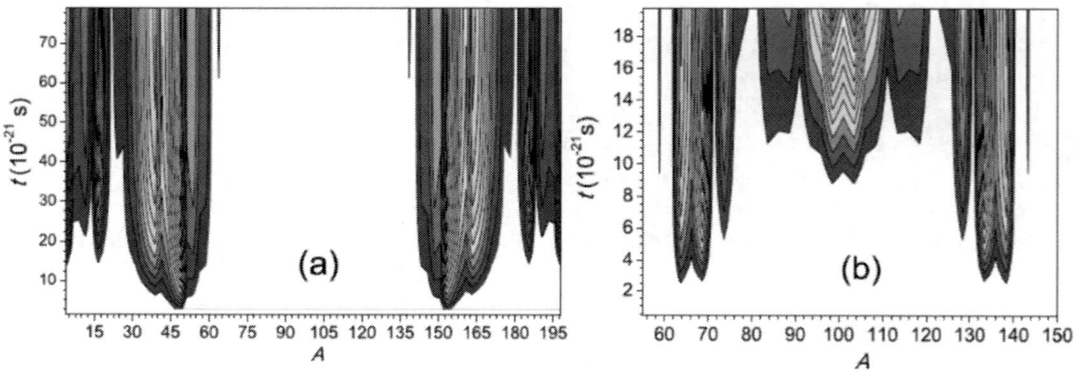

Figure 4. (a) The mass distribution of the quasifission products yield in the $^{48}Ca+^{154}Sm$ reaction at $E_{c.m.}$=140 MeV as a function of the lifetime of the dinuclear system formed at capture stage. (b) The mass distribution of the quasifission product yields in the $^{48}Ca+^{154}Sm$ reaction at $E_{c.m.}$=160 MeV as a function of the lifetime of the dinuclear system.

moreover, in figures we also present the cases of capture and DIC events by changing the beam energies and angular momentum values.

The lifetime of an excited DNS for a given reaction depends on the initial collision energy $E_{c.m.}$ and angular momentum distribution values (see formulae (6)-(8) and Fig.4 of our paper [12]). The nuclei constituting the DNS exchange mass and charge, as well as their shapes. The evolution of DNS depends on its excitation energy, orientation angles of the axial symmetry axes and shell structures of reacting nuclei. Therefore, the DNS during its evolution can evolve to complete fusion (fusion process) or can decay into two fragments (quasifission process). The competition between these two processes is related to the values of intrinsic fusion barrier $B^*_{\rm fus}$ and quasifission barrier $B_{\rm qf}$ [5–7] depending on the peculiarities of reacting nuclei, beam energy and angular momentum distribution.

The mass and angular momentum distributions of quasifission fragments in many cases and conditions can overlap with the mass and angular distributions of fusion-fission fragments, or/and with the ones of the fast fission fragments [10, 13], leading to a real difficulty in the experimental analysis in order to extract the true yields of fragments belonging to various processes (see for example [10, 14] and Fig. 3 of Ref. [15]).

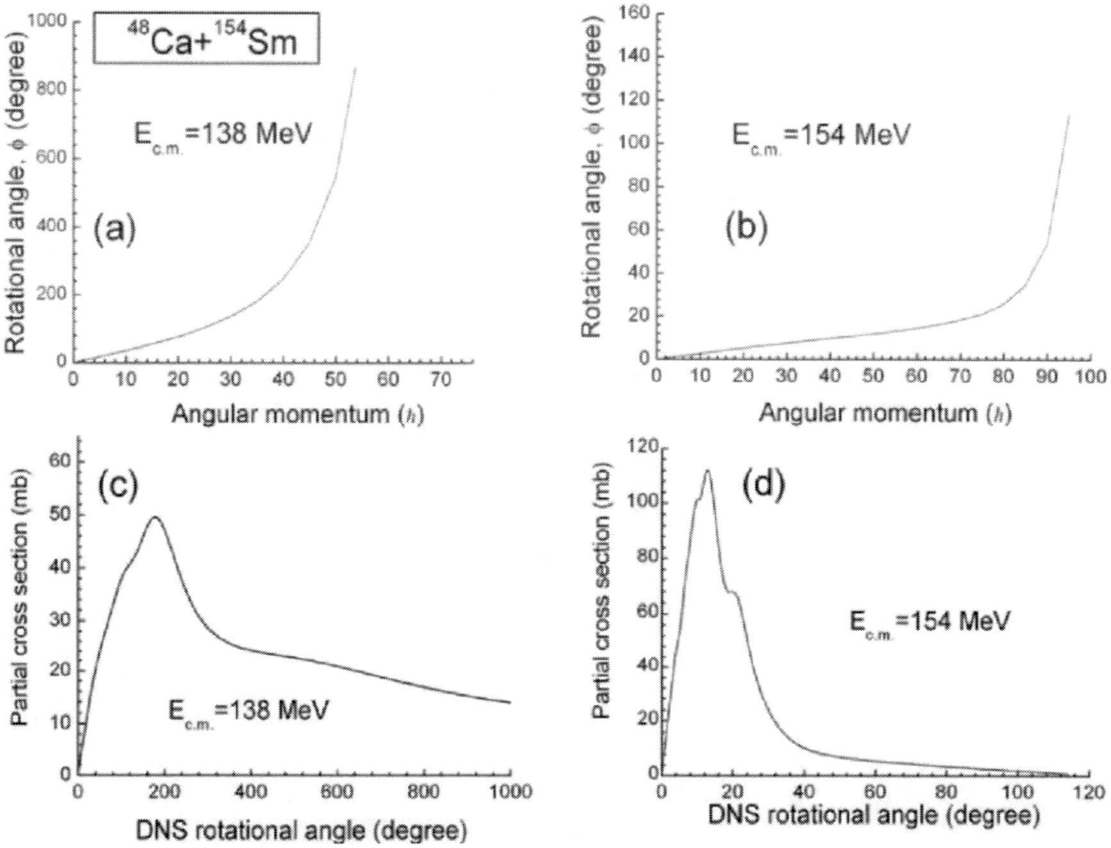

Figure 5. The rotational angle of the dinuclear system forming the ^{58}Cr+^{144}Ce fragments (by the ^{48}Ca+^{154}Sm) as a function of the orbital angular momentum (a) and (b), and angular distribution of the yield of quasifission fragments (c) and (d).

Usually, in reactions with actinides the maximum of the yield of the quasifission products is observed in the mass region between masses of the projectile-like products and symmetric masses. In the ^{48}Ca + ^{208}Pb reaction, the maximum of the quasifission products is very close to the region of projectilelike and targetlike fragments (see Fig. 4 of Ref. [10], as well as Fig. 1 of Ref. [16] and Fig. 8 of Ref. [17]). This is due to the fact that projectile or/and target are double magic nuclei. As a result the mass distributions of the capture, *i.e.* quasifission reactions and deep inelastic collisions are mixed. The reaction time of the deep inelastic collisions is much smaller than that of quasifission and fast fission reactions, because the last two reactions take place if capture occurs, *i.e.* projectile has been trapped into potential well (see Fig. 1).

It is difficult to separate without uncertainty products of these two processes in experiments. To demonstrate importance of this circumstance, the mass distributions of the deep-inelastic and capture reactions were calculated by the method developed on the base of the dinuclear system concept [18, 19]. The results are presented in Fig. 3. The reason of this result can be seen from the driving potential which has a deep valley around initial charge asymmetry due to the nuclear shell effects. Therefore, the sufficient part of the yields of quasifission and deep-inelastic products are mixed in the mass and charge distributions. The difference between them is in their angular distributions which are connected with the lifetime of the formed rotated dinuclear system. The overlap of the maximum of the mass distribution of quasifission with that of the deep inelastic collisions does not mean absent of the contribution of the quasifission process in the total reaction cross section. This explains the strong deviation of the theoretical values of the capture cross section at low energies from the experimental capture cross section which was determined by the symmetric fragmentation only.

So, the mixing of the mass distributions of the deep inelastic collisions and quasifission processes causes an ambiguous determination of the capture cross section from the experimental mass distributions in case of reactions with magic or double magic nuclei, like to the case of the reaction under discussion. As well as the mixing of the mass distributions of the quasifission, fast fission and fusion-fission reactions makes difficult the estimation of the complete fusion cross section leading to the genuine compound nucleus. Due to the strong dependence of the mass distribution on the peculiarities of nuclear shell structure, the proportion of mixing is changed from one reaction to another reaction. The theoretical study of this problem is useful to establish the reactions and beam energies which are favourable to obtain the maximum cross section of formation of the compound nucleus with the relatively low angular momentum and excitation energy.

We can state that, at the low energies, the products of the symmetric fragmentation are generally formed at usual fission of CN, and therefore they are related to the fusion cross section. At energies $E_{lab} > 212$ MeV the contribution of the fast fission appears and it increases by the beam energy.

3. Evolution of DNS : competition between Quasifission and Complete Fusion processes

During the lifetime of the excited DNS the composite system evolves by exchanging nucleons between the two nuclei constituting the DNS. For each event, during this stage of reaction, the DNS can reach the shape of a deformed mononucleus (complete fusion) or it can decay into two fragments (quasifission) before to reach the complete fusion stage. In the first case the nuclear system has to reach the statistical equilibrate shape of the CN, but the events which correspond to deformed mononucleus without barrier to provide stability ($B_{\rm f}$=0 for $\ell > \ell_{\rm f}$, where the $\ell_{\rm f}$ is characteristic of each CN with its structure of nucleons) cannot reach the shape of CN because the deformed complete fusion system immediately decays into two fragments(fast fission process). Therefore, the partial capture cross section is contributed by the following terms:

$$\sigma^{\ell}_{cap}(E_{\rm c.m.};\beta_P,\alpha_T) = \sigma^{\ell}_{qfiss}(E_{\rm c.m.};\beta_P,\alpha_T) + \sigma^{\ell}_{fus}(E_{\rm c.m.};\beta_P,\alpha_T) + \sigma^{\ell}_{fastfis}(E_{\rm c.m.};\beta_P,\alpha_T) \quad (3)$$

where

$$\sigma^{\ell}_{\rm fus}(E_{\rm c.m.}) = \sigma^{\ell}_{cap}(E_{\rm c.m.})P_{\rm CN}(E_{\rm c.m.},\ell), \quad (4)$$

is the partial fusion cross section and $P_{CN}(E_{\rm c.m.},\ell)$ is fusion probability or hindrance factor to complete fusion as a function of ℓ. The fusion cross section, for each orientation angles of the symmetry axes of the deformed reacting nuclei, is obtained by formula

$$\sigma_{fus}(E_{\rm c.m.};\beta_P,\alpha_T) = \sum_{\ell=0}^{\ell_f}(2\ell+1)\sigma_{cap}(E_{\rm c.m.},\ell;\beta_P,\alpha_T)P_{CN}(E_{\rm c.m.},\ell;\beta_P,\alpha_T). \quad (5)$$

It is clear that the fusion cross section in formula (4) includes the cross sections of evaporation residue and fusion-fission products. Therefore, taking into account the contributions of all configurations, the average value of the fusion cross section is obtained as

$$<\sigma_{fus}>_{\{\alpha_P,\alpha_T\}}(E_{c.m.}) = \int_0^{\pi/2}\sin\alpha_P\int_0^{\pi/2}\sin\alpha_T \times \sigma_{fus}(E_{c.m.};\alpha_P,\alpha_T)d\alpha_T d\alpha_P, \quad (6)$$

for deformed nuclei, or as

$$<\sigma_{fus}>_{\{\alpha_T\}}(E_{c.m.}) = \int_0^{\pi/2}\sin\alpha_T \times \sigma_{fus}(E_{c.m.};\alpha_T)d\alpha_T, \quad (7)$$

for spherical projectile and deformed target.

Obviously, the quasifission cross section is obtained as

$$\sigma_{qfis}(E_{\rm c.m.};\beta_P,\alpha_T)=\sum_{\ell=0}^{\ell_d}(2\ell+1)\sigma_{cap}(E_{\rm c.m.},\ell;\beta_P,\alpha_T)(1-P_{CN}(E_{\rm c.m.},\ell;\beta_P,\alpha_T)). \quad (8)$$

Another binary process which leads to the formation of two fragments similar to the ones of fusion-fission or quasifission is the fast fission. The fast fission cross section is calculated by summing the contributions of the partial cross sections related to the range $\ell_f \leq \ell \leq \ell_d$ (for which $B_{\rm f}$=0) leading to the formation of the mononucleus:

$$\sigma_{fastfis}(E_{\rm c.m.};\beta_P,\alpha_T)=\sum_{\ell_f}^{\ell_d}(2\ell+1)\sigma_{cap}(E_{\rm c.m.},\ell;\beta_P,\alpha_T)P_{CN}(E_{\rm c.m.},\ell;\beta_P,\alpha_T). \quad (9)$$

The fission and the evaporation residue cross sections are calculated by the advanced statistical code [4, 6, 10] that takes into account the damping of the shell correction in the fission barrier as a function of nuclear temperature and orbital angular momentum.

The fusion cross section is calculated from the branching ratio $P_{\rm CN}(Z)$ of the decay rates of overflowing the border of the potential well ($B^{(Z)}_{\rm qf}$) along R at a given mass asymmetry (decay of DNS in quasifission fragments) over the barriers on mass asymmetry axis $B^*_{\rm fus}$ for the complete fusion or $B^{(Z)}_{\rm sym}$ in opposite direction to the symmetric configuration of DNS (for details see Fig. 4 of Ref. [5]):

$$P^{(Z)}_{\rm CN}(E^*_{DNS})\approx\frac{\Gamma^{(Z)}_{\rm fus}(B^*_{\rm fus},E^*_{DNS})}{\Gamma^{(Z)}_{\rm (qf)}(B_{\rm qf},E^*_{DNS})+\Gamma^{(Z)}_{\rm (fus)}(B^*_{\rm fus},E^*_{DNS})+\Gamma^{(Z)}_{\rm sym}(B_{\rm sym},E^*_{DNS})}, \quad (10)$$

where $\Gamma_{\rm fus}$, $\Gamma_{\rm qf}$ and $\Gamma_{\rm sym}$ are corresponding widths determined by the level densities on the barriers $B^*_{\rm fus}$, $B^*_{\rm sym}$ and $B_{\rm qf}$ involved in the calculation of $P_{\rm CN}$ are used in the model [7, 20, 21] based on the dinuclear system concept [22]. Here $E^*_{\rm DNS}(Z_P,A_P,\ell)=E_{\rm c.m.}-V(Z_P,A_P,\ell,R_{\rm m})$ is the excitation energy of dinuclear system in the entrance channel, where Z_P and A_P are charge and mass numbers of the projectile nucleus. $V(Z,A,R_{\rm m},\ell)$ is the minimum value of the nucleus-nucleus potential well (for the DNS with charge asymmetry Z) and its position on the relative distance between the centers of nuclei is marked as $R=R_{\rm m}$. The value of $B_{\rm qf}$ for the decay of DNS with the given charge asymmetry of fragments is equal to the depth of the potential well in the nuclear-nuclear interaction. The intrinsic fusion barrier $B^*_{\rm fus}$ is connected with mass (charge) asymmetry degree of freedom of the dinuclear system and it is determined from the potential energy surface:

$$U(Z;R,\ell)=U(Z,\ell,\beta_1,\alpha_1;\beta_2,\alpha_2)=B_1+B_2+V(Z,\ell,\beta_1,\alpha_1;\beta_2,\alpha_2;R)-(B_{CN}+V_{CN}(\ell)). \quad (11)$$

Here, B_1, B_2 and $B_{\rm CN}$ are the binding energies of the nuclei in DNS and the CN, respectively, which were obtained from [23]; the fragment deformation parameters β_i are taken from the tables in [23–25] and α_i are the orientation angles of the reacting nuclei relative to the beam direction; $V_{\rm CN}(\ell)$ is the rotational energy of the CN. The distribution of neutrons between two fragments for the given proton numbers Z and Z_2 or ratios A/Z and A_2/Z_2 for both fragments were determined by minimizing the potential $U(Z;R)$ as a function of A for each Z.

The driving potential $U_{\rm dr}(Z)\equiv U(Z,R_m)$ is a curve linking minimums corresponding to each charge asymmetry Z in the valley of the potential energy surface from $Z=0$ up to $Z=Z_{\rm CN}$ (see Fig. 4 of Ref. [5]). We define the intrinsic fusion barrier for the dinuclear system with charge asymmetry Z as $B^*_{\rm fus}(Z,\ell) = U(Z_{\rm max}, R_{\rm m}(Z_{\rm max}),\ell) - U(Z, R_{\rm m}(Z),\ell)$, where $U(Z_{\rm max},\ell)$ is a maximum value of potential energy at $Z = Z_{\rm max}$ in the valley along the way of complete fusion from the given Z configuration. The $B^*_{\rm sym}(Z,\ell)$ is defined by the similar way as shown in Fig. 4 c of Ref. [5].

The masses and charges of the projectile and target nuclei are not constant during capture and after formation of the dinuclear system. The intense proton and neutron exchange between constituents of DNS is taken into account by calculation of the complete fusion probability $P_{\rm CN}$ as fusion from all populated DNS configurations according to the formula

$$P_{CN}(E^*_{\rm DNS}(Z,A,\ell);\{\alpha_i\}) = \sum_{Z_{\rm sym}}^{Z_{\rm max}} Y_Z(E^*_{\rm DNS}(Z,A,\ell))P^{(Z)}_{\rm CN}(E^*_{\rm DNS}(Z,A,\ell);\{\alpha_i\}) \qquad (12)$$

where $E^*_{\rm DNS}(Z,A,\ell) = E^*_{\rm DNS}(Z_P,A_P,\ell) + \Delta Q_{\rm gg}(Z)$ is the excitation energy of DNS with angular momentum ℓ for a given value of its charge-asymmetry configuration Z and $Z_{\rm CN}-Z$; $Z_{\rm sym} = (Z_1+Z_2)/2$; $\Delta Q_{\rm gg}(Z)$ is the change of $Q_{\rm gg}$-value by changing the charge (mass) asymmetry of DNS; $Y_Z(E^{*(Z)}_{\rm DNS})$ is the probability of population of the $(Z, Z_{\rm CN}\text{-}Z)$ configuration at $E^{*(Z)}_{\rm DNS}$ and given orientation angles (α_1,α_2). $Y_Z(E^*_{DNS},\ell,t)$ is the probability of population of the configuration $(Z, Z_{tot}-Z)$ at $E^*_{DNS}(Z)$ and ℓ. The evolution of Y_Z is calculated by solving the transport master equation:

$$\begin{aligned}\frac{\partial}{dt}Y_Z(E^*_Z,\ell,t) &= \Delta^{(-)}_{Z+1}Y_{Z+1}(E^*_Z,\ell,t) + \Delta^{(+)}_{Z-1}Y_{Z-1}(E^*_Z,\ell,t)\\ &\quad -(\Delta^{(-)}_Z + \Delta^{(+)}_Z + \Lambda^{qf}_Z)Y_Z(E^*_Z,\ell,t), \text{ for } Z = 2,3,...,Z_{tot}-2. \end{aligned} \qquad (13)$$

Here, the transition coefficients of multinucleon transfer are calculated as in [18]

$$\Delta^{(\pm)}_Z = \frac{1}{\Delta t}\sum_{P,T} |g^{(Z)}_{PT}|^2 \; n^{(Z)}_{T,P}(t)\;(1-n^{(Z)}_{P,T}(t))\;\frac{\sin^2(\Delta t(\tilde\varepsilon_{P_Z}-\tilde\varepsilon_{T_Z})/2\hbar)}{(\tilde\varepsilon_{P_Z}-\tilde\varepsilon_{T_Z})^2/4}, \qquad (14)$$

where the matrix elements g_{PT} describe one-nucleon exchange between the nuclei of DNS, and their values are calculated microscopically using the expression obtained in Ref. [26]. A non-equilibrium distribution of the excitation energy between the fragments was taken into account in calculations of the single-particle occupation numbers n_P and n_T as it was done in Ref. [[27]; $\tilde\varepsilon_{P_Z}$ and $\tilde\varepsilon_{T_Z}$ are perturbed energies of single-particle states. In Eq. 13, Λ^{qf}_Z is the Kramers rate for the decay probability of the dinuclear system into two fragments with charge numbers Z and $Z_{tot}-Z$ (details in Ref. [28]), and it is proportional to $\exp(-B_{qf}(Z)/(kT))$ where $B_{qf}(Z)$ is the quasifission barrier.

Eqs. (13) with the coefficients (14) and initial condition $Y_Z(E^*,0) = \delta_{Z,Z_P}$ are solved numerically and the primary mass and charge distributions are found for a given interaction time $t_{int} = 5\cdot10^{-21}$s (see Ref. [29]). In Eq. (12), we use the definition $Y_Z(E^*_Z,\ell) = Y_Z(E^*_Z,\ell,t_{int}(\ell))$. In (14) we use $\Delta t = 10^{-22}\text{s} << t_{int}$.

As an example of our calculation of the fusion probability $P_{\rm CN}$ versus the angular momentum ℓ, we present the results for the ^{48}Ca+^{154}Sm reaction at two different $E^*_{\rm CN}$ values (see Fig. 6 (a)) and the function $P_{\rm CN}$ versus the energy $E^*_{\rm c.m.} - E_{\rm B}$ for the reaction that lead to different isotopes of thorium (see Fig. 6 (b)). This last figure shows the effect of the entrance channel on the $P_{\rm CN}$ fusion probability.

Moreover, Fig. 7 shows the calculation of $P_{\rm CN}$ as a function of the energy $E^*_{\rm c.m.}$ for the two close ^{32}S+^{182}W and ^{32}S+^{184}W reactions for a wide range of excitation energies, in comparison with the calculation reported in Ref. [8]. For these two close reactions we observe two region

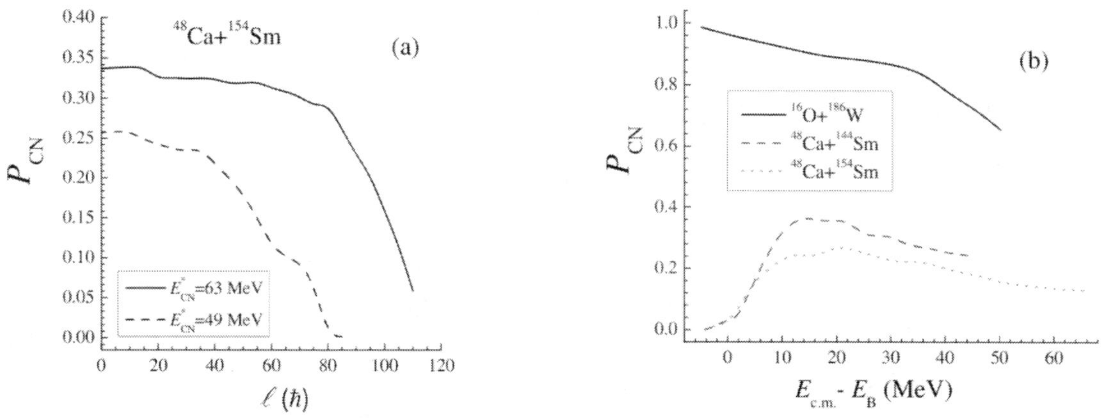

Figure 6. (a) The probability $P_{\rm CN}$ of compound nucleus formation as a function of the angular momentum of dinuclear system ℓ at energies $E_{\rm c.m.}$ =138 and 154 MeV, corresponding to the excitation energies of the compound nucleus $E^*_{\rm CN}$=49 and 63 MeV. (b) The DNS model results for the fusion probability P_{CN} for the ^{16}O+^{186}W, ^{48}Ca+^{144}Sm and ^{48}Ca+^{154}Sm reactions as a function of the collision energy relative to the interaction barriers corresponding to each of reactions.

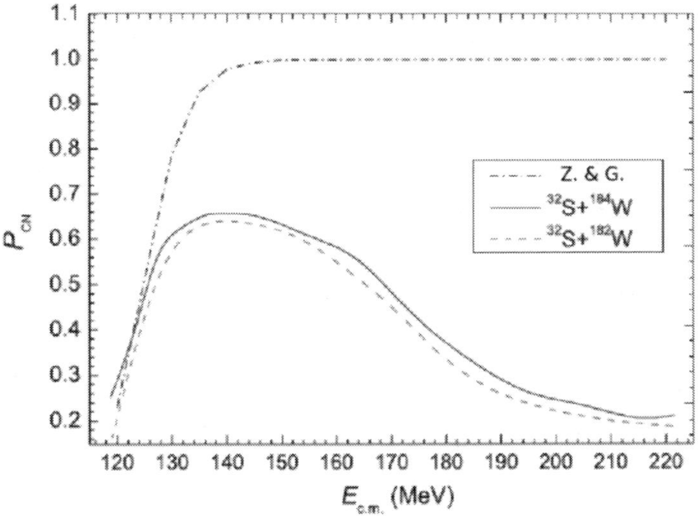

Figure 7. Our theoretical values of the fusion probability $P_{\rm CN}$ for the ^{32}S+^{184}W reaction as a function of collision energy $E_{\rm c.m.}$, and the one of other authors (Z. & G.) of Ref. [8].

of strong hindrance to fusion at very small and large values of the collision energy of $E^*_{\rm c.m.}$ for both reactions while the approximate formula used in Ref. [8] gives similar results only at small energies, but $P_{\rm CN}$ increase quickly reaching a maximum value equal to 1 at $E^*_{\rm c.m.}$ = 145 MeV. In addition, there is no difference between the values of $P_{\rm CN}$ calculated by the approximated formula for the ^{32}S+^{182}W and ^{32}S+^{184}W reactions.

3.1. Compound Nucleus and Fast Fission contributions from the complete fusion formation

In formula (3) the rates among cross sections of the quasifission, fusion-fission, and fast fission processes, which give fragments as reaction products, depend on the masses and charges, shell structure, mass asymmetry of projectile-target nuclei, as well as on the energy and impact parameter of the projectile. As it was noted in Section 1, the formation of quasifission and fast fission products bypasses the stage of the CN formation. The difference between the two mentioned processes is in their dependence on the orbital angular momentum: quasifission is

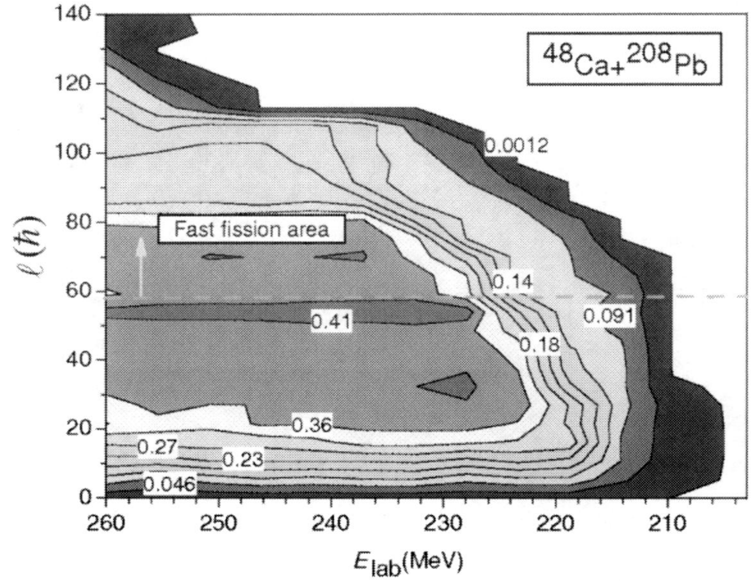

Figure 8. The results of calculation for the fusion angular momentum distribution of the ^{48}Ca+^{208}Pb reaction. The upper part of the figure, labelled by "Fast fission area", indicates the region where also the fast fission process appear.

possible at all angular momenta of collision as fast fission occurs only at $\ell > \ell_f$, where ℓ_f is the angular momentum value at which the fission barrier of the compound nucleus disappears. The mass distributions of products of the capture reactions can be mixed by different proportions in different mass regions. To avoid ambiguity, the authors of Ref. [17] referred to the registered products as ones of the symmetric fragmentation (capture) because the genuine fission process is the decay of the compound nucleus into two fragments. The mass distribution of products of the fast fission reactions can be alike to that of the fusion-fission reactions. But the angular distribution of products of the former reaction is more anisotropic than that of the latter reaction due to the difference in their reaction time. In order to estimate the contribution of the fast fission into the calculated capture cross section, we use the value $\ell_f = 58$ obtained in Ref. [17] for the ^{48}Ca+^{208}Pb reaction.

In Fig. 8 the results of calculation for the fusion angular momentum distribution of this reaction is presented. One can see that maximum of the partial fusion cross section is around $\ell = 50$. The fast fission contribution for $\ell > \ell_{\rm f}$ is calculated by formula (11), while the fusion cross section of CN must include the evaporation residues and fusion-fission cross sections only. In our calculation of evaporation residues, the values of the angular momentum $\ell < 58$ were used.

3.2. Evaporation Residue

The probability of the formation of a evaporation residue nucleus surviving fission with mass number $A = A_{CN} - (\nu(x) + y(x) + 4k(x))$ and charge number $Z = Z_{CN} - (y(x) + 2k(x))$ from a heated and rotated compound nucleus $^{A_{CN}}Z_{CN}$ after emissions of ν neutrons, y protons, and k α-particles at the xth step of the de-excitation cascade is determined by the formula [6, 10]:

$$\sigma_{ER(x)}(E^*_x) = \sum_{\ell=0}^{\ell_f}(2\ell+1)\sigma^{\ell}_{(x-1)}(E^*_x)W_{\text{sur}(x-1)}(E^*_x, \ell), \tag{15}$$

where $\sigma^{\ell}_{(x-1)}(E^*_x)$ is the partial cross section of the intermediate nucleus formation at the $(x-1)$th step and $W_{\text{sur}(x-1)}(E^*_x, \ell)$ is the survival probability of the $(x-1)$th intermediate nucleus against fission along the de-excitation cascade of CN; ℓ_f is the value of angular momentum ℓ at which the fission barrier for a compound nucleus disappears completely [30]; E^*_x is the excitation energy of the nucleus formed at the xth step of the de-excitation cascade. It is clear that $\sigma^{\ell}_{(0)}(E^*_0) = \sigma^{\ell}_{\rm fus}(E^*_{CN})$ at

$$E^*_{CN} = E^*_0 = E_{\rm c.m.} + Q_{gg} - E_{rot}, \tag{16}$$

where $E_{\rm c.m.}$, Q_{gg}, and E_{rot} are the collision energy in the center of mass system, the reaction Q_{gg}-value, and rotational energy of the compound nucleus, respectively. The numbers of the emitted neutrons, protons, α-particles and γ-quanta, $\nu(x)$n, $y(x)$p, $k(x)\alpha$, and $s(x)\gamma$, respectively, are functions of the step x. The emission branching ratios of these particles depend on the excitation energy E^*_A and angular momentum ℓ_A of the cooling intermediate nucleus.

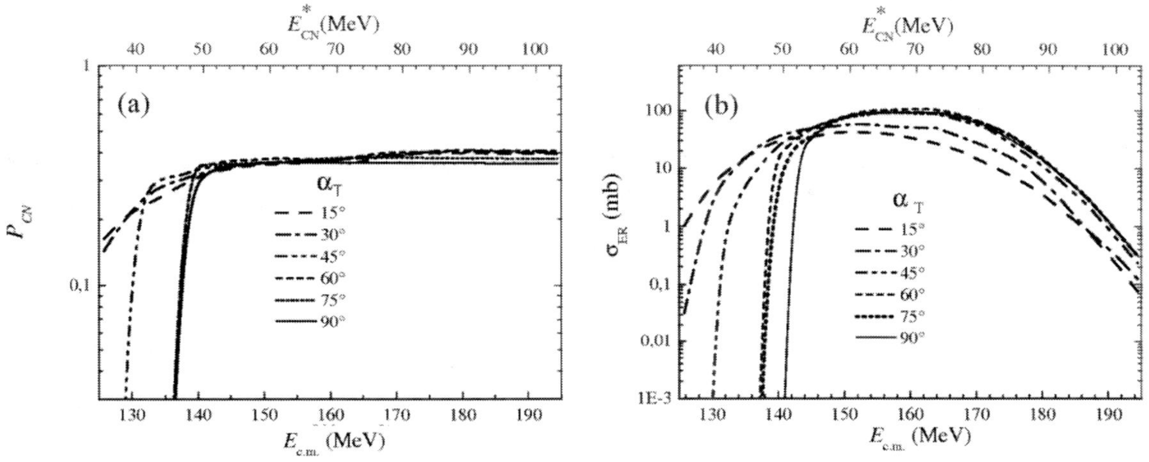

Figure 9. Fusion probability P_{CN} (a) versus the collision energy $E_{\rm c.m.}$ for different values of the α_T angle. Evaporation residue cross sections (b) versus the $E_{\rm c.m.}$ energy for various orientation angles α_T of the ^{154}Sm target.

The de-excitation cascade, characterized by the emission of the above-mentioned particles, starts from the compound nucleus $^{A_{CN}}Z_{CN}$. Its formation probability is the partial cross section of complete fusion $\sigma^{\ell}_{\rm fus}(E^*_{CN})$ corresponding to the orbital angular momentum ℓ. The fusion cross section is equal to the capture cross section for light systems or light projectile-induced reactions, while for reactions with massive nuclei, it becomes a model-dependent quantity. Concerning estimation of the fusion cross section from the experimental fragments data, its value is sometimes an ambiguous quantity because of difficulties in separating the fusion-fission fragments from the quasifission fragments in the case of overlap of mass and angular distributions.

As an example we present the results of our calculation for the ^{48}Ca+^{154}Sm reaction regarding the fusion probability $P_{\rm CN}$ (see Fig. 9 (a)) and evaporation residue cross sections $\sigma_{\rm ER}$ (see Fig. 9 (b)) as a function of the collision energy $E_{\rm c.m.}$ for different values of the target orientation angles $\alpha_{\rm T}$. The dependence of the quasifission-fusion competition during the evolution of the dinuclear system and the sensitivity of the fission-evaporation competition during the de-excitation cascade of the compound nucleus on the values of orientation angle α_T are demonstrated. The analysis of the dependence of the compound nucleus and evaporation residue formation cross sections on α_T shows that the observed yield of evaporation residues in the ^{48}Ca+^{154}Sm reaction at the low energies ($E_{\rm c.m.}$ <137 MeV) is formed in the collisions $\alpha_T < 45°$. Because the initial beam energy is enough to overcome the corresponding Coulomb barrier for the collisions with these orientation angles α_T. Only in this case it is possible formation of dinuclear system which evolves to compound nucleus or breaks up into two fragments after

Journal of Physics: Conference Series **282** (2011) 012006 doi:10.1088/1742-6596/282/1/012006

multinucleon exchange without formation of the compound nucleus. At larger energies (about $E_{c.m.}$ =140–180 MeV) all orientation angles of the target-nucleus can contribute to σ_{ER} and its values are in the 10–100 mb range. At the larger collision energies ($E_{c.m.}$ >158 MeV) the complete fusion still increases but the evaporation residue cross section σ_{ER} goes down and its values are in the 1–0.1 mb range due to the strong decrease of the survival probability of the heated compound nucleus along de-excitation cascade. This is connected by the decrease of the fission barrier for a compound nucleus by increasing its excitation energy [3, 4] and angular momentum [30].

Another phenomenon leading to decrease of σ_{ER} at higher energy is the fast fission process which is the splitting of the mononucleus into two fragments due to absence of the fission barrier at very high the angular momentum $\ell > \ell_f$.

4. Mass distribution for quasifission and fusion-fission fragments at different energies

To show the evidence of the meaningful overlap of mass distribution of quasifission and fusion-fission process products, and to demonstrate the difficulty in experiments to solve unambiguously the selection of events belonging to true fusion-fission process from the ones of quasifission process, we present in Fig. 10 and discuss the results of investigation on the ^{48}Ca+^{248}Cm reaction.

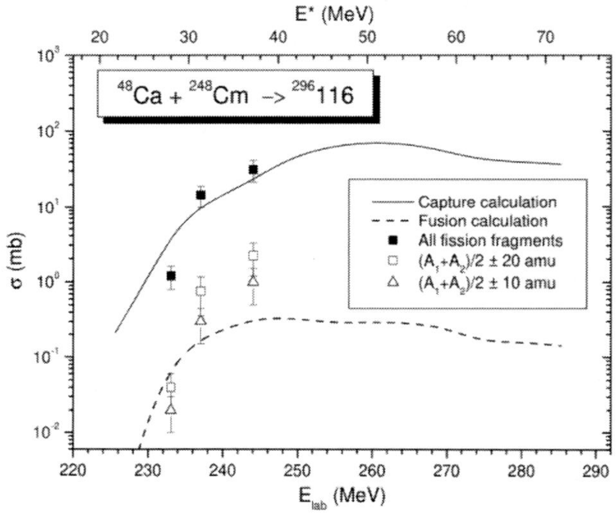

Figure 10. The capture and fusion cross section calculation, in comparison with the experimental data [31, 32], for the ^{48}Ca + ^{248}Cm reaction leading to the ^{296}116 superheavy CN. The difference between the fusion cross section and more symmetric fragment yield ($(A_1 + A_2)/2 \pm 10$ amu) at a higher excitation energy is related to the contribution of the quasifission process yielding more symmetric fragments.

In accordance with the above-mentioned considerations, our calculation of the capture cross section (solid line in Fig.10) is in complete agreement with the experimental data [31] for the production of all fragments (full squares), while the fusion cross section (dashed line) is not in agreement with the data (open squares) for the symmetric mass fragments ($(A_1 + A_2)/2$) when a large mass interval of ±20 amu is assumed.

Such disagreement is connected with the contribution of the quasifission process in the range of the more symmetric fragments in which the fusion-fission process also contributes. If it is assumed that the experimental fusion - fission events are in the $(A_1 + A_2)/2 \pm 10$ amu interval (almost close to the $\sqrt{(A_1 + A_2)/2}$ value), the calculated fusion cross sections (the dashed line) will be closer to the new set of the experimental data (open triangles [32] in Fig.10). Indeed, in this case there is an appreciable contribution of the quasifission process (or a contribution which cannot be neglected), in addition to the fusion-fission fragment formation. Therefore, the estimated experimental fusion cross section, connected with the new set of the experimental events of fission fragments, still appears to be a little larger than the calculated fusion excitation

Journal of Physics: Conference Series **282** (2011) 012006 doi:10.1088/1742-6596/282/1/012006

function at higher excitation energies. A preliminary calculation of the mass distribution of quasifission fragments for a fixed reaction time t_{reac} of a DNS performed in the framework of the model developed on the basis of the dinuclear system concept [33, 34], indicates that the fragments of the quasifission process also appear in the mass-symmetric region and are mixed with the fragments coming from the fusion-fission process.

5. Study of the reaction dynamics of two close mass asymmetric reactions by the analysis of the fusion-evaporation and fusion-fission processes

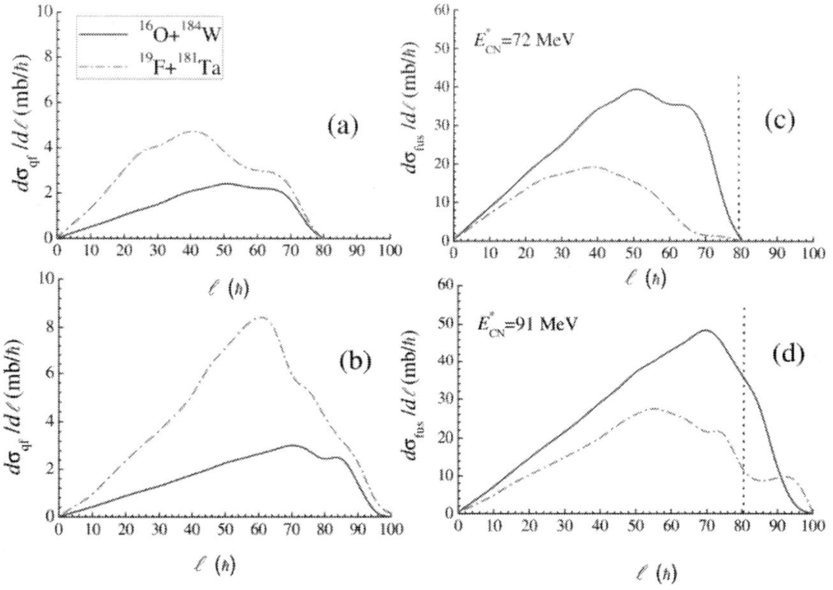

Figure 11. Partial quasifission ((a) and (b) panels) and fusion ((c) and (d) panels) cross sections as a function of the angular momentum *ell* for the $^{16}O+^{184}W$ (solid line) and $^{19}F+^{181}Ta$ (dashed line) reactions. In the panel (a) the beam energies of reactions induced by ^{16}O and ^{19}F are $E_{c.m.}$ =96.2 MeV and 95.7 MeV, respectively. These energies correspond to the same excitation energy of CN E^*_{CN} =72 MeV for the both systems shown in the panel (c). Analogously, in the panel (b) the beam energies are $E_{c.m.}$ =115.2 MeV and $E_{c.m.}$ =114.7 MeV, respectively. The corresponding excitation energy of CN is 91 MeV shown in the panel (d). In (c)and (d), the vertical dashed line at $\ell_f = 80\hbar$ separates compound nucleus and fast fission contributions.

There are three main processes causing hindrances to ER formation in reactions with massive nuclei: quasifission, fusion-fission, and fast fission [9]. All of these processes produce binary fragments in different stages of reaction. Moreover, the angular and mass distributions of some parts of their products can overlap [9,35]. Ignoring this mixing may lead to ambiguity at analysis of the experimental data connected with the binary fragments. This problem should be studied carefully.

The ER formation process is often considered as the third stage of the three-stage process. The first stage is capture–formation of the DNS after full momentum transfer into the deformation energy of nuclei, their excitation energy, and rotational energy from the initial relative motion of the colliding heavy ions in the center-of-mass system. Capture takes place if the initial energy of the projectile in the center-of-mass system is enough to overcome the interaction barrier (Coulomb barrier + rotational energy of the entrance channel) [20]. The study of the dynamics of heavy ion collisions at energies near the Coulomb barrier shows that

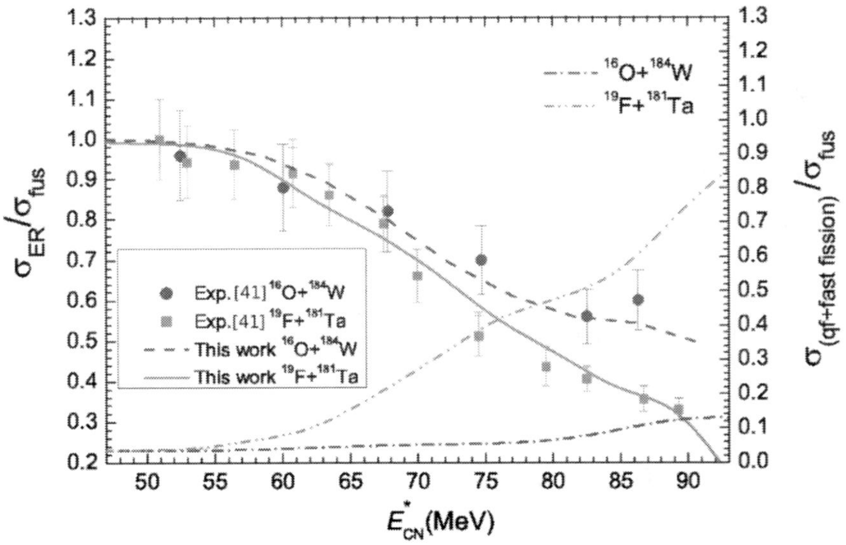

Figure 12. Comparison of the experimental values of the evaporation residue cross sections normalized with respect to the capture cross sections for the $^{16}O+^{184}W$ (solid circles) [41] and $^{19}F+^{181}Ta$ systems (solid squares) [41] with the corresponding theoretical results (dashed and solid lines, respectively) as a function of the excitation energy $E^*_{\rm CN}$ of CN (left axis). Theoretical results of the sum of the quasifission and fast fission cross sections (normalized with respect of the fusion cross sections) for the $^{16}O+^{184}W$ (dot dashed line) and $^{19}F+^{181}Ta$ (dot-dot dashed line) systems are presented versus $E^*_{\rm CN}$ and compared on the right axis.

complete fusion does not occur immediately in collisions of massive nuclei [4,9,21,36,37]. After formation of the DNS, the quasifission process competes with the formation of CN. Quasifission occurs when the DNS prefers to break down into fragments instead of being transformed into a fully equilibrated CN. The number of events contributing to quasifission increases drastically by increasing the sum of the Coulomb interaction and rotational energy in the entrance channel [1,7,9].

Another reason for the decreasing yield of ER with increasing excitation energies is the usual fission of a heated and rotating CN that was formed in competition with quasifission. The stability of a massive CN decreases due to the decrease in the fission barrier by increasing its excitation energy E^*_{CN} and angular momentum ℓ [38–40]. The theoretical values of the quasifission partial cross sections for the $^{19}F+^{181}Ta$ and $^{16}O+^{184}W$ reactions are presented in the left panels of Fig. 11. It is seen from these figures that the quasifission takes place at all values of ℓ leading to capture. The angular momentum distributions of CN formed in these reactions at the excitation energies $E^*_{\rm CN}$ =72 and 91 MeV are presented in the right panels of Fig. 11. The spin distributions of CN formed in each of these reactions differ mainly by the probability but not by the values of the angular momentum ranges. This means that the number of CN formed in both reactions under discussion are different, but they have a similar range of the angular momentum ℓ. The vertical dotted lines at $\ell_f = 80\hbar$ in these panels separates the complete fusion ($\ell_f < 80\hbar$) and fast fission ($\ell_f \geq 80\hbar$) regions of the angular momentum.

The quasifission and fast fission processes produce binary fragments which can overlap with those of the fusion-fission channel and the amount of mixed detected fragments depends on the mass asymmetry of entrance channel, as well as on the shell structure of the reaction fragments being formed. The suggestion for the experimental studies of the difference between characteristics of the fusion-fission, quasifission and fast fission products can be made when their mass (charge), kinetic energy and angular distributions are explored in detail by dynamical calculations allowing to obtain the relaxation times of these processes. Therefore, the correct estimation of the CN formation probability in the reactions with massive nuclei is a difficult task for both experimentalists and theorists. Different assumptions about the fusion process are

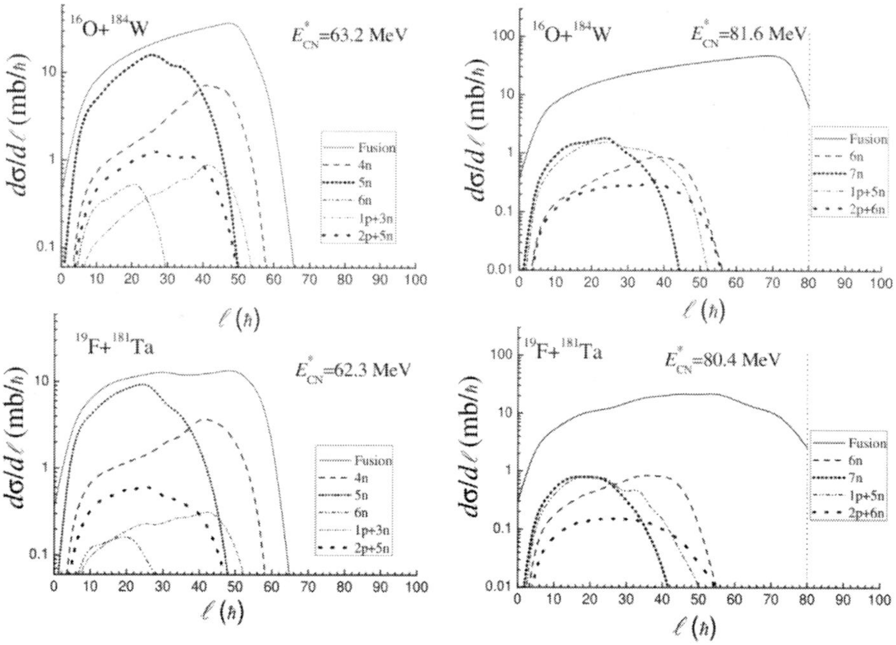

Figure 13. Spin distribution of evaporation residue cross sections as a function of the angular momentum ℓ. The upper part is for the ^{16}O+^{184}W reaction, the lower part is for ^{19}F+^{181}Ta reaction, both at two $E^*_{\rm CN}$ energies: approximately 62-63 MeV and 80-81 MeV.

used in different theoretical models which can give different cross sections. The experimental methods used to estimate the fusion probability depend on an unambiguous identification of the complete fusion products among the quasifission products. The difficulties arise when the mass (charge) and angular distributions of the quasifission and fusion-fission fragments strongly overlap, depending on the reaction dynamics. As a result, the complete fusion cross sections may be overestimated [9].

We confirm that the compared ratios of the cross sections between evaporation residues and complete fusion σ_{ER}/σ_{fus} for the ^{16}O+^{184}W and ^{19}F+^{181}Ta reactions discussed in [41] are not free from the above-mentioned ambiguity in the determination of the fusion cross section σ_{fus}. Theoretical values of the fusion cross section include only evaporation residues and fusion-fission cross sections

$$\sigma_{\rm fus} = \sigma_{\rm ER} + \sigma_{\rm ff}. \tag{17}$$

The experimental values of fusion cross section reconstructed from the detected fissionlike fragments and evaporation residues [41]:

$$\sigma^{\rm (exp)}_{\rm fus} = \sigma_{\rm ff} + \sigma_{\rm qf} + \sigma_{\rm fast\,fis} + \sigma_{\rm ER}, \tag{18}$$

where $\sigma_{\rm ff}$, $\sigma_{\rm qf}$, and $\sigma_{\rm fast\,fis}$ are the contributions of fusion-fission, quasifission and fast fission processes, respectively, and $\sigma_{\rm ER}$ is the ER contribution. According to the statement of the authors of Ref. [41], the complete fusion cross sections are obtained by adding fission cross sections [42] to the measured data of the evaporation residue cross sections [43]. Therefore, we can state that the definition of the experimental fusion cross section is similar with the definition of capture [9]: $\sigma^{\rm (exp)}_{\rm fus} = \sigma_{\rm cap}$.

In Ref. [42], the complete fusion cross section is derived from a statistical model where only neutron evaporation and fission are included. We think that the fission data from Ref. [42] contain quasifission fragments and, at larger beam energies, also fast fission contributions, which appear as hindrances to complete fusion. This argument is confirmed by our results obtained in the framework of the DNS model. We calculate the total ER and fusion-fission excitation functions in the framework of the advanced statistical model [38–40].

6. Study for the synthesis of the ^{297}117 element by the ^{48}Ca+^{249}Bk reaction

By using the above-mentioned method we calculated the capture, quasifission, fast fission and fusion (formation of CN) cross sections for the ^{48}Ca induced reaction on the ^{249}Bk target (see Fig. 14 (b)). In the $E_{c.m.}$ <200 MeV energy range the fusion cross section is about three orders

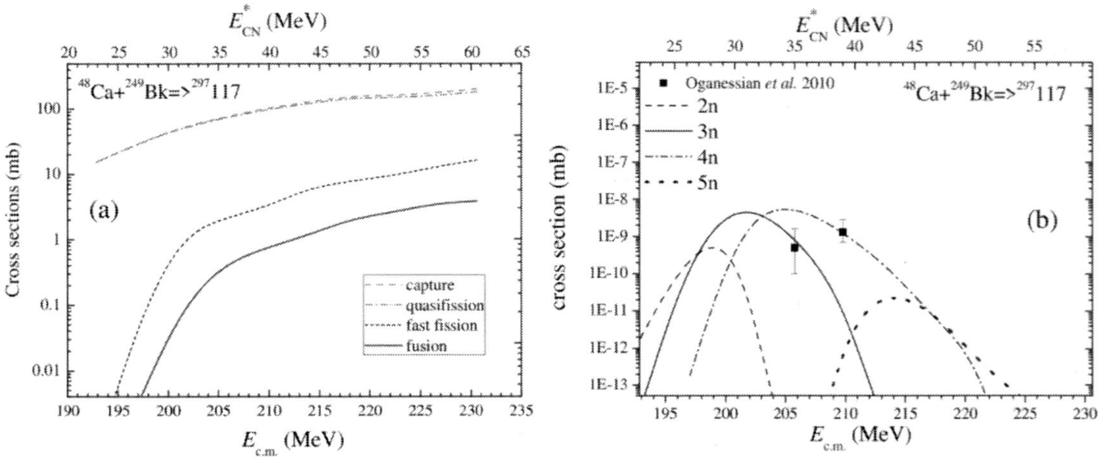

Figure 14. (a)Capture, quasifission, fast fission, and fusion cross sections for the ^{48}Ca+^{249}Bk reaction leading to ^{297}117 compound nucleus as a function of the colliding energy $E_{c.m.}$. (b) Evaporation residue cross sections, after neutron emission only, calculated for the ^{48}Ca+^{249}Bk reaction leading to ^{297}117 compound nucleus as a function of the colliding energy $E_{c.m.}$, in comparison with the experimental data (full square) presented in Ref. [47].

of magnitude lower than that of quasifission (which is close to the capture cross section in the whole explored energy range), and it is about one order of magnitude lower than the one of fast fission. At higher $E_{c.m.}$ energies, the fusion cross section is about two orders of magnitude lower than that of quasifission, while it is about five times lower than the one of fast fission. In this reaction the quasifission is the completely dominant process in comparison to complete fusion, and the fast fission is also a strongly relevant process in comparison to the fusion formation (CN). In this case, fast fission process takes place starting from the ℓ_f value equal to 30 $\hbar$.

The shell corrections for the odd-odd and odd-even nuclei in the 292-297 mass region of the 117 element were obtained from the static fission barrier B_f (for ℓ=0) calculated by Sobiczewski [44], which correspond to the realistic values of shell corrections, as an extension of the results presented in the paper of Kowal *et al.* [45] for the even-even superheavy nuclei. We calculated the average effective fission barriers and evaporation residue cross sections for all nuclei along the de-excitation cascade of the ^{297}117 compound nucleus by using such values of shell corrections (for ℓ=0) and taking into account their fade-out (see Refs. [4,6,10,46] and references therein) as a function of the temperature and angular momentum. In Fig. 14(b), we compare the individual excitation functions of the evaporation residue for emission of neutron only along the cascade of

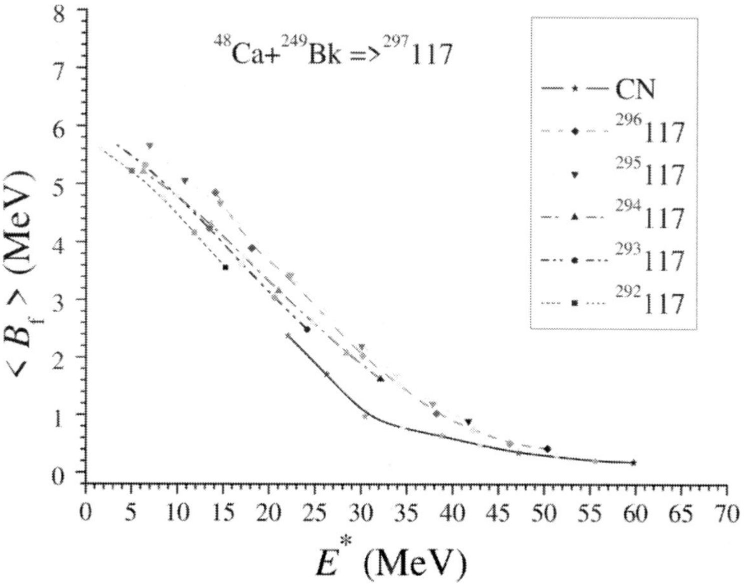

Figure 15. The effective fission barrier $< B_{\mathrm{f}} >$ as a function of the excitation energy for the $^{48}Ca+^{249}Bk$ reaction leading to $^{297}117$ compound nucleus.

CN with ER cross sections after 3n and 4n emission from CN measured by Oganessian *et al.* in their experiment [47]. As one can see our result are in good agreement with the experimental data and we also find that the peaks of such excitation functions for 3n and 4n emission are present at about 30.6 and 34.0 MeV, respectively. The maximum yield of the 2n contribution (5×10^{-1}pb) is lower than the ones of the 3n and 4n contributions. The maximum yield of the 5n contribution is about 2.3×10^{-2} pb at E^*_{CN}= 43.2 MeV. Moreover, in Fig. 15 we report the effective fission barrier $< B_{\mathrm{f}} >$ distribution as a function of E^* along the cascade of CN at different excitation energies and neutron number of the intermediate excited nuclei. In Table 1 are reported such $< B_{\mathrm{f}} >$ values obtained for the chain of nuclei, at three different $E*$ values of the considered excited nuclei. In Table 1 the rows show the evolution of the effective fission

E^* (MeV)	A= 297 N= 180	A= 296 N= 179	A= 295 N= 175	A= 294 N= 177	A= 293 N= 176	A= 292 N= 175
17	....	4.2	4.3	3.8	3.6	3.3
25	1.9	2.9	3.0	2.6	2.4	....
33	0.8	1.7	1.8	1.6	....	....

Table 1. Effective fission barrier $< B_{\mathrm{f}} >$ for the $^{48}Ca+^{249}Bk \rightarrow ^{297}117$ reaction leading to intermediate excited nuclei with masses A=297, 296, 295, 294, 293 and 292, at excitation energies of nuclei E^*= 17, 25 and 33 MeV.

barrier $< B_{\mathrm{f}} >$ values for different neutron number of excited nuclei when they are formed with a defined excitation energy E^*(for example 17, 25, 33 MeV). Of course, such conditions for the intermediate nuclei are reached starting from different excitation energy values of E^*_{CN}. In details, to form the nuclei with masses A of 297, 296, 295, 294 and 293 u at the same excitation energy, for example E^*=25 MeV, it is necessary to start from CN with excitation energy of 25.0, 33.5, 41.7, 51.6 and 60.8 MeV, respectively. The changing of the $< B_{\mathrm{f}} >$ values along the rows are related to the structure effects of nuclei by changing the neutron number. Such changes of $< B_{\mathrm{f}} >$ are similar for the three presented cases of excitation energies. Instead, the $< B_{\mathrm{f}} >$ values along a column show the change of the effective fission barrier for a defined intermediate nucleus by changing its excitation energy E^*. For all intermediate nuclei with different neutron number, the changes have the same trend by changing the excitation energy E^*. It is clear that the $< B_{\mathrm{f}} >$ value decreases increasing E^*.

7. Conclusions

In a large number of capture reactions with massive nuclei the quasifission process with its peculiarities is the main subject for the understanding of the reaction dynamics. Therefore, by using beams of ^{124}Sn and ^{132}Sn (neutron-rich) in collisions with targets heavier than ^{92}Zr , and for reactions with the Coulomb parameter $z = \frac{Z_1 \times Z_2}{A_1^{1/3}+A_2^{1/3}}$ (connected with the intense Coulomb repulsion between reacting nuclei) higher than 200, for which the quasifission yield is at least two orders of magnitude higher than the complete fusion yield (and then the fusion probability $P_{\rm CN}$ is lower than about 10^{-2}, as for example for the reactions listed in Table 2), are produced

Reaction	$Z_{\rm tot}$	$z = \frac{Z_1 \times Z_2}{A_1^{1/3}+A_2^{1/3}}$	$P_{\rm CN}$
^{86}Kr+^{136}Xe	90	204	$\sim 4\times10^{-2}$
^{92}Zr+^{132}Sn	90	209	$\sim 5\times10^{-2}$
^{136}Xe+^{136}Xe	108	284	$<10^{-10}$
^{64}Ni+^{208}Pb	110	232	$<10^{-5}$
^{208}Pb+^{70}Ge	114	262	$<10^{-7}$
^{139,149}La+ ^{139,149}La	114	306, 317	$<10^{-10}$
^{86}Kr+^{208}Pb	118	286	$<10^{-7}$
^{238}U+^{58}Ni	120	256	$<10^{-6}$
^{132}Sn+^{174}Yb	120	327	$<10^{-10}$
^{238}U+^{70}Ge	124	286	$<10^{-7}$
^{58}Fe+^{249}Cf	124	251	$<10^{-3}$
^{132}Sn+^{186}W	124	343	$<10^{-11}$
^{132}Sn+^{208}Pb	132	373	$<10^{-13}$
^{160}Gd+^{186}W	138	431	$<10^{-16}$
^{132}Sn+^{249}Cf	148	430	$<10^{-15}$
^{238}U+^{248}Cm	188	707	- - - -

Table 2. Reactions, total Z of the system, z parameter of the Coulomb repulsion, and fusion probability $P_{\rm CN}$ are reported.

intense yields of quasifission fragments. Therefore, all above-mentioned reactions are excellent reactions for the study of the quasifission process. For the last reaction (^{238}U+^{248}Cm) in Table 2 the $P_{\rm CN}$ value is not reported because for this reaction capture cross section does not exist.

Moreover, at energies higher than the Coulomb barrier, the fast fission gives a relevant contribution to decay of the complete fusion system into two fragments. For these reactions contribution of the complete fusion cross section to the capture cross section is very small. In such a context the ER cross section is very small, or in many cases is lower than the limit (about 0.2 pb) of the present possibility of the experimental setup. Therefore, before to plan new long beam-time and expensive experiments it is necessary to estimate, by a hopeful analysis of the reaction dynamics, the quasifission and complete fusion cross sections, and then the fast fission and CN cross sections because only the fusion-fission and fusion-evaporation yields contribute to the CN cross section. Only for the two above-mentioned reactions (^{92}Zr+^{132}Sn , ^{86}Kr+^{136}Xe) it is possible to observe evaporation residues. For all other reactions listed in Table 2 it is not possible to observe meaningful fusion-fission contributions and least of all evaporation residues. The ratios of fusion-fission fragment to quasifission fragment yields for the all other mentioned reactions are much lower than 10^{-3} and therefore it is impossible to separate the fragments of fusion-fission process from the fragments of the huge amount of quasifission process. Our theoretical model is a powerful predictive method of calculation of reaction products and also to describe the processes of the reaction dynamics. Moreover, also for reactions leading to synthesis of superheavy elements our model is able to calculate the effective fission barriers along the various steps of the de-excitation cascade of the compound nucleus at various excitation energy and number of neutrons of intermediate excited nuclear systems .

References

[1] Giardina G *et al.* 2000 *Eur. Phys. J. A* **8** 205
[2] Giardina G *et al.* 2000 *Nucl. Phys A* **671** 165
[3] Fazio G *et al.* 2003 *J. Phys. Soc. Jpn.* **72** 2509
[4] Fazio G *et al.* 2004 *Eur. Phys. J. A* **19** 89
[5] Zhang H Q *et al.* 2010 *Phys. Rev C* **81** 034611
[6] Fazio G *et al.* 2004 *Eur. Phys. J. A* **22** 75
[7] Fazio G *et al.* 2005 *Phys. Rev C* **72** 064614
[8] Zagrebaev V and Greiner W *et al.* 2008 *Phys. Rev C* **78** 034610
[9] Oganessian Yu. *et al.* 2009 *Phys. Rev C* **79** 024608
[10] Fazio G *et al.* 2005 *Mod. Phys. Lett. A* **20** 391
[11] Fazio G *et al.* 2008 *J. Phys. Soc. Jpn.* **77** 2509
[12] Nasirov A K *et al.* 2009 *Int. J. Mod. Phys. E* **18** 841
[13] Nasirov A K *et al.* 2010 *Phys. Rev C* **79** 024606
[14] Nasirov A K *et al.* 2010 *Phys. Lett. B* **686** 72
[15] Nasirov A K *et al.* 2010 *J. Phys: Conf. Ser.* **205** 012018
[16] Itkis M G *et al.* 2004 *Acta Phys. Hung. A* **19** 9
[17] Bock R *et al.* 1982 *Nucl Phys. A* **388** 334
[18] Jolos R V *et al.* 1986 *Sov. J. Nucl. Phys.* **44** 228
[19] Nasirov A K *et al.* 2004 *Acta Phys. Hung. A* **19** 109
[20] Nasirov A K *et al.* 2005 *Nucl. Phys. A* **759** 342
[21] Adamian G G *et al.* 2003 *Phys. Rev. C* **68** 034601
[22] Volkov V V in Contributed Papers of Nucleus-Nucleus Collision II, Visby, 1985, edited by B. Jakobson and K. Aleclett (North-Holland, Amsterdam, 1985), Vol.1, p.54; Izv. Akad. Nauk SSSR Ser. Fiz. 50, 1879 (1986); in Proceedings of International School-Seminar on Heavy Ion Physics, Dubna, 1986, D7-87-68 (Dubna, 1987), p. 528; in Proceedings of the 6th International Conference on Nuclear Reaction Mechanisms, Varenna, 1991, edited by E. Gadioli (Ricerca Scientifica ed Educazione Permanente Supplemento n.84, 1991), p.39. on Heavy Ion Physics
[23] Moller P and Nix J R 1988 *Atomic Data and Nuclear Data* **39** 213
[24] Raman S *et al.* 1987 *Atomic Data and Nuclear Data* **36** 1
[25] Spear R H 1989 *Atomic Data and Nuclear Data* **42** 55
[26] Adamian G G *et al.* 1992 *Sov. J. Nucl. Phys.* **55** 660
[27] Adamian G G *et al.* 1994 *Z. Phys. A* **347** 203
[28] Adamian G G *et al.* 2003 *Phys. Rev. C* **68** 034601
[29] Tõke J *et al.* 1985 *Nucl. Phys. A* **440** 327
[30] Sierk A J 1986 *Phys. Rev. C* **33** 2039
[31] Itkis M G *et al.* 2002 Proc. of the International Conference on Nuclear Physics at Border Lines, May 21-24, 2001 Lipari (Messina), Italy, eds. G.Fazio, G.Giardina, F.Hanappe, G.Imme, N.Rowley. World Scientific, Singapore, p.146-156
[32] Kozulin E 2002 private communication
[33] Dzholos R V *et al.* 1989 *Sov. J. Nucl. Phys.* **50** 382
[34] Fazio G *et al.* 2003 Proc. of the International Symposium on New Projects and Lines of Research in Nuclear Physics, October 24-26, 2002, Messina, Italy, eds. G.Fazio, F.Hanappe. World Scientific, Singapore, p.258-271
[35] Nasirov A K *et al.* 2007 *Eur. Phys. J. A* **34** 325
[36] Back B B 1985 *Phys. Rev. C* **31** 2104
[37] Antonenko N A *et al.* 1995 *Phys. Rev. C* **51** 2635
[38] D' Arrigo A *et al.* 1992 *Phys. Rev. C* **46** 1437
[39] D' Arrigo A *et al.* 1994 *J. Phys. G* **20** 365
[40] Sagaidak R N *et al.* 1998 *J. Phys. G* **24** 611
[41] Shidling P D *et al.* 2008 *Phys. Lett. B* **670** 99
[42] Forster J S *et al.* 1987 *Nucl. Phys. A* **464** 497
[43] Hinde D J *et al.* 1982 *Nucl. Phys. A* **385** 109
[44] Sobiczewski A 2011 private communication
[45] Kowal M *et al.* 2010 *Phys Rev C* **82** 014303
[46] Drozdov V *et al.* 2003 *J. Phys. Soc. Jpn.* **72** 2118
[47] Oganessian Yu Ts *et al.* 2010 *Phys. Rev. Lett* **104** 142502

Transfer reaction studies in the region of heavy and superheavy nuclei at SHIP

S Heinz[1], V Comas[1], S Hofmann[1,2], D Ackermann[1], J Heredia[1], F P Heßberger[1], J Khuyagbaatar[1], B Kindler[1], B Lommel[1] and R Mann[1]
[1] GSI Helmholtzzentrum für Schwerionenforschung GmbH, 64291 Darmstadt, Germany
[2] Institut für Physik, Goethe-Universität Frankfurt, 60054 Frankfurt, Germany

E-mail: s.heinz@gsi.de

Abstract. We studied multi-nucleon transfer reactions in the region of heavy and superheavy nuclei. The goal was to investigate these reactions as possibility to create new superheavy neutron-rich isotopes, which cannot be produced in fusion reactions. The experiments have been performed at the velocity filter SHIP at GSI. At SHIP we can detect and identify the heavy, target-like, transfer products. Due to the low background at the focal plane detector and the isotope identification via radioactive decays, the setup allows to reach an upper cross-section limit of 10 pb/sr within one day of beamtime. We investigated the systems ^{58,64}Ni + ^{207}Pb and ^{48}Ca + ^{248}Cm at beam energies below and up to 20% above the Coulomb barrier. At all energies we observed a massive transfer of protons and neutrons, where transfer products with up to eight neutrons more than the target nucleus could be identified.

1. Introduction

Isotopes in the region of superheavy nuclei with $Z > 100$ are usually produced in fusion-evaporation reactions applying Pb and Bi targets (cold fusion) or actinide targets (hot fusion) [1, 2]. Several isotopes with $Z \leq 101$ have also been made in nucleon transfer reactions using heavy and intermediate heavy beams on actinide targets like ^{238}U + ^{248}Cm or ^{48}Ca + ^{248}Cm [3, 4]. By comparing the same reaction product, fusion reactions lead usually to (10 – 50) times higher yields than transfer reactions. However, the isotopes resulting from fusion-evaporation reactions are located in the rather neutron-deficient region of the nuclear chart. This is due to the limited number of stable isotopes which can be applied as beams and target materials. Therefore, other reactions than fusion have to be considered in order to extent the nuclear chart towards more neutron-rich nuclei. In this context, multi-nucleon transfer reactions, which have been already extensively studied in the late 1970s up to the 1990s are brought into discussion again. Especially from theoretical side many contributions have been published in recent years concerning the creation of superheavy nuclei in transfer reactions with medium-heavy and very heavy beams [5, 6], while new experimental data in the region around $Z =$ 100 and beyond are not available. Concerning the experiments performed some decades ago only those which applied chemical methods for particle identification were able to distinguish isotopes up to $Z = 101$ [3, 4]. These isotopes were still located within the region of nuclei which can also be

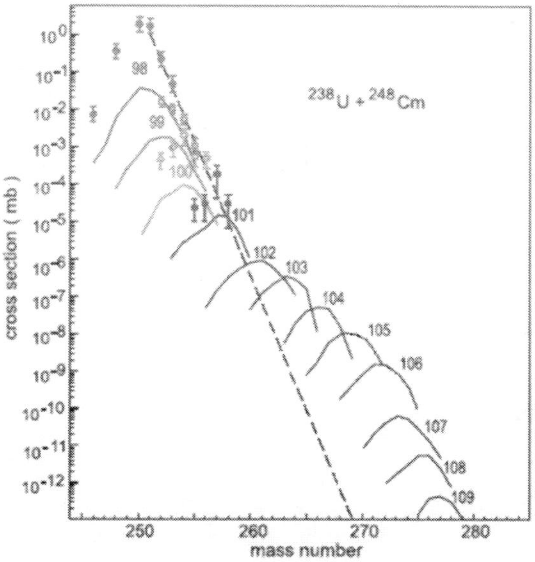

Figure 1. Calculated (solid lines) [6] and measured (symbols) [3] cross-sections for transfer products from ^{238}U + ^{248}Cm collisions. The dashed line represents the expected yields assuming a diffusion process.

reached in fusion reactions. In collisions of ^{238}U + ^{248}Cm numerous isotopes have been observed, where the heaviest one was Z = 101, N = 157 [3]. These data are compared with model calculations for the same system at a centre-of-mass energy of E_{cm} = 800 MeV (6.6×A MeV) [6] in figure 1. The calculations are represented by the solid lines, while experimental data are represented by the different symbols. The calculations reveal a pronounced shoulder in the yield distributions at Z = 106 and a steep drop in the yield for isotopes with Z > 106. The shoulder appears due to shell effects which govern the break-up of the di-nuclear system (DNS) U+Cm after nucleon exchange. The strong shell closures at Z = 82 and N = 126 in ^{208}Pb lead to a pronounced minimum in the potential energy surface and to a preferred production of isotopes close to ^{208}Pb in the transfer reactions. In this case the heavier transfer product is located close to ^{278}Sg (Z = 106, N = 172). For proton numbers of Z < 82 and Z > 106, respectively, the potential increases steeply, which leads to the steep drop in the yields for isotopes with Z < 82 and Z > 106. For comparison, the dashed line indicates the cross-sections, which are expected according to the diffusion model, without assuming the influence of the ^{208}Pb shells. In this case, the cross-sections drop exponentially with increasing number of transferred protons. The data in figure 1 reveal that the maxima of the experimental distributions rather follow the yields expected from the diffusion model. This is especially visible for the distributions for Z = 98, 99.
Experimental data from medium heavy beams on heavy targets are available from collisions of $^{40,44,48}Ca$ with ^{248}Cm targets [4]. Here, the heaviest measured isotope, Z = 100, N = 156, was observed with the most neutron rich beam, ^{48}Ca, with a total cross-section of ≈1 μb. By comparing the experimental data for the isotopes up to Z = 101 for medium heavy and very heavy projectiles, similar cross-sections were measured with ^{48}Ca and ^{238}U beams. From the technical point of view intermediate heavy beams are more favourable, because they are available at about 100 times higher intensities than the heavy beams like uranium. This is an important point regarding the expected low cross-sections of nanobarn and below for nuclei with Z > 100.

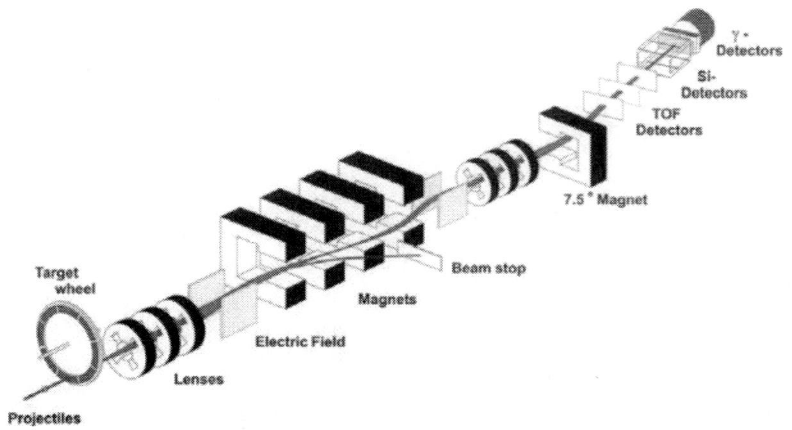

Figure 2. Sketch of the experimental setup: velocity filter SHIP and detection system.

At the velocity filter SHIP we started the investigation of transfer reactions in heavy systems, applying very sensitive detection techniques, which are also used for the identification of superheavy nuclei produced in fusion evaporation reactions. The identification of the isotopes is performed via their radioactive alpha-, beta- or spontaneous fission (SF) decays. This method together with the low background on the detection system allows to reach the cross-section limit of 10 pb/sr for a certain isotope, within one day of beam time. In a first experiment we investigated transfer reactions with ^{207}Pb targets applying ^{58,64}Ni beams. The application of lead targets leads to the population of a broad region of alpha emitters with proton numbers between Z = 82 and Z = 92. This allowed us to investigate the transfer from the projectile to the target nucleus in the region of heavy nuclei and its dependence on the projectile neutron number. Further, we could investigate the applied detection techniques and their limitations. In a follow-up experiment we studied transfer reactions using ^{248}Cm targets (Z = 96) and ^{48}Ca beams, where we explored the population of isotopes in the region around Z = 100. All experiments were performed at rather low beam energies, starting from about 10% below the Coulomb barrier up to about 20% above the barrier. At these low energies the (primary) transfer products are created at low excitation energies. This reduces the losses due to fission of the primary nuclei, which is an important point for the heavy nuclei around Z = 100, which have low fission barriers. Moreover, for the primary isotopes surviving sequential fission low excitation energies reduce the number of evaporated neutrons and lead to more neutron rich secondary nuclei.

2. Experimental setup and method

A sketch of the experimental setup is shown in figure 2. The beams of ^{48}Ca and ^{58,64}Ni were delivered by the UNILAC accelerator of GSI. The UNILAC accelerator delivers beams with a pulsed structure. The 5 ms long beam-on periods are followed by 15 ms long beam-off periods. Eight targets were mounted on a rotating target wheel, which is installed at a distance of ≈ 500 mm in front of the first quadrupole triplet of SHIP. Reaction products which leave the target at forward angles of (0 ± 2) degree with respect to the beam direction are accepted by the entrance aperture of SHIP. The respective fraction of the solid angle is $\Delta\Omega$ = 10 msr. The electric and magnetic fields are chosen such, that the relatively light projectiles and projectile-like reaction products, which have about ten times larger velocities than the transfer products, are deflected to the beam stop. Only the slow ($v \approx 0.01c$) target-like nuclei can pass SHIP and reach the focal plane detector. All reaction products which pass the velocity filter are implanted in a position sensitive silicon detector in the focal plane of SHIP where their time of implantation, position, kinetic energy and radioactive decays are registered. Especially the alpha decay properties allow for an unambiguous identification of the respective isotope. In order to reach the focal plane the nuclei must have half-lives of at least 1 μs which corresponds to the flight time of the reaction products through the 11 m long setup. For the identification via radioactive decay chains their half-life has to be ≥ 20 μs which corresponds to the conversion time plus dead time of the data acquisition system. For nuclei with shorter half-lives pile-up of the signals from the implanted reaction product and the particle emitted during its successive decay occurs.

For the identification of the reaction channel (transfer, fusion-evaporation etc.) we measured the velocity spectra of the produced isotopes. To obtain the spectra we varied the electric and magnetic field values, E and B, of the velocity filter accordingly. The ratio E/B determines at which velocity a particle can pass SHIP. The accepted velocity width at a given setting is $\Delta v/v = 0.1$ (FWHM).

The SHIP setup allows the identification of single nuclei. This results from the combination of strong background suppression and the isotope identification via the radioactive decays of the nuclei. This is valid for isotopes produced in fusion-evaporation reactions as well as for isotopes produced in nucleon transfer reactions. The suppression of the primary beam depends on the velocity difference between the projectile- and the target-like nucleus. This difference increases with increasing mass asymmetry between projectile and target. In the here-discussed reactions of $^{58,64}Ni + ^{207}Pb$ and $^{48}Ca + ^{248}Cm$ the factor of beam suppression is on the order of $10^{11} - 10^{12}$. The remaining projectile-like particles which still reach the focal plane can be distinguished from target-like particles by a time-of-flight (TOF) and energy measurement. Such, projectile- and target-like particles are located on different branches in the energy-versus-TOF spectrum. The obtained mass resolution is $A/\Delta A \approx 10$. The decisive point for identification of a certain isotope is its radioactive decay or decay chain, respectively, which is registered in the focal plane detector as described above. However, this method is only applicable for nuclei with sufficiently short half-lives. The upper limits for the correlation times are given by the number of background events which have identical signatures like true events [7]. The here described method of identification allows to reach cross-section limits of 10 pb within one day of beam time by applying usual beam intensities of several 10^{12} particles per second. For more details concerning the experimental setup and identification see ref [1, 8].

Concerning the transmission of SHIP there is a significant difference between fusion-evaporation residues and nuclei produced in transfer reactions (the term transmission comprises the angular acceptance and the transport efficiency through SHIP). The full momentum transfer in fusion reactions leads to an emission of the compound nuclei in the direction of the primary beam. Scattering angles, which differ from zero degree result predominantly from small angle scattering in the target. The effect of small angle scattering increases with increasing mass asymmetry between projectile and target nucleus. This causes the drop in transmission for very asymmetric combinations of projectile and target mass [9]. Usual transmissions obtained for fusion-evaporation reactions in the region of superheavy nuclei are (30 – 50)%. In transfer reactions, only a fraction of the incident momentum is transferred to the heavier transfer product, while the remaining fraction is carried by the lighter one. This leads to relatively broad angular distributions in the centre-of-mass as well as in the laboratory frame. For quasi-elastic transfer like pickup or stripping reactions, the angular distributions peak at the grazing angle. But for deep inelastic multi-nucleon transfer with long lifetimes of the underlying DNS the distributions broaden significantly and can be isotropic in the utmost case. Therefore, the transmission of SHIP for transfer products depends strongly on the angular distribution of the transfer products from a certain reaction, which is usually not fully known. In order to estimate the lower limit of the SHIP transmission for transfer products with $Z > 100$ we assumed the extreme case of an isotropic angular distribution in the centre-of-mass frame. In the laboratory, the heavy target-like product is still emitted within a rather forward directed cone with opening angles up to about 40°. In this case we obtain a transmission of ≈5%. Even with this relatively small transmission one can detect one event per day for reactions with total cross-sections of 200 pb. The small angle scattering is not dominant for transfer products since they have three times higher kinetic energies than fusion products created at the same beam energy.

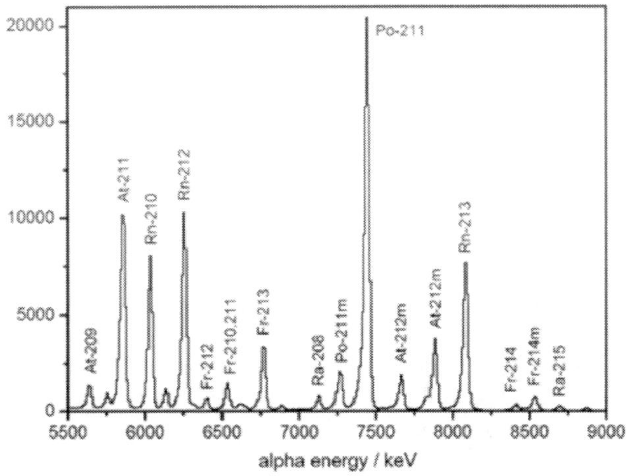

Figure 3. Alpha spectrum measured in ^{64}Ni + ^{207}Pb collisions during beam-off periods. All observed alpha lines can be attributed to isotopes, which have been created in transfer reactions.

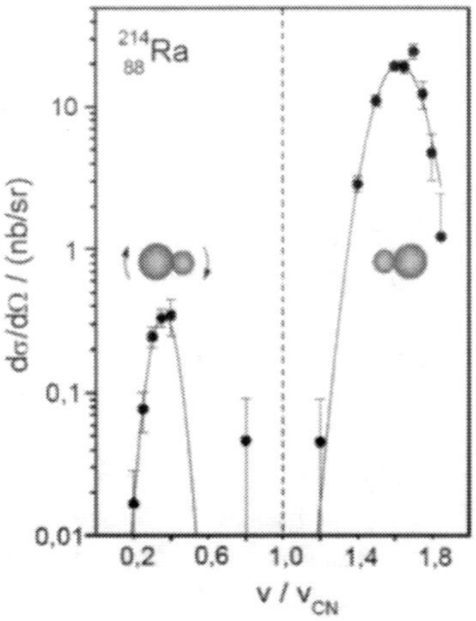

Figure 4. Velocity spectrum for ^{214}Ra isotopes created in transfer reactions of ^{64}Ni + ^{207}Pb at 5.9×A MeV (for description see text).

3. Results and discussion

1.1. Transfer reactions in the systems ^{58}Ni + ^{207}Pb and ^{64}Ni + ^{207}Pb

Transfer reactions with Pb targets lead to the population of a broad region of alpha emitters between proton numbers Z = 82 and Z = 92. This allowed us to investigate nucleon transfer from the projectile to the target nucleus in the region of heavy nuclei as well as the available identification and detection methods. We made experiments with the relatively neutron-rich ^{64}Ni and the more neutron-deficient ^{58}Ni beams. The energies of the ^{64}Ni beam were in the range (4.80 – 5.92)×A MeV, corresponding to (0.95 – 1.17) times the Bass barrier (V_{Bass} = 5.05×A MeV). The respective grazing angles are (180 – 96)° in the centre-of-mass system, which correspond to (0 – 42)° in the laboratory system for the target-like transfer product. For the ^{58}Ni beam we applied energies of (4.85 – 6.50)×A MeV, corresponding to (0.90 – 1.20) times the Bass barrier (5.4×A MeV). The grazing angles were (180 – 91)° in the centre-of-mass and (0 – 45)° in the laboratory system. The targets were prepared from enriched ^{207}Pb. Layers of 380 μg/cm^2 thickness of the chemical compound ^{207}PbS have been evaporated on carbon foils of 40 μg/cm^2 thickness and covered by 10 μg/cm^2 carbon. Eight targets were mounted on a rotating wheel. At each beam energy we scanned the velocity setting of SHIP in the interval (0.2 – 2.0) times the compound nucleus velocity v_{CN} in steps of 0.05×v_{CN}.

Figure 3 shows the alpha spectrum measured in reactions of ^{64}Ni + ^{207}Pb. It contains the sum over all beam energy and velocity settings. We detected isotopes of elements with $84 \leq Z \leq 88$, which were directly produced in the reaction, i. e. they were not populated by the decay of mother nuclei. Nuclei with $Z \leq 83$ could not be identified because they are not alpha emitters or have too long half-lives. With ^{58}Ni beams the same elements were populated, but the isotopic distributions peak at the more neutron-deficient side. All observed alpha lines can be attributed to nuclei produced in transfer reactions.

The velocity spectrum of ^{214}Ra (Z = 88) is shown in figure 4 representatively for the isotopes created in transfer reactions. The velocity of the detected reaction product was normalized on the compound nucleus velocity. For the different velocity settings limits for the differential cross-sections of (dσ/dΩ) = (10 – 50) pb/sr were reached, depending on the respective beam dose and measurement time.

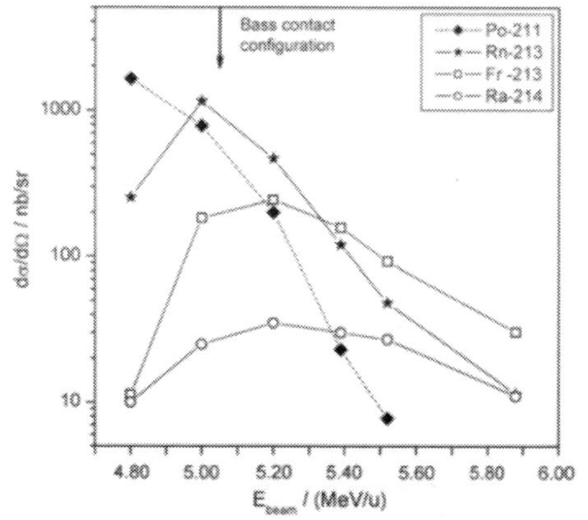

Figure 5. Excitation functions for the isotopes ^{211}Po, ^{213}Rn, ^{213}Fr and ^{214}Ra measured in ^{64}Ni + ^{207}Pb transfer reactions. The arrow marks the energy corresponding to the Bass barrier.

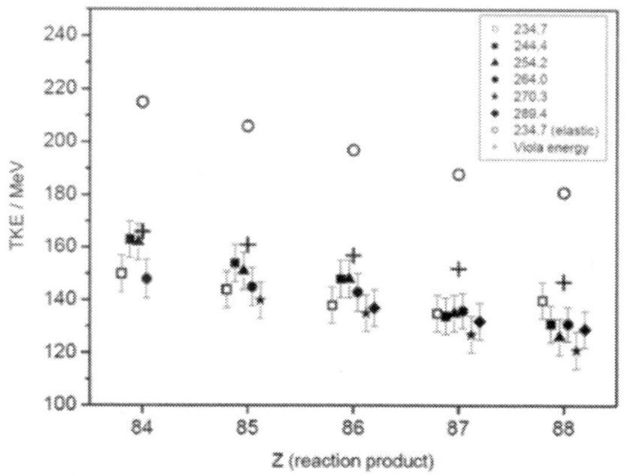

Figure 6. TKE values as a function of the proton number Z of the detected heavy reaction products for different beam energies (the respective beam energies in the centre of mass are given in the inset). For details see text.

For all isotopes we found pronounced maxima around velocities of v/v_{CN} = 1.6 and further weak maxima around v/v_{CN} = 0.4. The maxima are located symmetrically with respect to the compound nucleus velocity. This structure is caused by the narrow angular acceptance of SHIP of $(0 \pm 2)^{\circ}$. Only transfer products which are created in central collisions can enter the separator. In this case the lighter, projectile-like nucleus is scattered in backward direction and the heavier, target-like nucleus in beam direction with a velocity of $1.6 \times v_{CN}$. This peak position cannot be explained by quasi-elastic transfer, which would lead to velocities of v/v_{CN} = 1.9. Rather, the DNS which is formed after capture re-separates with the Coulomb-barrier energy of the exit channel,, which means that the isotopes originate from deep inelastic transfer processes (see discussion below). The low-velocity component is consistent with this scenario if we assume that the DNS can rotate around its centre of gravity after a non-central collision. The rotation brings the system again to an orientation parallel to the beam axis where, however, projectile and target have changed their positions with respect to the above described case of a central collision. Therefore, if the re-separation occurs in this orientation, the lighter, projectile-like fragment will be emitted in forward direction and the heavier, target-like in backward direction in the centre-of-mass frame. In the laboratory frame the target-like nucleus is still emitted in beam direction but with reduced velocity [10].

As an example, the excitation functions for the isotopes ^{211}Po (Z = 84), ^{213}Rn (Z = 86), ^{213}Fr (Z = 87) and ^{214}Ra (Z = 88) measured in ^{64}Ni + ^{207}Pb reactions are shown in figure 5. One has to note, that the detected nuclei are secondary reaction products which result from the excited primary products after evaporation of neutrons and, at sufficiently high excitation energies, also protons. The excitation functions reveal a systematic broadening and shift of the maximum with increasing number of transferred protons. For ^{211}Po and ^{213}Rn the maxima are located at $E \approx 0.95\ V_{Bass}$ and at $E \approx V_{Bass}$, respectively. At these beam energies the grazing angle is 180° in the centre-of-mass. This corresponds to a scattering angle of the target-like nucleus of 0° in the laboratory frame, which is also the detection angle of SHIP. This explanation holds in the case of direct reactions. However, the excitation functions for ^{213}Fr and ^{214}Ra do no longer fit into this scheme. For a possible explanation we took into account the Q_{gg} values of the ^{64}Ni + ^{207}Pb reactions leading to the different isotopes. In the case of polonium and radon the least negative Q_{gg} values are obtained when ^{211}Po and ^{213}Rn are created. This means that the detected ^{211}Po and ^{213}Rn nuclei are very close to the primary fragments and must have been created mainly at low excitation energies or beam energies, respectively. For the creation of

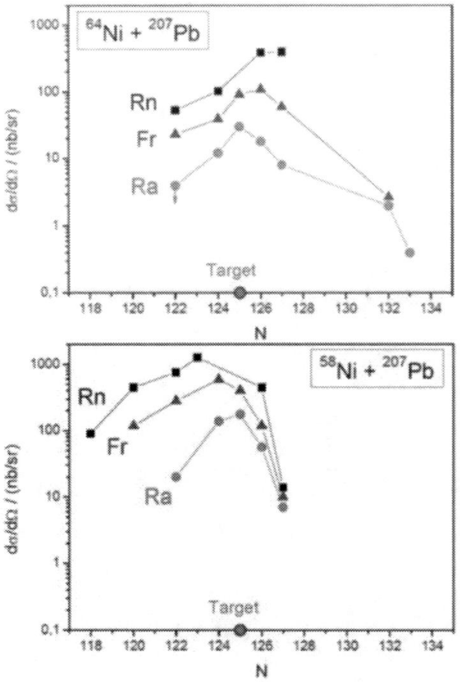

Figure 7. Isotopic yields of transfer products from ^{64}Ni + ^{207}Pb (top) and ^{58}Ni + ^{207}Pb reactions (bottom) as a function of the neutron number N of the reaction products. The neutron number of N = 125 of the target nucleus ^{207}Pb is indicated by the green dot.

francium and radium isotopes the optimum Q_{gg} values are obtained for mass numbers A = 216 and A = 218, respectively. Therefore, ^{213}Fr and ^{214}Ra result after the evaporation of three to four neutrons. Larger beam energies are needed to obtain the required excitation energies of the primary fragments for the evaporation of several neutrons. The broadening of the curves we interpret as a hint that the correlation between incident energy and grazing angle becomes weak like it is observed in deep-inelastic transfer reactions with relatively long lifetimes of the underlying DNS.
Indeed, the total kinetic energies (TKE) in the exit channel reveal deep inelastic kinematics for all detected isotopes (figure 6). The TKE values were reconstructed from the measured kinetic energy of the target-like transfer product by assuming two-body breakup. The data points in figure 6 which belong to the same value of Z but have been created at different beam energies have been plotted with an offset for better discrimination. The values expected from the Viola systematic (crosses) are also shown. The TKE values of the reaction products are, within error bars, independent of the beam energy, which points to a deep inelastic transfer process. In comparison, the open circles represent the TKE values, which are expected if the detected isotopes would originate from elastic kinematics at 4.80×A MeV (the Q_{gg} values have been taken into account). The maximum available excitation energy of the DNS is the difference between the centre-of-mass energies in the entrance and exit channels. This is (50 – 60) MeV for the lowest beam energy of 4.80×A MeV (see figure 6). For more details on the ^{64}Ni + ^{207}Pb reactions see ref [11].
The observations described for the ^{64}Ni beams concerning excitation functions and TKE values are also true for the reactions with ^{58}Ni beams. The only exception is ^{211}Po, which we will not discuss here.

The isotopic yield distributions obtained with ^{64}Ni and ^{58}Ni beams are compared in figure 7 for the Rn, Fr and Ra isotopes (Z = 86, 87, 88). For these graphs all beam energies have been added. A common feature for both projectile-target combinations is the drop of the cross-sections by roughly a factor of five to ten with every transferred proton. This is true for all detected isotopes. Further, two characteristic differences between ^{58}Ni and ^{64}Ni reactions can be observed. With ^{58}Ni beams rather neutron-deficient isotopes are populated and the maximum of the distributions shift to the more neutron-rich side with increasing proton number of the transfer product (see figure 7, bottom). A steep drop in the yields can be observed for N > 126. For the more neutron-rich ^{64}Ni beams the yield distributions extend noticeably to the neutron-rich side. The isotope with the largest neutron number is ^{221}Ra (N = 133), which has eight neutrons more than the target nucleus. The corresponding differential cross-section is 500 pb/sr, which is 100 times lower than the differential cross-section of ^{213}Ra (N = 125) which is located in the maximum of the yield curve. Contrary to the ^{58}Ni reactions, the maxima of the distributions shift to the neutron deficient side with increasing proton transfer. Concerning the yields, we found about two times larger cross-sections for ^{58}Ni reactions. These results are in agreement with observations from previous experiments described in ref [4] with ^{40,44,48}Ca beams on ^{248}Cm targets. The largest neutron transfer observed in the experiments of ref [4] was $\Delta N = 4$. The reason is probably, that the half-lives of more neutron-rich isotopes are on the scale of milliseconds or seconds and are therefore not accessible with chemical methods, which require several minutes. In our experiment the observation of the most neutron-rich isotopes with N > 128 was only possible by searching for alpha decay chains correlated in time with a preceding implanted recoil nucleus. This method, which is in the same way applied at SHIP for the identification of superheavy nuclei produced in fusion reactions, allows the identification of single events and gives access to the short-living isotopes with half-lives down to 20 μs. Further isotopes with still larger neutron numbers could not be identified, because they have decay chains with short-living members with half-lives in the region of microseconds and below which cannot be distinguished with the present data acquisition system of SHIP. For the same reason also no Rn isotopes with neutron numbers larger than N = 127 could be identified.

1.2. Transfer reactions in the system ^{48}Ca + ^{248}Cm

The cross-sections of transfer products drop by up to one order of magnitude with every transferred proton. Therefore, targets with the largest available proton numbers must be applied for the population of superheavy nuclei. Moreover, in order to reach out to the neutron-rich side, the neutron numbers of the projectile as well as of the target nucleus should also be as large as possible. One of the heaviest available target materials is the radioactive ^{248}Cm with a half-live of 3.4×10^5 y (Z = 96, N = 152). With these targets we investigated transfer reactions using the neutron-rich ^{48}Ca beams. ^{48}Ca is a doubly magic nucleus and ^{248}Cm has a subshell closure at N = 152, which leads in both cases to high binding energies. In this case we expect, in analogy to fusion reactions, relatively low excitation energies of the primary transfer products. ^{248}Cm targets are applied at SHIP since June 2010. The first application was from June 25 to July 26, 2010, for the production of different isotopes of element 116 in the fusion reactions of ^{48}Ca + ^{248}Cm $\rightarrow$ ^{296}116*. This experiment, where a total of six decay chains of isotopes ^{292}116 and ^{293}116 were observed as well as the applied target technique are discussed in detail in ref [12]. After the setting for fusion-evaporation residues we applied two days of beam time for the investigation of transfer reactions at the beam energy of 5.63×A MeV. This energy is 15% larger than the Coulomb barrier, which is obtained by assuming spherical Cm nuclei. Data were taken at five different velocity settings, 1.70, 1.80, 1.85, 1.90 and 1.95 times the compound nucleus velocity v_{CN}. The analysis of the transfer reactions from ^{48}Ca + ^{248}Cm is still in progress. Therefore, more detailed or improved information than discussed below might be available after finishing the analysis.

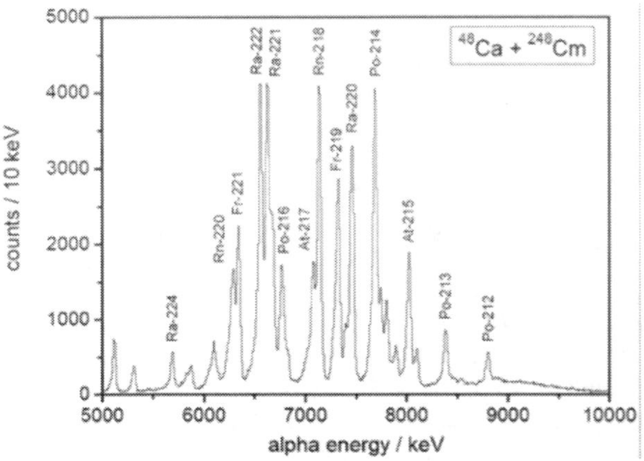

Figure 8. Alpha spectrum measured in reactions of ^{48}Ca + ^{248}Cm at 5.63×A MeV.

We identified numerous isotopes in the region of $Z < 96$ via alpha decays. The respective alpha energy spectrum is shown in figure 8. Like in the case of ^{58,64}Ni + ^{207}Pb reactions the detected nuclei are secondary reaction products which result after evaporation of neutrons from the excited primary products. Figure 8 reveals alpha lines which belong to isotopes with proton numbers $Z = 84$ (polonium) to $Z = 88$ (radium). We observed the maximum production yield for isotopes with $Z = 88$, which result after the transfer of 8 protons from the target to the projectile nucleus. The respective differential cross-sections are $d\sigma/d\Omega \approx (100 - 500)$ nb/sr related to the angular range of $(0 \pm 2)^{o}$ accepted by SHIP. According to the present status of analysis we can identify isotopes down to $Z = 86$ as directly produced, i.e. not populated by the decay of mother isotopes. The main contribution to the alpha line intensities for $Z \leq 86$ nuclei comes from the population due to mother decays of $Z = 88$ and $Z = 87$ isotopes. If also the $Z = 84$ lines contain a contribution from direct production is subject to the ongoing analysis. It is noteworthy, that we observed a sudden steep drop in the yield by more than a factor of ten for nuclei with proton numbers $Z \geq 89$. We identified these isotopes by their alpha decay chains. In figure 8 they give only a small contribution to the much more dominant alpha lines of isotopes with lower proton numbers. The only exception is some isotopes around the ^{248}Cm target nucleus, which were populated in quasi-elastic nucleon stripping or pickup reactions. They were produced with differential cross-sections of up to ≈10 μb/sr. We identified these nuclei due to their beta-decays, which lead to excited states of the respective daughter nuclei. The gamma rays following the de-excitation were detected in coincidence with the electron from the preceding beta-decay. The gamma-ray energies allowed for the assignment of the respective mother isotope.

We are not yet sure about the cross-section maximum observed for $Z = 88$. Actually, we would expect that the isotopes closer to the target are populated stronger and that the cross-sections decrease with every transferred proton, like observed in the reactions of ^{58,64}Ni + ^{207}Pb. A possible reason for the large cross-section for radium isotopes might be connected with the fission barriers, which decrease gradually with increasing proton number in the region of actinide nuclei. Since the measured nuclei are secondary reaction products the observed alpha spectra reflect also the loss of primary fragments due to sequential fission or nucleon evaporation, where especially the probability for sequential fission increases for nuclei closer to ^{248}Cm, while the nuclei closer to lead are more resistive. This might lead to the yield maximum at $Z = 88$ for the secondary isotopes, while the yield of the primary isotopes still decreases gradually with increasing number of transferred protons. Theoretical studies are presently undertaken to investigate this scenario [13].

Besides the alpha activity described above, we observed also numerous SF decays during the settings for transfer products. After the beam time the registration of SF events was continued and will run for

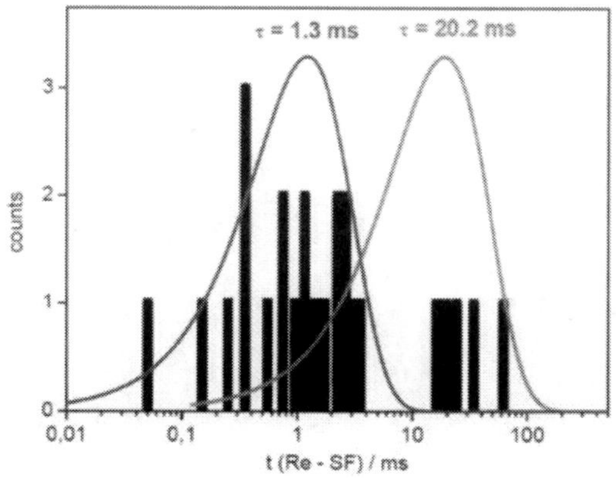

Figure 9. Correlations between fission events and preceding recoil nuclei within a time window of 100 ms. The decays were observed in ^{48}Ca + ^{248}Cm transfer reactions.

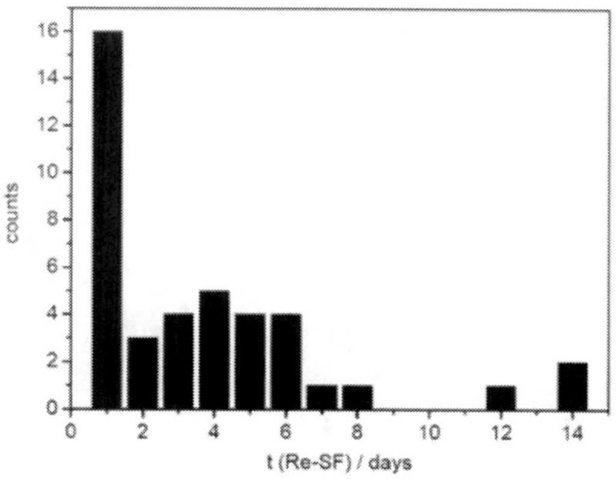

Figure 10. Long-living spontaneous fission events created in ^{48}Ca + ^{248}Cm transfer reactions. Each bar represents the number of fissions observed within one day. The counting was started just after the beam time ended.

about half a year more in order to see also possible long-living activities (see below). Since the energies of spontaneous fission events are no clear fingerprint of the decaying nucleus, especially not at small numbers of events, an assignment can only be made via recoil-fission correlations. This method is, however, restricted to sufficiently short-living isotopes. The maximum possible correlation time is given by the counting rate of possible recoils. In our experiment it allowed non-ambiguous correlations within time windows of 100 ms. The observed correlations between SF events and preceding recoil nuclei within a time of 100 ms are shown in figure 9. Two groups of SF events with different half-lives can be distinguished. For the group of events with correlation times below 10 ms we obtain a mean correlation time of t(Re-SF) = (1.63 ± 1.14) ms and for the group with correlation times larger than 10 ms we obtain t(Re-SF) = (25.26 ± 14.88) ms. We attributed these events to fission isomers in ^{242}Am ($T_{1/2}$ = 14 ms) and ^{244}Am ($T_{1/2}$ = 0.9 ms) with mean lifetimes of 20.2 ms and 1.30 ms, respectively. The americium isotopes (Z = 95) can be reached by the exchange of one proton and the exchange or/and evaporation of several neutron. The calculated envelopes of the time distributions, which we expect for events from ^{242}Am and ^{244}Am are also drawn in figure 9. The respective differential cross-sections are ≈700 pb/sr for ^{244m}Am and ≈100 pb/sr for ^{244m}Am. The activity with the shorter half-live would also fit to an isomeric state in ^{240}Am with a lifetime of 1.36 ms ($T_{1/2}$ = 0.94 ms). However, within error bars, we can not decide, if the SF events with τ = 1.63 ms originate from ^{240m}Am or from ^{244m}Am or if there are contributions from both. One has still to note that for Am nuclei we can only identify the isomeric but not the ground states, because they are too long-living.

The time distribution of the long-living SF events is shown in figure 10 for the first 15 days after the beam time. The respective measurement was started just after the end of the beam time. Each bar contains the events which were observed within one day. The activity measured within the first day after beam time is significantly higher than on the following days. Within statistical fluctuations, the activity decreases from the second day on. The missing events for days nine and ten are due to a stop of the measurement during these two days, while on days 11, 13 and 15 no decays occurred. The shape of the distribution suggests, that there are contributions from several activities with different half-lives. A more detailed analysis is in progress.

For the long-living fission events a clear assignment to a certain mother nucleus via time correlations is no longer possible. Different time components can only be roughly distinguished. Figure 11 shows the isotopes, which are the most probable candidates for the observed SF decays. The location of the target nucleus, ^{248}Cm, is also shown for orientation. The chart contains only isotopes which have half-lives shorter than one year and sufficiently large SF branches of > 0.5%. Nuclei, with longer half-lives

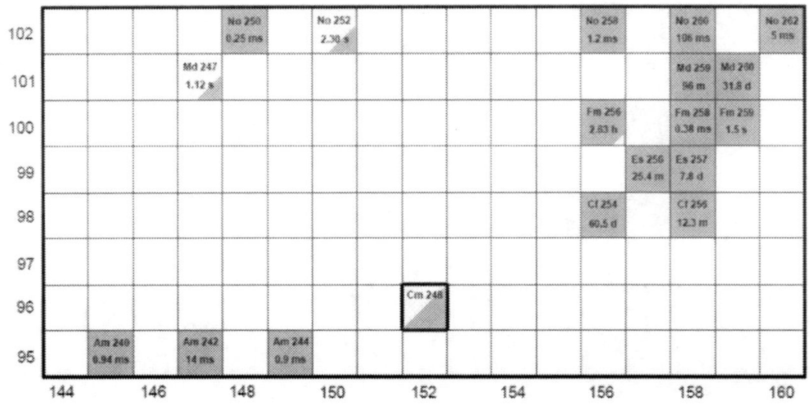

Figure 11. Isotopes which are the most probable candidates for the observed long-living SF events from $^{48}Ca + ^{248}Cm$ transfer reactions at 5.63×A MeV.

and smaller fission branches are not detectable at SHIP taking into account the expected small production cross-sections. The half-lives of several isotopes are too similar and cannot be unambiguously attributed to a certain nucleus. However, one can deduce from these results that the region of nuclei with $Z \approx 100$ was populated during this experiment with high probability.

In order to reach the neutron-rich nuclei in the region of $Z \geq 100$, low beam energies should be favourable. The nuclei in this region have relatively low fission barriers. As a consequence, the probability that the excited primary transfer product undergoes sequential fission is larger than for lighter nuclei and is increasing with increasing excitation energy or beam energy, respectively. Secondly, the nuclei which survive sequential fission de-excite by evaporating neutrons. The lower the excitation energy, the lower is the number of evaporated neutrons and the more neutron-rich are the secondary nuclei. Similar to fusion-evaporation reactions, the cross-sections for the production of a certain transfer product can be written as the product of three factors: $\sigma = \sigma_{capture} \times P_{transfer} \times P_{survival}$. The factor $\sigma_{capture}$ is the capture cross-section and $P_{transfer}$ is the probability to create a certain primary transfer product. These two values increase when the beam energy exceeds the Coulombbarrier. The factor $P_{survival}$ is the probability that the primary transfer product survives sequential fission and that it evaporates just the number of neutrons which lead to the desired secondary product. $P_{survival}$ decreases with increasing excitation energy. Especially the influence of sequential fission on $P_{survival}$ becomes more critical for very heavy nuclei with decreasing fission barriers. The extreme case is superheavy nuclei, where the fission barriers are only determined by the shell correction energies. These qualitative considerations are also found by theoretical model calculations [13]. Figure 12 shows the excitation functions for the isotopes ^{259}No ($Z = 102$), ^{264}Lr ($Z = 103$) and ^{271}Sg ($Z = 106$) created in transfer reactions in the collision system $^{48}Ca + ^{248}Cm$. These three isotopes are already located towards the neutron-rich side of the nuclear chart. ^{264}Lr and ^{271}Sg are also new isotopes in the sense, that they have not been produced directly, so far. According to the model calculations, the maximum cross-sections for ^{264}Lr and ^{271}Sg are located at energies, which are only 2% above the Coulomb barrier.

4. Summary and outlook

We studied multi-nucleon transfer reactions in the collision systems $^{58,64}Ni + ^{207}Pb$ and $^{48}Ca + ^{248}Cm$. In both cases we observed a massive transfer of protons and neutrons already at beam energies below the Coulomb barrier. The transfer from the projectile to the target nucleus was of special interest, since we investigated the possibility to create new neutron-rich nuclei in the region of superheavy elements

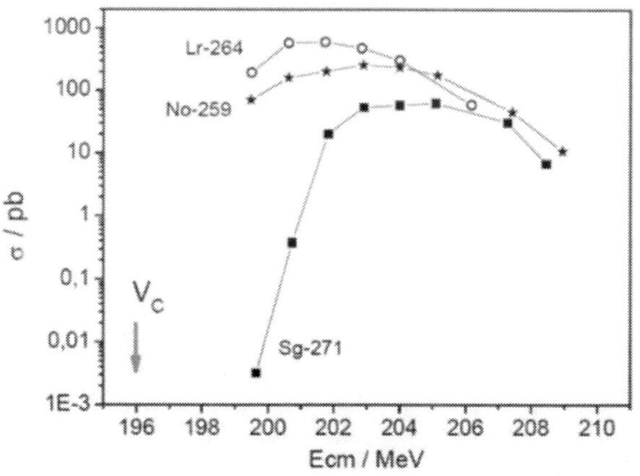

Figure 12. Calculated excitation functions for the isotopes ^{259}No, ^{271}Sg and ^{264}Lr produced in nucleon transfer reactions in the collision system ^{48}Ca + ^{248}Cm [13]. The arrow denotes the Coulomb barrier energy.

with $Z \geq 100$, which cannot be reached in fusion-evaporation reactions. We performed the studies at the velocity filter SHIP. The setup allows the detection of the heavy, target-like transfer products and their identification via radioactive decays. Such, even single nuclei can be unambiguously identified. With this detection method, which is also applied for superheavy nuclei produced in fusion reactions, in combination with the low background on the focal plane detector a cross-section limit of 10 pb can be reached within one day of beam time applying usual beam intensities of 5×10^{12} particles per second. Therefore SHIP is presently the most sensitive apparatus for the study of such transfer reactions for which cross-sections of nanobarn and below are expected. A further advantage is that SHIP allows the detection of the transfer products, which are emitted in forward direction. This is essential if the reactions are performed at beam energies below and close to the Coulomb barrier. Such low beam energies seem favourable for the production of very heavy nuclei with large neutron numbers, since excitation energies are low in this case, which prevents the excited primary transfer products from sequential fission or from the evaporation of too many neutrons.

In collisions of ^{58}Ni and the more neutron-rich ^{64}Ni with ^{207}Pb nuclei we observed the transfer of up to six protons from the projectile to the target nucleus. With ^{64}Ni beams we reached out to the neutron-rich side with respect to the target neutron number N = 125. We detected isotopes resulting from the transfer of up to eight neutrons (^{221}Ra). Still more neutron-rich nuclei were not observed. They would have been accessible concerning the cross-sections, but their radioactive decay chains comprised short-living members (<20 μs), which were not accessible with the data acquisition system. Also in the ^{48}Ca + ^{248}Cm collisions we observed numerous transfer products with proton numbers below as well as above the target proton number. The transfer products with $Z < Z_{Cm}$ can be identified via their alpha decays. But for the transfer products with $Z > Z_{Cm}$ the limitations of the present detection technique become visible. The nuclei in this region are very long-living and can no longer be identified via their decay chains. Further, the nuclei on the neutron-rich side undergo SF or beta-decays, which allow, in contrary to alpha decay chains, no unambiguous identification of the mother nucleus. In our experiment we observed several SF activities which can be roughly attributed to isotopes with different half-lives. However, an unambiguous attribution to certain mother nuclei was not possible. Therefore, for long-living isotopes and those which undergo SF- or beta-decays other detection methods have to be applied. Presently we are investigating the possibility to identify them via precision mass measurements. Penning traps, like SHIPTRAP at GSI [14], and time-of-flight mass spectrometers, like the MR-TOF-MS developed at the University of Gießen [15], have sufficiently large resolution of $m/\Delta m > 10^5$ to perform also an isobaric separation for most of the isotopes. Part of our program at GSI is to make feasibility studies of this method for the application to cross-sections below nanobarn. Especially the efficiency of the buffer gas cell has to be improved, which is presently on the level of 1%. The buffer gas cell is needed for stopping the reaction products, which have initial energies of ≈100 MeV, for the injection to a Penning trap or time-of-flight mass spectrometer, which requires ion energies on the order of 100 eV. The application of alternative detection methods is also essential to study the development of production cross-sections with increasing proton and neutron transfer. With intermediate heavy beams on heavy targets, like $^{48}Ca + ^{248}Cm$, neutron-rich isotopes with proton numbers up to $Z \approx 105$ are predicted by model calculations with total cross-sections of several ten up to several 100 pb. This would make transfer reactions a feasible means to create these nuclei if the predicted cross-sections can be verified experimentally.

Acknowledgments
We would like to thank especially our colleagues from theory, G. Adamian, N. Antonenko, A. Nasirov (Bogoliubov Laboratory for Theoretical Physics, Dubna) and V. Zagrebaev (Flerov Laboratory, Dubna) for their continuous interest in our data. Their accompanying model calculations and related discussions are very fruitful and valuable for the interpretation of our data and the progress of the experiments.

[1] Hofmann S and Münzenberg G 2000 *Rev. Mod. Phys.* **72** 733
[2] Oganessian Yu Ts 2007 *J. Phys. G* **34** R165
[3] Schädel M et al. 1982 *Phys. Rev. Lett.* **48** 852
[4] Türler A, von Gunten H R, Leyba J D 1992 *Phys. Rev. C* **46** 1364
[5] Adamian G G, Antonenko N V and Zubov A S 2005 *Phys. Rev. C* **71** 034603
[6] Zagrebaev V I, Oganessian Yu Ts, Itkis M G and Greiner W 2006 *Phys. Rev. C* **73** 031602(R)
[7] Schmidt K H, Sahm C C, Pielenz K, Clerc H G 1983 *Z. Phys. A* **316** 19
[8] Armbruster P 2000 *Annu. Rev. Nucl. Part. Sci.* **50** 411
[9] Mazzocco M 2007 *Eur. Phys. J. Special Topics* **150** 363
[10] Heinz S et al. 2008 *Eur. Phys. J. A* **38** 227
[11] Heinz S et al. 2010 *Eur. Phys. J. A* **43** 181
[12] Hofmann S et al. to be published in *Eur. Phys. J. A*
[13] Adamian G and Antonenko N *private communication*
[14] Block M et al. 2005 *Eur. Phys. J. A* **25 s01** 49
[15] Plaß W R et al. 2008 *Nucl. Instr. Meth. B* **266** 4560

Symmetric and asymmetric quasifission in reactions with heavy ions

E.M. Kozulin[1], **M.G. Itkis, G.N. Knyazheva**
FLNR, Joint Institute for Nuclear Research, Dubna, Russia

E-mail: kozulin@jinr.ru

Abstract. The main aim of the present study is to evaluate the fusion probabilities and investigate competing quasifission process in the reactions with heavy ions leading to the formation of superheavy composite systems. The mass-energy distributions of binary fragments as well as their cross sections have been measured for a wide range of composite systems with Z=82-122 formed in the reactions with ^{22}Ne, ^{26}Mg, ^{48}Ca, ^{58}Fe and ^{86}Kr ions at energies around the Coulomb barrier. The experiments were carried out using a double-arm time-of-flight spectrometer of binary reaction products CORSET. The results of the experimental investigation of the influence of the entrance channel properties on the competition between fusion-fission and quasifission for the "warm" fusion reactions is discussed.

1. Introduction

The collision of two massive nuclei takes a special place in nuclear reaction studies due to the large number of interacting nucleons. In this type of reactions a drastic change of the reaction partners may occur that leads to different reaction mechanisms. In reactions with heavy ions complete fusion and quasifission (QF) are competing processes [1, 2, 3, 4]. The relative contribution of QF to the capture cross section becomes dominant for superheavy composite systems and compound nucleus (CN) formation is hindered by QF process. It strongly depends on the entrance channel properties, such as mass-asymmetry, deformation of interacting nuclei, collision energy and the Coulomb factor Z_1Z_2.

In the context of the Liquid Drop Model (LDM) taking into account the nuclear interaction in the form of a proximity potential, calculations of the potential interaction energy were performed [5] for a large number of reactions involving heavy ions. These calculations show that the fusion barrier for the systems with $Z_1Z_2 \geq 1800$ has a double-humped shape. The internal barrier becomes predominant for the systems with $Z_1Z_2 \geq 2300$. The existence of a minimum in the potential energy after a contact of two nuclei, which is bounded by the internal barrier, is likely to be responsible for an onset of QF. These calculations, however, did not take into account the influence of the shell effects playing an important role in the interaction between two heavy nuclei.

A realistic description of the mass, energy and angular distributions of the reaction fragments formed in deep inelastic scattering, QF and compound nucleus fission (CNF) processes in low-energy heavy ion collisions was performed in [6] by using Langevin type dynamic equations of motion. It was shown that the multi-dimensional adiabatic potential energy surface (calculated within the two-centre shell model) plays the most important role in such processes. Fig. 1 shows the potential energy surface

[1] E. M. Kozulin, 6, st. Joliot-Currie, 141980 Dubna, Russia.

as a function of the mass-asymmetry and elongation for the superheavy nuclear system consisting of 116 protons and 180 neutrons. This potential energy surface is strongly modulated by shell effects and leads to the appearance of deep valleys corresponding to the formation of well bound magic nuclei. In accordance with these calculations, at least three paths leading to the formation of fission-like fragments can be distinguished: (1) asymmetric QF (QFasym in Fig. 1) caused by the influence of proton shells with Z = 28, 82 and neutron shells with N = 50 and 126; (2) symmetric QF (QFsym in Fig. 1) determined by the shells with Z = 50 and N = 82; (3) CNF (CNF path in Fig. 1) leading to the formation of symmetric fragments.

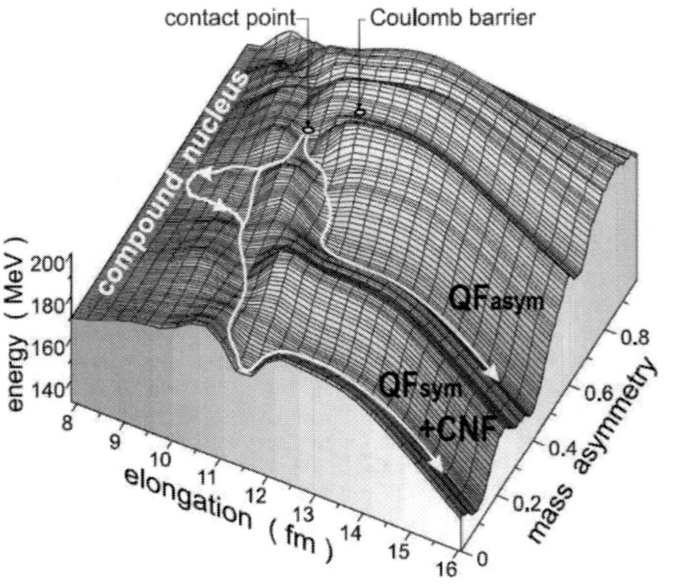

Figure 1. Example of a driving potential energy surface as a function of elongation and mass asymmetry for a heavy ion induced reaction leading to the superheavy composite system [6].

2. Asymmetric quasifission

While QF is a process occurring without a CN stage the QF fragment properties are close to CNF. The main experimental difference between asymmetric QF and asymmetric fission is connected with angular distribution of their fragments.

At comparative study of mass-energy distributions of fission-like fragments obtained in the reactions $^{12}C+^{204}Pb$ and $^{48}Ca+^{168}Er$ at excitation energy of about 40.5 MeV leading to the formation of the same CN $^{216}Ra^*$ the large contribution ~30% of the asymmetric fission mode in the case of the reaction $^{48}Ca+^{168}Er$ is immediately evident, manifesting itself in the form of wide ''shoulders''. The symmetric fission component is described by a Gaussian shape for the both reactions, but it is wider in the ^{48}Ca induced reaction than that for the ^{12}C reaction. This increase in variance is consistent with the results of Ref. [7], where it is shown that dispersion of mass distribution increases approximately linearly with angular momentum. According to [8] the critical angular momentum is 31ħ for the $^{12}C+^{204}Pb$ reaction and 54 ħ for $^{48}Ca+^{168}Er$. Thus the observed increase of dispersion agrees with the expectations for normal symmetric fission of the excited CN. The shape of the curve obtained for <TKE> in the case of $^{48}Ca+^{168}Er$ is far from parabolic, and is much wider than the parabola of the $^{12}C+^{204}Pb$.

Thus in the ^{48}Ca reaction we observe a strong increase (by a factor of ~20) in the contribution of asymmetric fission compared with the ^{12}C reaction. This increase is also reflected in the energy distributions of the fragments. It is interesting to note that this effect is observed at all energies around the fusion barrier in the ^{48}Ca reaction though its importance decreases sharply with increasing excitation energy.

Angular distributions for symmetric (98-118 u) and asymmetric (68-88 u) fragment masses were derived attempting to find experimental evidence for the QF nature of the mass-asymmetric "shoulders" observed in the reaction ^{48}Ca+^{168}Er (see Fig.3). A significant forward-backward asymmetry is observed in the angular distribution of fragments with masses 68-88 u (the region where asymmetric "shoulders" is observed), while symmetric fragments show a typical for CN symmetrical with respect to 90° in the centre-of-mass system angular distribution. Forward-backward asymmetry in angular distribution indicates that the composite system lives less than the time of one turn of composite system (~10^{-21}s) and this time is not sufficient for the formation of the fully equilibrated CN.

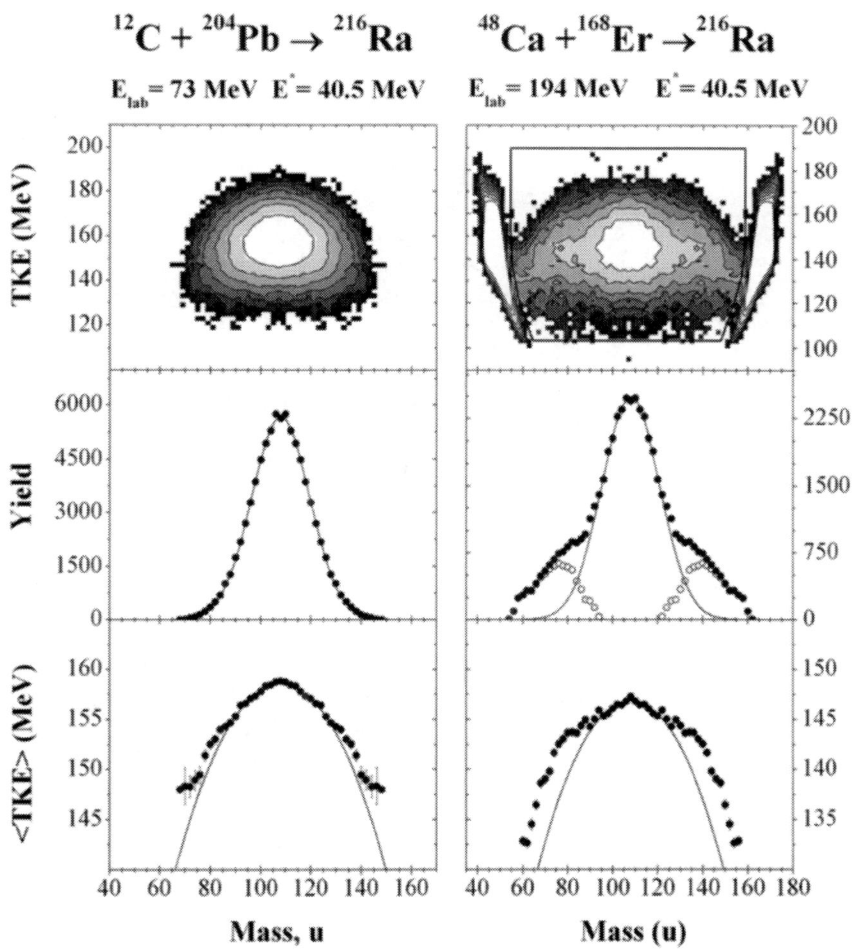

Figure 2. From top to bottom: two-dimensional matrix (TKE, Mass) of binary fragments obtained in the reactions ^{12}C+^{204}Pb and ^{48}Ca+^{168}Er; mass distributions and average TKE's as functions of fragment masses.

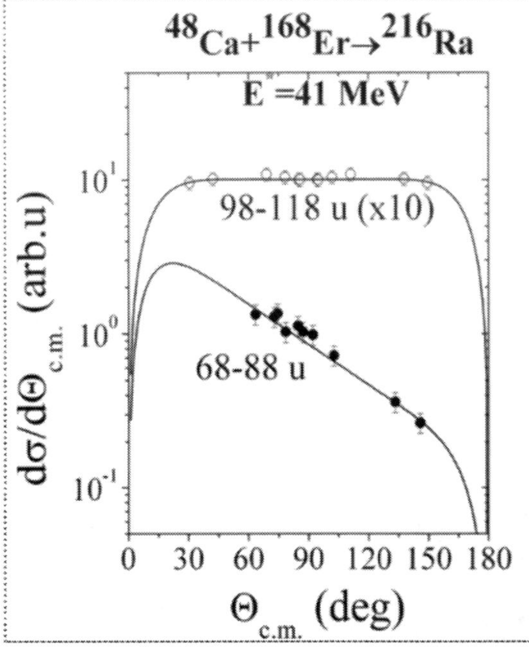

Figure 3. Differential cross sections for fission-like fragments in the reaction $^{48}Ca+^{168}Er$ for asymmetric and symmetric fragment masses.

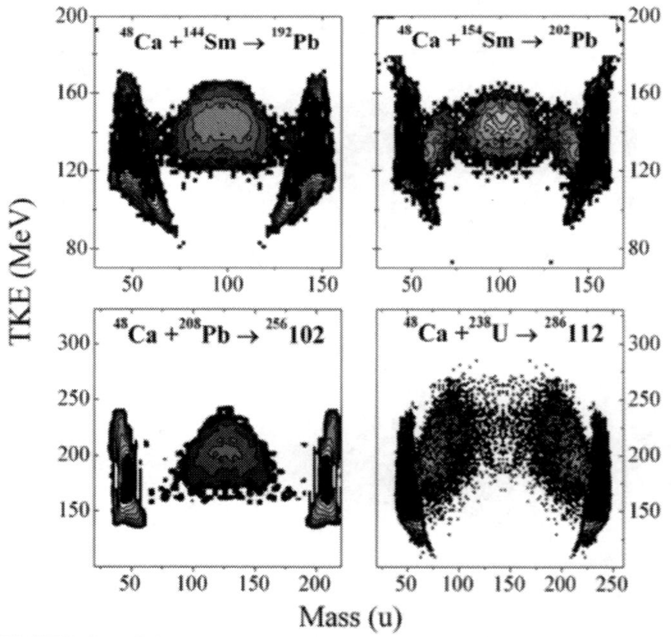

Figure 4. Two-dimensional matrices (TKE, Mass) for the reactions $^{48}Ca+^{144}Sm$ and $^{48}Ca+^{208}Pb$ (targets and projectile are spherical) in comparison with the $^{48}Ca+^{154}Sm$ and $^{48}Ca+^{238}U$ (targets are strong deformed) at the energy near the Coulomb barrier.

The QF process has been found for the reactions ^{48}Ca+^{168,170}Er [9], ^{40,48}Ca+^{154}Sm [10] which involve deformed targets and lead to moderately fissile compound nuclei. For these reactions along with a maximum corresponding to the symmetric mass division around $A_{CN}/2$, reveals two additional "asymmetric shoulders" corresponding to QFasym. The major part of the asymmetric component fits into the region of the Z=28 and N=82 shells, and its maximum yield is a "compromise" between them. Thus, we have an indication that the shell structure of the fragments formed in the ranges of the light M=60-75 u and heavy M=130-145 u masses, respectively, strongly favour the QF process, the shells in both light and heavy fragments playing an important role.

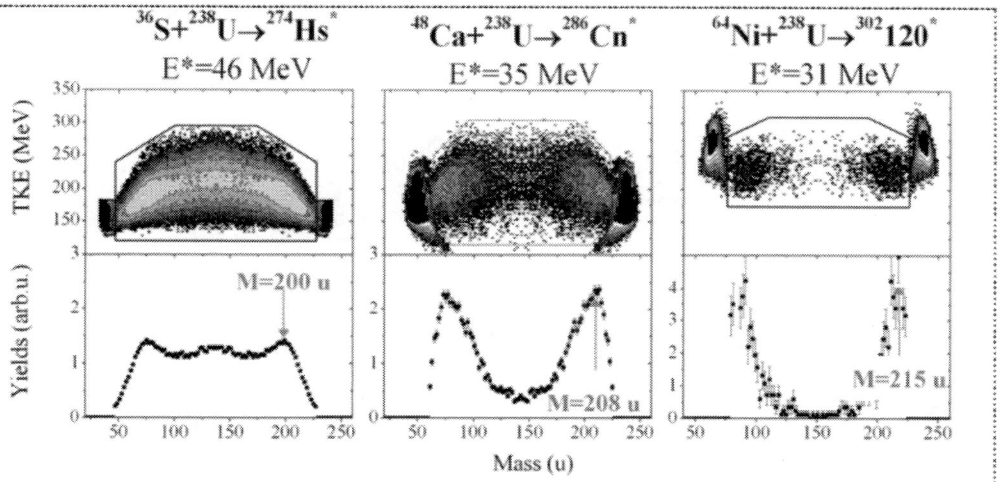

Figure 5. Mass-energy distributions for the reactions ^{36}S, ^{48}Ca, ^{64}Ni +^{238}U at energies close to the Coulomb barrier.

However, for the ^{48}Ca+^{144}Sm reaction with spherical nuclei the QFasym is not observed, the closed shells with N=50 and Z=50 fall into the region of the symmetric mass distribution, i.e., in the region of "normal" CNF. The same trend may be expected in the reactions leading to the superheavy nuclei. In the case when both reaction partners are spherical the contribution of QF process may be expected relatively small (for example, ^{48}Ca+^{208}Pb [11]).

The yields of evaporation residues, fusion-fission, and quasifission fragments in the ^{48}Ca + ^{144,154}Sm and ^{16}O + ^{186}W reactions have been analyzed in the framework of the combined theoretical method based on the dinuclear system concept and advanced statistical model [12]. According to these calculations the decrease in the measured yield of QF fragments in ^{48}Ca + ^{154}Sm at the large collision energies and the lack of QF fragments in the ^{48}Ca + ^{144}Sm reaction are explained by the overlap in mass angle distributions of the QF and CNF fragments.

In Fig. 4 the mass-energy distributions for the reactions ^{48}Ca+^{144,154}Sm and ^{48}Ca+^{208}Pb, ^{238}U at the energy near the Coulomb barrier are shown. Deformation of the interacting nuclei favours the QF process which mainly leads to the formation of the clusters in the exit channel of the reaction.
The contribution of asymmetric QF component in the capture cross section increase with increasing the reaction Coulomb factor Z_1Z_2 and becomes dominant for the composite systems with Z_1Z_2>1800 [13]. Fig. 5 shows the mass energy distributions of binary fragments obtained in the reactions of ^{16}S, ^{48}Ca, ^{64}Ni ions with an uranium target. The Coulomb factors are 1472, 1840 and 2576 for the ^{16}S+^{238}U, ^{48}Ca+^{238}U and ^{64}Ni+^{238}U, respectively.

Some noteworthy features of QFasym component of fragment mass distributions for the studied reactions can be highlighted at this point. Generally, in heavy-ion induced reactions the formation of QFasym fragments is connected with the strong influence of the nuclear shell at Z = 82 and N = 126 (doubly magic lead). In fact, as it was shown in Ref. [14], for the ^{48}Ca+^{238}U reaction the maximum yield corresponds to fragments with masses 208 u. However, in reactions with lighter projectiles on an uranium target, the asymmetric QF peak shifts to more symmetric masses [15]. By contrast, for the heavier projectile ^{64}Ni the maximum yield of QFasym fragments corresponds to the heavy mass 215u [14]. This trend is illustrated in Table 1 where the positions of light and heavy QF fragments for these reactions are presented. Notice, that even in the case of the ^{36}S+^{238}U and ^{26}Mg+^{248}Cm leading to the formation of the same composite nucleus ^{274}Hs* the position of the asymmetric QF peaks changes [16] (see Fig. 6). This trend is contrary to the multimodal fission [17] where the position of each mode determined by the nuclear shells is constant for specific compound nucleus and only the relative contribution of each mode varies in dependence of excitation energy. Thus, for more asymmetric reactions the Coulomb repulsion is expected to be smaller. For lighter projectiles this may lead to longer reaction times before separation for asymmetric QF and thus to larger numbers of exchange nucleons.

Table 1. Positions of heavy and light peaks in the mass distributions of QFasym fragments.

Reaction	Z_1Z_2	M_L	M_H	Exchange nucleons	Ref.
^{30}Si + ^{238}U	1288	90	178	60	Nishio et al. [15]
^{36}S + ^{238}U	1472	74	200	38	Itkis et al [16], Nishio et al. [15]
^{40}Ar + ^{238}U	1656	74	204	34	Nishio et al. [15]
^{48}Ca + ^{238}U	1840	78	208	30	Kozulin et al. [14]
^{64}Ni + ^{238}U	2576	87	215	23	Kozulin et al. [14]
^{26}Mg + ^{248}Cm	1152	94	180	68	Itkis et al [16]

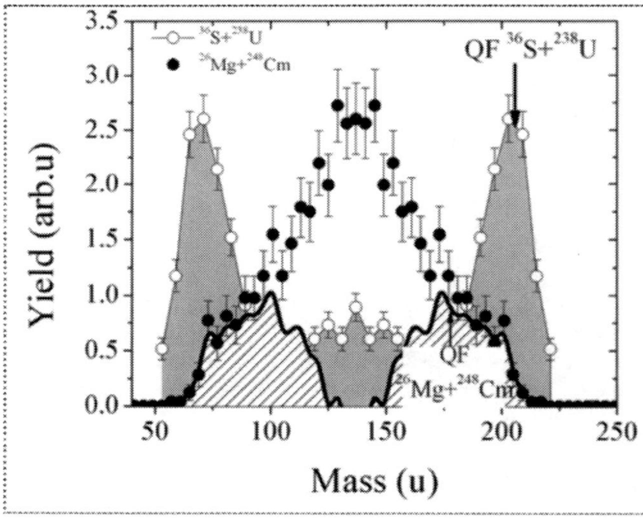

Figure 6. Mass distributions of fission-like fragments of the reactions ^{36}S+^{238}U and ^{26}Mg+^{248}Cm leading to the formation of the same CN of ^{274}Hs at excitation energies of the CN of about 35 MeV.

3. Symmetric quasifission

Besides asymmetric also symmetric component may be affected by the presence of QF process. Consequently the question of whether the symmetric fragments originate from CNF or QF processes arises. On the one hand the angular distribution for these fragments is symmetric with respect to 90° in the centre-of-mass system and the estimated reaction time is about 10^{-20} s typical for CNF processes. On the other hand mass drift is rather slow process and long reaction time is needed to reach the mass equilibration. But this condition is not enough for CN formation, in which all degrees of freedom have to come to equilibration. Thereby, the possible reaction channel for superheavy composite systems is a process occurring without a CN stage. This process is characterized by long reaction times sufficient for mass equilibration and resulting in the formation of symmetric fragments (symmetric quasifission (QFsym) in Fig. 1).

Figure 7 displays the obtained distributions of binary fragments obtained in the reaction $^{48}Ca+^{238}U$ at different excitation energies, namely (from top to bottom): the two-dimensional matrix of events as a function of the mass and total kinetic energy; the mass distribution for fission events framed into the contour line drawn on the two-dimensional mass-energy plot. Reaction products lying between elastic peaks can be identified as totally relaxed events, i.e., as fission (or fission-like) fragments. We have outlined them by solid lines in the panels. Henceforth we consider the properties of these events only.

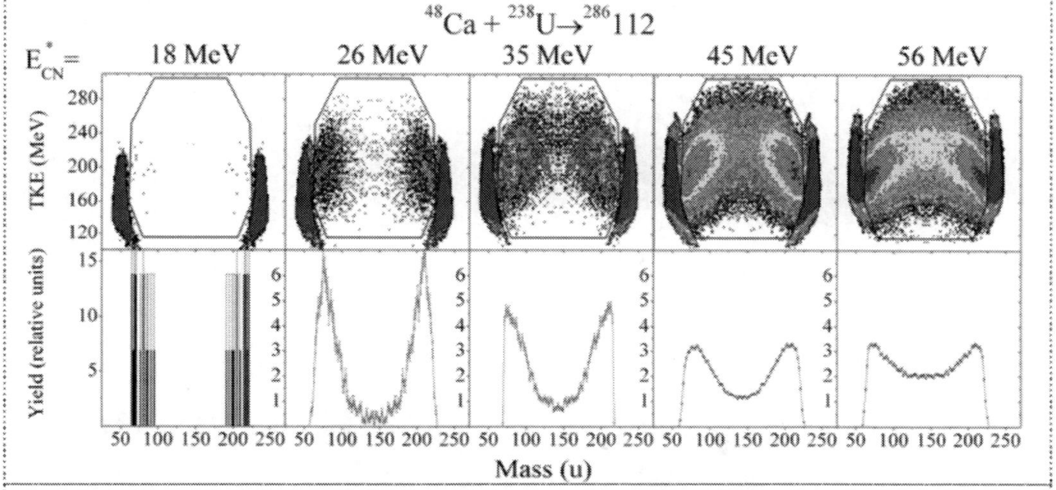

Figure 7. The two-dimensional TKE–M matrices (upper panels) and yields of fragments inside the contour lines on TKE-M matrices (bottom panels) in the $^{48}Ca+^{238}U$ reaction at projectile energies 212, 222, 232, 244 and 258 MeV corresponding to excitation energies of the CN of 18, 26, 35, 45 and 56 MeV, respectively.

Mass-energy distributions for the reaction $^{48}Ca+^{238}U$ have the wide two-humped shape caused by QFasym process. In spite of domination of asymmetric QF fragments at energies below the Coulomb barrier, especially, the yield of symmetric fragments increases with increasing excitation energy. Figure 8 presents the TKE distribution of fragments with masses $A_{CN}/2\pm20$ u for the reaction $^{48}Ca+^{238}U$ at energy 232 MeV. It is readily seen that TKE distribution has a complex structure which is not consistent with only CNF process. In fact, it is known that in CNF process the average TKE of the partner fragments is substantially independent on the excitation energy and shows a typical Gaussian-like shape. The TKE distribution was described by the sum of the three Gaussians with the next values of mean energies and dispersions: 188±3.0 MeV and 213±5 MeV^2; 228.5±1.5 MeV and 420±5 MeV^2; 265.9±1.9MeV and 74±4 MeV^2.

From the Viola systematics [18] we infer that the average TKE is in a first approximation a linear function of the Coulomb barrier $Z^2/A^{1/3}$ whereas from the systematics in Ref. [7] we can estimate the variance of the TKE distribution. For the $^{286}112$ CN the variance of the TKE distribution is about ~450 MeV^2 and the TKE is 226 MeV. Given the considerable good agreement with the systematics, we associate the middle component of the TKE distribution with the CNF process. Since the asymmetric fragments have lower TKE that the symmetric ones, the low energy component of the experimental TKE distribution may be associated with fragments originating from the asymmetric QF process. The high energy part may arise instead from the symmetric mode of the QF. Furthermore, we note that the mean TKE from this mode is about 40 MeV higher that the mean TKE for the CNF process. Considering that both processes give rise to symmetric mass fragments, the difference in mean TKE can be taken as an evidence that in the QF process a complete dissipation of the entrance channel energy does not occur. As a consequence, the symmetric fragments with high TKE do not originate from complete fusion because the final fragments retain part of the entrance channel total kinetic energy.

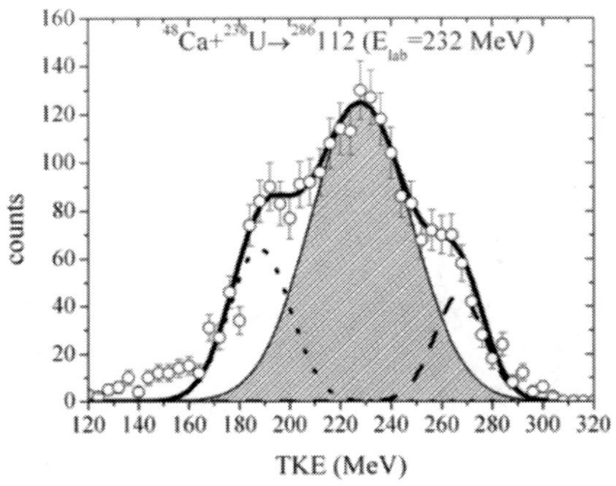

Figure 8. TKE distribution of fragments with masses $A_{CN}/2\pm20$ u for the reaction $^{48}Ca+^{238}U$ at energy 232 MeV. The open circles are the experimental points, the hatched region corresponds to CN fission with energy taken from the Viola systematic, dashed and dotted curves represent high and low energy components of the TKE distribution.

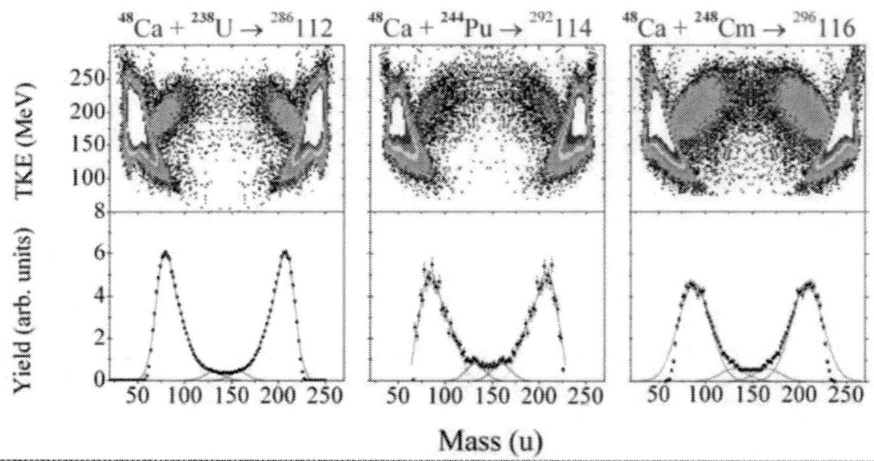

Figure 9. The two-dimensional TKE–M matrices (upper panels) and yields of fission-like fragments (bottom panels) in the $^{48}Ca+^{238}U$, ^{244}Pu, ^{248}Cm reactions at the CN excitation energies around 33 MeV.

At the transition from ^{238}U to the heavier ^{244}Pu and ^{248}Cm targets the mass-energy distributions of binary reaction products do not change practically. Figure 9 displays the TKE-M matrices for these reactions at CN excitation energy of about 33 MeV. The typical wide two-humped shape with small contribution to the symmetric mass region is observed for these reactions. It is important to note that the relative contribution of the symmetric fragments into the capture cross section does not drop dramatically with increasing of charge of composite system and equal 5-10% at CN excitation of 33 MeV. The evaporation residues cross sections are of the picobarn level for these reactions [19]. Consequently, the reaction mechanism is practically the same for the reactions with the ^{48}Ca ions with actinide targets.

To explore the relative contributions of CNF and QF to symmetric splitting we have investigated binary reaction channels of the composite systems with Z = 108 produced in reactions with ^{22}Ne, ^{26}Mg, ^{36}S and ^{58}Fe ions at energies below and above the Bass barrier. The entrance channel properties of these systems vary strongly: for the reaction ^{58}Fe+^{208}Pb entrance channel mass-asymmetry η= 0.571, for ^{36}S+^{238}U - 0.737, for ^{26}Mg+^{248}Cm - 0.810, and for ^{22}Ne+^{249}Cf - 0.838. It is important to note that all reaction partners, except ^{208}Pb, are well deformed nuclei. In the reactions with deformed nuclei the potential energy surface strongly depends on the relative orientation of the reaction partners. Except for reactions with strong mass-asymmetry in the entrance channel the dominance of tip configurations at energies below the barrier leads to the increase of QF contributions. As demonstrated in Fig. 10 the mass-energy distributions change with decreasing asymmetry η in the entrance channel from symmetric for incoming ^{22}Ne-ions to strongly asymmetric for incoming ^{58}Fe-ions. These changes are understood as reflecting the relative contributions from FF and QF to the fission process of Hs depending on the reaction studied.

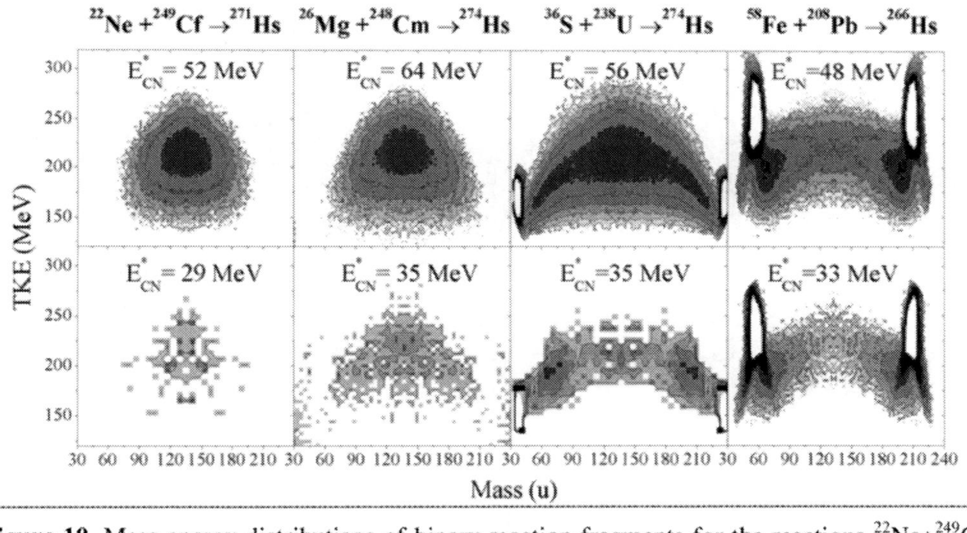

Figure 10. Mass-energy distributions of binary reaction fragments for the reactions ^{22}Ne+^{249}Cf, ^{26}Mg+^{248}Cm, ^{36}S+^{238}U, ^{58}Fe+^{208}Pb leading to the formation of elements with Z=108 at energies above (top panel) and below (bottom panel) the Coulomb barrier.

It is clearly seen that even at similar CN excitation energies the mass-energy distributions are vastly different for these reactions. In the case of the reactions ^{22}Ne+^{249}Cf and ^{26}Mg+^{248}Cm the mass distributions have a near Gaussian shape with no evidence for asymmetric fission. The reactions are considered to be mainly CNF process. For the ^{36}S+^{238}U reaction the mass distributions of the fission-like fragments change markedly. This difference in mass distributions for the ^{26}Mg+^{248}Cm and ^{36}S+^{238}U reactions is connected with an increasing contribution of the QF process for the ^{36}S-induced reaction. At low excitation energies QF is the dominant process for the reaction ^{36}S+^{238}U. At higher excitation energies the mass distribution becomes symmetric and similar to the reaction ^{26}Mg+^{248}Cm though - due to a remaining trace of QF - slightly wider.

In the case of the ^{58}Fe + ^{208}Pb reaction the mass-energy distribution has a wide two-humped shape even at 48 MeV excitation energy. For this reaction the QF process dominates at energies below and above the Bass barrier. The strong overlap between QF fragments, quasi-elastic and deep-inelastic events is observed due to the fact that one of the partners is doubly magic lead.

The TKE of symmetric fragments is also higher than predicted. In Fig. 11 the TKE distributions are shown for symmetric fragments with masses $A_{CN}/2 \pm 20$ u obtained in the $^{22}Ne + ^{249}Cf$ and $^{26}Mg + ^{248}Cm$ reactions at two excitation energies of the compound nuclei (below and above the Bass barrier). It is readily seen that the TKE distributions have a complex structure at low excitation energy, while at high excitation energy the TKE distributions are well described by single Gaussians with parameters coming from the LDM.

At low excitation shell effects show up in the CNF process giving rise to a structure in the mass-energy distributions of fission fragments. The structural features of the TKE distributions at low excitation may arise due to the fact that in symmetric fission both fragments are close to the spherical neutron shells with N = 82. A similar behaviour could be observed when both fission fragments are close to the spherical proton shell Z = 50. In fact, the phenomenon of bimodality has been disclosed in the case of spontaneous and low energy fission of nuclei in the Fm-Rf (Z = 100-104) region [20].
One can see in Fig. 11 that they may be deconvoluted into two Gaussians, the constituent peaks lying near ~214 MeV and ~230 MeV. Only the lower energy corresponds to the established linear dependency of the TKE on the Coulomb parameter $Z^2/A^{1/3}$ in the LDM. Hence, in this case bimodal fission is observed.

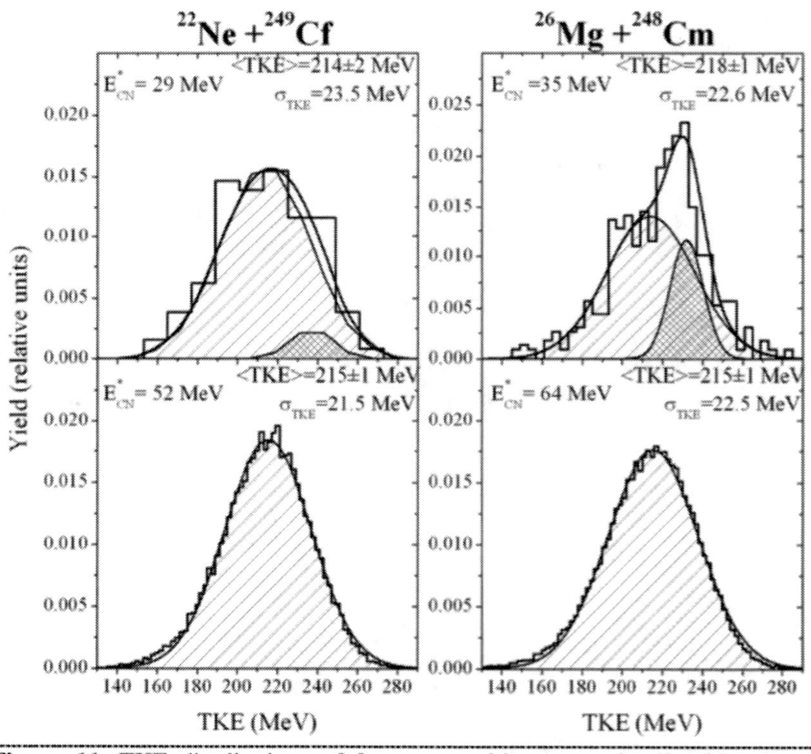

Figure 11. TKE distributions of fragments with masses $A_{CN}/2 \pm 20$ u for the reactions $^{22}Ne + ^{249}Cf$ (left panel) and $^{26}Mg + ^{248}Cm$ (right panel). Mean values and standard deviations of the experimental TKE distributions are indicated in the top right corner of each plot. High and low kinetic energy components are given as densely and sparsely hatched regions, respectively.

In the case of the $^{22}Ne + ^{249}Cf$ reaction we have some evidence for bimodal fission at an excitation energy of 29 MeV. The standard deviation of the experimental distribution is higher for the lower energy. It is important to note the enhanced relative contribution of the high energy component in the case of $^{274}Hs^*$-fission ($^{26}Mg + {}^{248}Cm$ reaction) compared to $^{271}Hs^*$-fission ($^{22}Ne + {}^{249}Cf$ reaction), while the excitation energy is higher for the former reaction. At symmetric fission of $^{274}Hs^*$ both fragments have N = 83, while in the case of ^{271}Hs the neutron number N is 81.5. According to the systematics for pre- and post-scission neutron emission accompanying fission [21], the values of 1.1 and 0.6 pre-scission neutrons are expected for $^{274}Hs^*$ ($E^* = 35$ MeV) and $^{271}Hs^*$ ($E^* = 29$ MeV), respectively. Thus the formation of two spherical fragments with neutron number N = 82 is more favourable in the case of $^{274}Hs^*$ compared to $^{271}Hs^*$.

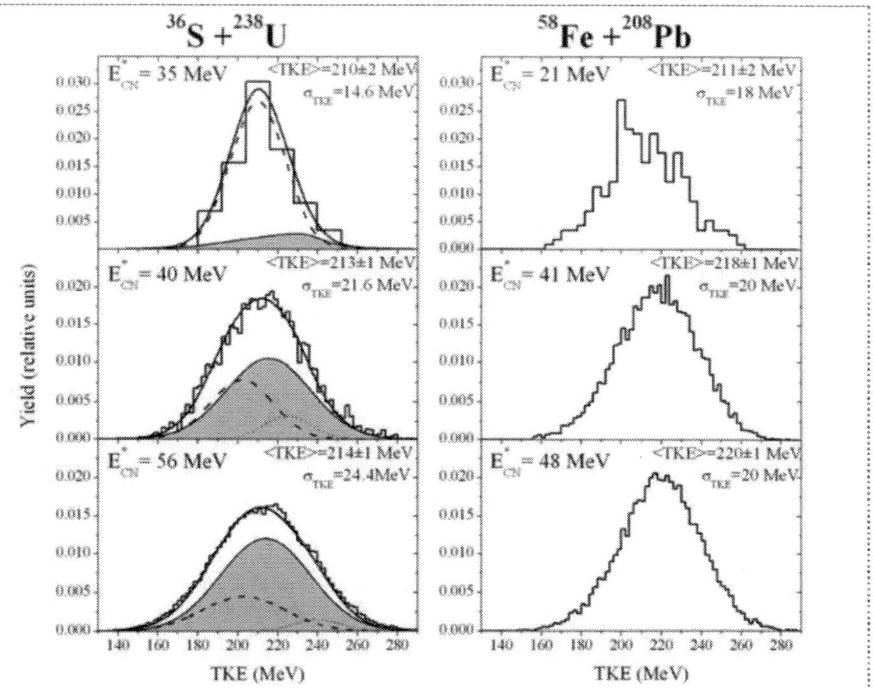

Figure 12. TKE distributions of fragments with masses $A_{CN}/2 \pm 20$ u for the reactions $^{36}S + {}^{238}U$ (left panel) and $^{58}Fe + {}^{208}Pb$ (right panel). Mean values and standard deviations of the experimental TKE distributions are indicated in the top right corner of each plot. In the case of the $^{36}S + {}^{238}U$ reaction the filled region corresponds to the TKE distribution for the CNF process with parameters taken from the $^{26}Mg + {}^{248}Cm$ reaction. The dashed and dotted curves are associated with asymmetric and symmetric QF, respectively.

As it was mentioned above the reaction $^{36}S + {}^{238}U$ leads to the formation of a similar CN as in the reaction $^{26}Mg + {}^{248}Cm$. But, the TKE distributions for these reactions are different, in particular, at the lowest excitation energy of 35 MeV studied. In the case of the S induced reaction the TKE distribution is narrower and shifted to lower energies, the mean value of the TKE being 210 MeV instead of 218 MeV for the Mg induced reaction. Since in CNF the properties of the same CN formed in different reactions at the same excitation energy do not change, this shift of the TKE is due to a large contribution of QF to the fragment region with masses $A_{CN}/2 \pm 20$ u. Thus, this points to the fact that the main part of fission-like fragments originates from the asymmetric QF process and the contribution of other processes is relatively small.

The mean value and dispersion of TKE increase with increasing energy of ^{36}S ions and exceed the LDM predictions at the highest energy. The TKE distributions of fragments formed in the reaction with ^{36}S ions at energies above and below the Coulomb barrier for symmetric mass splits are shown in the left panel of Fig. 12. As just mentioned, at the excitation energy of 35 MeV the main component of the TKE distribution is connected with QFasym. For the higher energies each TKE distribution is decomposed as a sum of three Gaussians. One of them is associated with the CNF process. We fix the mean value and variance of this component to the values predicted from the systematics [18] and [7], respectively. The low energy component in Fig. 12 is attributed to QFasym while the high energy one is connected with QFsym.

Thus we can estimate the fusion probability using the measured mass-energy distributions as the ratio between the number of events attributed to CNF in the frame of the present analysis and all fission-like fragments. The fusion probabilities as a function of the energy above the barrier are presented as open symbols in Fig. 13 for these reactions. As it follows from the present experimental data, the properties of entrance channels strongly affect the reaction dynamics. At the excitation energy near the barrier the estimated values of P_{CN} are about 70% in the case of the Mg-induced reaction and ~25% in the S-induced reaction.

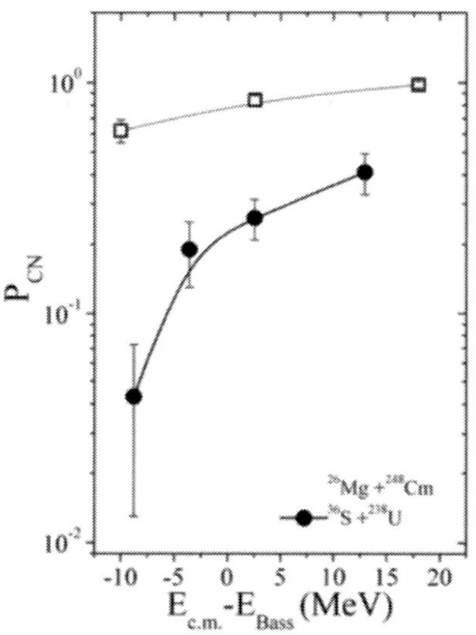

Figure 13. The fusion probabilities in the reactions ^{26}Mg + ^{248}Cm and $^{36}S+^{238}U$ obtained from the present analysis of mass-energy distributions of fission-like reaction products.

Unfortunately, the transition to heavier projectile results in the large increase in the Coulomb factors Z_1Z_2 that is crucial in the competition between CNF and QF. The mass-energy distributions for the reactions $^{58}Fe+^{244}Pu$ and $^{64}Ni+^{238}U$ (leading to the formation of the same composite system with Z=120 and N=182) at the CN excitation energies about 45 MeV are presented in the figure 14. At first glance the distributions are similar for the both reactions: the wide two-humped shape with large QF component.

However, at the same CN excitation energy the mass drift to the symmetry (estimated as a distance between masses corresponding to the maximum and half maximum of QF yields) is 22 nucleons in the case of the ^{58}Fe reaction and only 11 nucleons in the case of the ^{64}Ni ions. It is significant that the mass drift to the symmetry is about 34 nucleons for the $^{48}Ca+^{238}U$ at the same CN excitation energy. The contribution of the symmetric fragments with masses $A_{CN}/2\pm20$ u to all fission-like events is about 8% and 4% for Fe and Ni-ions, respectively. The average TKE's are similar for asymmetric QF fragments for the both reactions, while in the symmetric mass region the TKE for the Fe+Pu reaction is higher than for the Ni+U. The solid line corresponds to the parabolic dependence following from the Liquid Drop Model and the Viola systematics for TKE. The dispersions of the TKE are different for the reactions: it is bigger in the case of Fe-ion induced reaction. In the reaction $^{64}Ni+^{238}U$ the dispersion does not depend on a fragment mass practically and its mean value is about

350 MeV2, while for the reaction $^{58}Fe+^{244}Pu$ dispersion increases for symmetric fragments and its mean value is about 610 MeV^2. The extrapolation of the experimental dependence on dispersion of TKE distributions from [7] gives the value of about 550 MeV^2 for the fission of $^{302}120$. Hence, for the reaction with Ni-ions TKE dispersion is lower than this value, whereas it is higher in the case of Fe-ions.

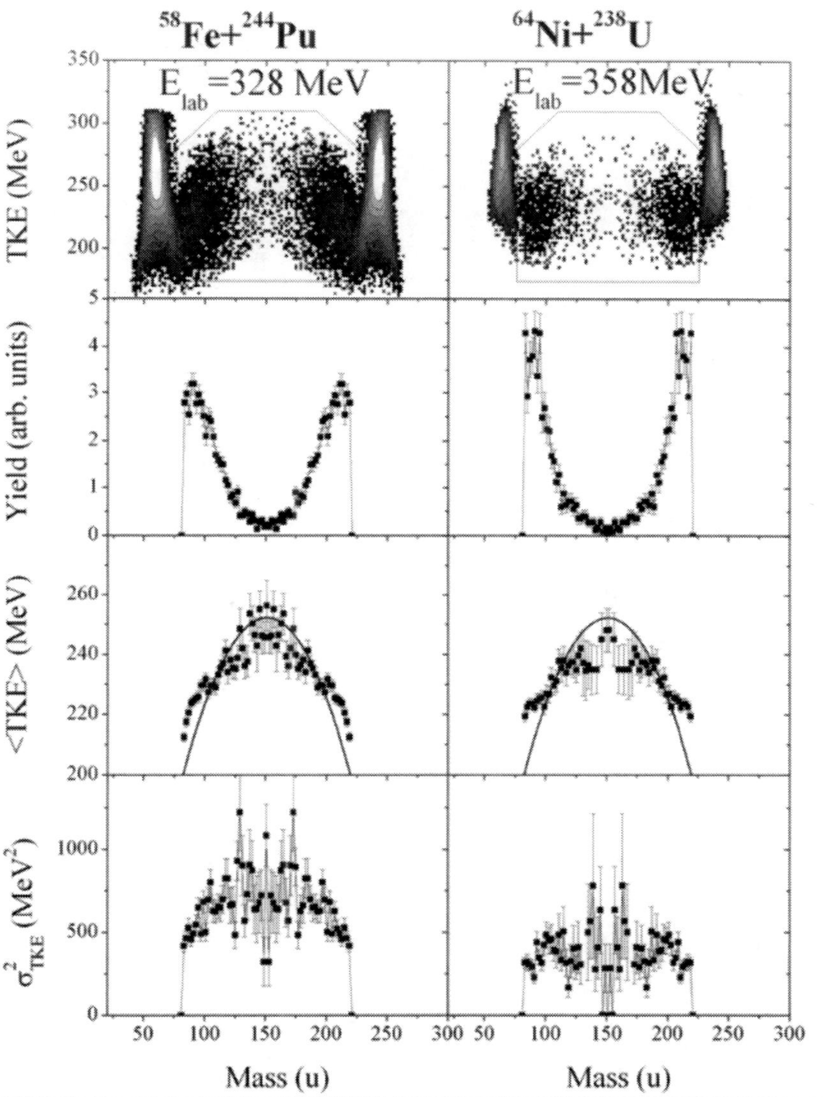

Figure 14. From top to bottom: the two-dimensional TKE–M matrices, yields, average TKEs and TKE dispersions of fragments inside the contour lines on TKE-M matrices (bottom panels) in the $^{58}Fe+^{244}Pu$ and $^{64}Ni+^{238}U$ reactions at the CN excitation energies about 45 MeV.

In the figure 15 the TKE distributions of fission-like fragments in the mass region $A_{CN}/2\pm20$ u for the reactions ^{58}Fe+^{244}Pu and ^{64}Ni+^{238}U are presented. It is seen that both TKE distributions have a complex structure as in the case of the ^{48}Ca+^{238}U reaction (see Fig. 8).
In contrast to the ^{58}Fe+^{244}Pu, for the reaction ^{64}Ni+^{238}U the TKE distribution has more pronounced low and high energy components (see fig. 14b), while the component with average value of 252 MeV (corresponding to the Viola systematics) is highly hindered. Because of the low statistics, only an upper value for the relative yield of the CN-fission component can be reasonably given. Table 2 gives the relative contribution of the all symmetric fragments in the mass range $A_{CN}/2\pm20$ u and symmetric fragments with TKE corresponding to the Viola systematics.

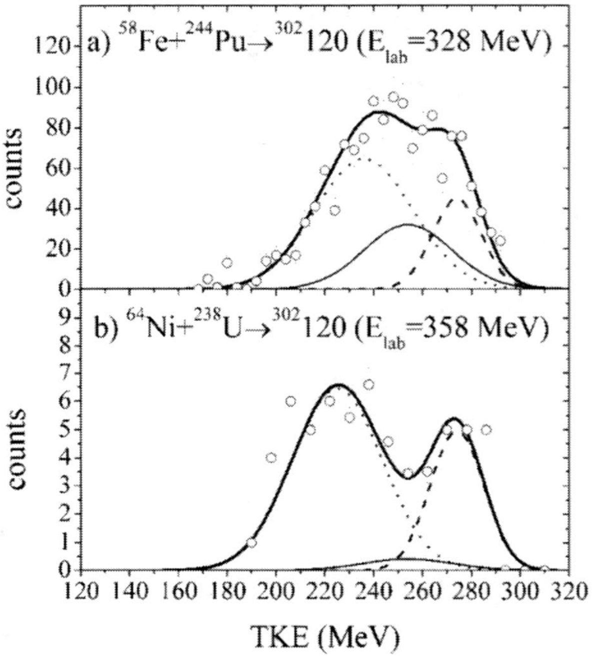

Figure 15. TKE distributions of fragments with masses $A_{CN}/2\pm20$ u. for the reactions ^{58}Fe+^{244}Pu and ^{64}Ni+^{238}U reaction at the CN excitation energies about 45 MeV.

The obtained captures cross sections as well as cross section for formation of symmetric fragments with mass $A_{CN}/2\pm20$ u are presented in the figure 16 for the reaction ^{64}Ni + ^{238}U. The capture cross sections are about a few hundred millibarns for Ca and Ni induced reactions, whereas the formation of symmetric fragments is one order of magnitude less for the reaction ^{64}Ni + ^{238}U. Yet, in the case of the Ca+U at the highest energy, approximately 70% of the events have the TKE expected for the CN fission process, whereas in the case of the Ni reaction only a few percent of symmetric fragments have the TKE compatible with the Viola prediction for the ^{302}120 CNF. The CNF cross-section in the ^{64}Ni+^{238}U→^{302}120 case drops three orders of magnitude with respect to the ^{48}Ca+^{238}U→^{286}112 case. This is unfortunately a limiting factor. Furthermore, the relative contribution of the CNF from ^{64}Ni+^{238}U is much lower than in the case of ^{58}Fe+^{244}Pu. Resently the experiments aimed at the synthesis of isotopes of element Z=120 have been performed using the ^{244}Pu(^{58}Fe,xn)$^{302-x}$120 reaction [22] and ^{238}U(^{64}Ni,xn)$^{302-x}$120 reaction [23]. A cross-section limit of

0.4pb at E^*=44.7 MeV for the former reaction and 0.09pb at E^*=36.4 MeV for the latter reaction were obtained. In the case of ^{48}Ca+^{238}U reaction the evaporation residue cross-section for 3n, 4n channels is about a few pb. Thereby in the transition from Ca to Fe and Ni ions, the evaporation residue cross-section drops down at least one and two order of magnitude, respectively.

According to the calculations performed in the frame of dinuclear system concept [12] the estimated evaporation residue cross sections are 10^{-3}pb in the case of the ^{244}Pu(^{58}Fe,xn)$^{302-x}$120 and 10^{-6}pb in the case of the ^{238}U(^{64}Ni,xn)$^{302-x}$120.

Table 2. The relative contributions of all symmetric fragments ($\sigma_{ACN/2\pm20}$) and symmetric fragments with TKE corresponding to the older Viola systematics (σ_{CNF}) to the capture cross section (σ_{cap}) for the reactions ^{48}Ca+^{238}U, ^{58}Fe+^{244}Pu and ^{64}Ni+^{238}U at CN excitation energy of around 45 MeV.

Reaction	$\sigma_{ACN/2\pm20}/\sigma_{cap}$ (%)	$\sigma_{CNF}/\sigma_{ACN/2\pm20}$ (%)	$\sigma_{CNF}/\sigma_{cap}$ (%)
^{48}Ca+^{238}U	12±2	68±3	8±4
^{58}Fe+^{244}Pu	8±3	≤25	≤2
^{64}Ni+^{238}U	4±1	≤5	≤0.2

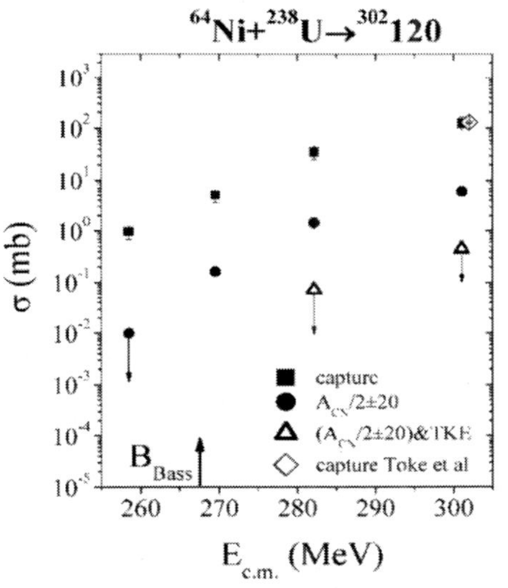

Figure 16. Capture cross section (squares), cross section for the formation of fragments with masses $A_{CN}/2\pm20$ (circles) and the fragments, but with TKE corresponding to the Viola systematic (open triangles) for the ^{64}Ni+^{238}U reaction. Rhomb is the capture cross section from [2].

4. Summary

Mass-energy distributions of binary fragments obtained in the reactions ^{20}Ne+^{249}Cf, ^{26}Mg+^{248}Cm, ^{36}S+^{238}U, ^{48}Ca+^{144,154}Sm,^{168}Er,^{208}Pb,^{238}U,^{244}Pu,^{248}Cm, ^{58}Fe+^{244}Pu, ^{64}Ni+^{238}U are analysed in the terms of symmetric and asymmetric quasifission and CN-fission. The asymmetric QF shows up even in the case of the ^{26}Mg+^{248}Cm reaction (Z_1Z_2=1152) at energy below the Coulomb barrier. The contribution of asymmetric QF component in the capture cross section increase with increasing the reaction Coulomb factor Z_1Z_2 and becomes dominant for the composite systems with Z_1Z_2>1800 formed in the reactions with deformed nuclei. Generally, in heavy-ion induced reactions the formation of QFasym fragments is mainly connected with the strong influence of the nuclear shell at Z = 82 and N = 126 (doubly magic lead). But present analysis shows that the position of asymmetric QF peak shifts to more symmetric masses for the reactions with lighter projectiles (with smaller value of Z_1Z_2). Even in the case of the ^{36}S+^{238}U and ^{26}Mg+^{248}Cm leading to the formation of the same composite nucleus ^{274}Hs* the position of the asymmetric QF peaks changes from ~200 u to ~180 u, respectively. This trend is contrary to the multimodal fission where the position of each mode determined by the nuclear shells is constant for specific compound nucleus and only the relative contribution of each mode varies in dependence of excitation energy. Thus, for more asymmetric reactions the Coulomb repulsion is expected to be smaller. For lighter projectiles this may lead to longer reaction times before separation for asymmetric QF and thus to larger numbers of exchange nucleons.

For the symmetric mass splitting the presence of all there processes, namely, QFasym, QFsym and CNF was found from the analysis of the TKE distributions for these events. In the case of the $^{48}Ca+^{238}U$ the contribution of the CNF process in TKE distribution for fragment masses $A_{CN}/2\pm20u$ was estimated as ~65%, while for the case of the more symmetric reaction $^{64}Ni+^{238}U$ this ratio drop to ~5%.

At the low excitation energy of 35 MeV, when shell effects should become more effective in fission, the TKE distribution of symmetric fragments obtained in the reaction $^{26}Mg + ^{248}Cm$ differs strongly from a Gaussian shape. Besides a low energy, a high energy component not foreseen in the LDM arises. This is attributed to the fact that both fission fragments are close to the spherical neutron shell N = 82. It means that for the compound nucleus Hassium formed in the reaction $^{26}Mg + ^{248}Cm$ the phenomenon of bimodal fission was discovered. In contrast to the $^{26}Mg + ^{248}Cm$ reaction where both fission fragments are close to the spherical neutron shell, in the case of the $^{58}Fe + ^{208}Pb$ reaction only one fragment is spherical (at symmetric mass split both fragments have only about 79 neutrons) and consequently bimodal fission is not expected for the latter reaction. But for the reaction $^{58}Fe + ^{208}Pb$ a high energy component was also observed in the TKE distribution of symmetric fragments. Its contribution increases with increasing excitation energy. Since shell effects fade away when the excitation energy is increasing, this component should be associated with the symmetric QF process because, quite generally, the QF process is colder than CNF. Therefore shell effects still play a dominant role in QF even at relatively high excitation energies.

Acknowledgments

The authors are grateful to Prof. G.Giardina, Prof. B.B. Back and Prof. E.Vardaci for the interesting and fruitful discussions. The authors are grateful to the Organizing Committee International Symposium "Quasifission Process in Heavy Ion Reactions"and Prof. Giorgio Gardina for the support and warm hospitality during stay at the University of Messina. The work has been also supported by the Alexander von Humboldt-foundation.

References

[1] Bock R. *et al.* 1982 *Nucl. Phys. A* **388** 334
[2] Toke J. *et al.*1985 *Nucl. Phys. A* **440** 327
[3] Shen W.Q. *et al.* 1987 *Phys. Rev. C* **36** 115.
[4] Itkis M.G. *et al.* 2007 *Internat. J. Modern Phys. E* **16** 957
[5] Royer G.and Remaundm B.1985 *Nucl. Phys. A* **444** 477
[6] Zagrebaev V.I. and Greiner W. 2005 *J. Phys. G* **31** 825
[7] Itkis M.G. and Rusanov A.Ya. 1998 *Fiz. Elem. Chastits At. Yadra* **29** 389; *Phys. Part. Nucl.* **29** 160
[8] Bass R.1977 *Phys. Rev. Lett.* **39** 265
[9] Chizhov A.Yu. *et al.* 2003 *Phys. Rev. C* **67** 011603, R.N. Sagaidak et al. 2003 *Phys. Rev. C* **68** 014603
[10] Knyazheva G.N. *et al.* 2007 *Phys. Rev. C* **75** 064602; A. M. Stefanini *et. al.* 2005 *Eur. Phys. J. A* **23** 473
[11] Prokhorova E.V. *et al.* 2008 *Nucl. Phys. A* **802** 45
[12] Nasirov A.K. *et al.* 2009 *Phys. Rev. C* **79** 024606
[13] Itkis M.G. *et al.* 2007 *Nucl Phys. A* **787** 150c

[14] Kozulin E.M. *et al.* 2010 *Phys. Lett. B* **686** 227
[15] Nishio K. *et al.*, *in Proceedings of International Symposium on Exotic Nuclei (EXON 2009)*, edited by Yu.E. Penionzhkevich and S.M. Lukyanov, (AIP Conference Proceedings, Volume 1224) p.301
[16] Itkis Yu. M. *et al.* 2011 to be published.
[17] Brosa U., Grossman S. and Muller A. 1990 *Phys. Rep.* **197** 167
[18] Viola V.E. 1966 *Nucl. Data Tables A* **1** 391
[19] Oganessian Yu.Ts. 2007 *J. Phys. G: Nucl. Part. Phys.* **34** R165–R242
[20] Hulet E.K. *et al.* 1986 *Phys. Rev. Lett.* **56** 313
[21] Kozulin E.M., Rusanov A.Ya., and Smirenkin G.N. 1993 *Phys. At. Nucl.* **56** 166
[22] Oganessian Yu. Ts. *et. al.* 2009 *Phys. Rev. C* **79** 024603
[23] Hofman S. *et al.* 2008 *GSI Scientific Reports* 131

Discrimination between fission and quasi-fission from reaction time measurements

M. Morjean[1], **D. Jacquet**[2], **M. Laget**[2] **and J.F. Berger**[3]

[1] GANIL, CEA-DSM and IN2P3-CNRS, B.P. 55027, F-14076 Caen Cedex, France
[2] IPNO, CNRS/IN2P3, Université Paris-Sud 11, F-91406 Orsay Cedex, France
[3] CEA/DAM/DIF, Bruyères-le-Châtel, F-91297 Arpajon, France

E-mail: morjean@ganil.fr

Abstract. The formation of compound nuclei with Z = 120 and Z = 124 has been evidenced from their very long fission times measured by the blocking technique in single crystals. A possible explanation for the long measured fission times might be found in the temperature dependence of the fission barriers, as predicted by Hartree-Fock-Bogolubov calculations at finite temperature in this super-heavy nucleus domain.

1. Introduction

A compound nucleus formed by fusion of two heavy nuclei can decay either by particle emission or by fission. The latter becomes more and more probable when the mass of the compound nucleus increases. However, in the 80's, it has been shown [1, 2, 3] that the detection of fission-like fragments does not constitute an evidence for fusion: quasi-fission processes occur, characterized by the formation of composite systems that do not reach the fully equilibrated compound nucleus before splitting in two fission-like fragments. Numerous works have been devoted to quasi-fission studies. They have investigated the differences with true compound nucleus fission, usually concluding to broader mass distributions or higher most probable kinetic energies for fission-like fragments arising from quasi-fission processes. A few experiments have tried to disentangle the quasi-fission cross-section from the true fusion-fission one. However, the definition of quasi-fission in these experiments is often somewhat arbitrary, based for instance on selections justified by rough extrapolations in mass-energy plots.

The quasi-fission processes can actually arise from quite different physical origins. It can arise from vanishing fission barriers at high angular momenta, fusion being still present for the most central collisions [4]. It can also result from trajectories of the system in the potential landscapes that lead to fast separations of the partners [5]. Depending on the actual multi-dimensional potential landscape, these trajectories can be more or less favored with respect to the ones driving the system to the minimum corresponding to the compound nucleus. Finally, quasi-fission can also be due to the disappearance in the potential of the compound nucleus minimum, making thus fusion obviously impossible. Whatever the physical origin of the quasi-fission process is, the associated dynamical separation of the two fission-like fragments takes place in a very short time scale that has been indeed early measured in a quite direct way from the anisotropic associated angular distributions [3]. It was found to be less than a rotational period of the system, corresponding to typical times between 10^{-21} and 10^{-20}s.

Journal of Physics: Conference Series **282** (2011) 012009 doi:10.1088/1742-6596/282/1/012009

In the present experimental work, our goal was to determine for very heavy systems, possibly leading to compound nuclei with atomic numbers larger than 110, wether the true fusion process still exists, even with rather small cross-sections as compared to quasi-fission. We have therefore searched for evidences for composite systems surviving against fission at times longer by orders of magnitude than the quasi-fission time scale, only compatible with fusion followed by fission.

2. Experimental procedure and results

For very heavy highly excited nuclei, the fission time distributions present tails arising from the competition between particle (even maybe γ-ray) emission and fission at each of the step of the nucleus decay [6, 7]. Such a tail for excited uranium nuclei can extend with sizable probabilities up to very long times, of the order of 10^{-16}s [8, 9]. The blocking technique in single crystal [10] has been applied to get direct evidence for such a tail at times longer than about 10^{-18}s for composite systems with $Z \geqslant 114$.

The principle of the blocking technique as applied in our experiments is to measure the fragment angular distributions around the direction of a major axis of a single crystal that is used as a target. If the composite system splits rapidly, a fission-like fragment emitted in the direction of the major axis will be deflected from its initial direction by its atomic interactions with all the atoms of the crystal row. Therefore, for short splitting times of the composite systems, the measured angular distribution presents a dip with a minimum precisely in the

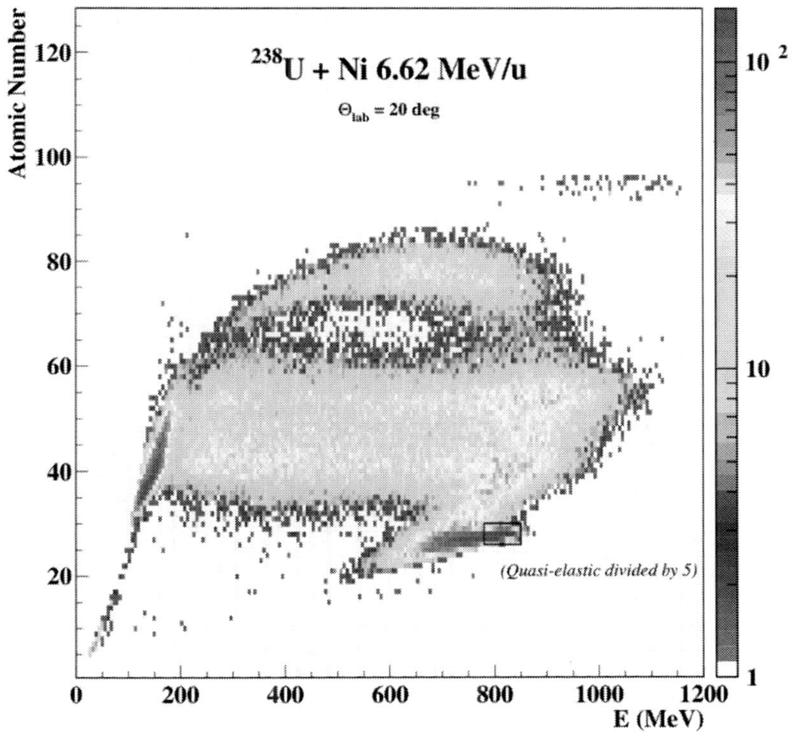

Figure 1. Atomic number versus total kinetic energy.

direction of the crystal axis. When the composite system lifetime becomes longer, the potential felt by the emitted fragments is lower resulting in a less marked dip. However, for the shortest fission times, the splitting can take place within the thermal vibrations of the crystal atoms, leading thus to a probability to detect a fragment precisely in the direction of the crystal axis that does not depend anymore on the splitting time. It has been shown that to first order, the width of the blocking dips evolves as the square root of Z/E [10], but the blocking technique has a lower time sensitivity time limit that has a pure geometrical origin, the amplitude of the thermal vibrations. Any filling of the dip in the direction of the crystal axis is a model independent signature that the composite system had a fission time longer than the time it needs to recoil out of the thermal amplitude area.

Three systems have been studied [11] at Ganil for which the possibly formed compound nuclei have excitation energies between 70 and 80 MeV. The first system was ^{208}Pb + Ge at 6.2 MeV/A that would lead to a compound nucleus with Z = 114. The second system was ^{238}U + Ni at 6.6 MeV/A, leading to compound nuclei with Z = 120, and the last one was ^{238}U + Ge at 6.1 MeV/A, leading to compound nuclei with Z = 124.

For the three systems studied, the lower sensitivity limit given by the thermal vibrations of the atoms is $\sim 10^{-18}$s, at least two orders of magnitude longer than the time scale associated to quasi-fission. The blocking effects were measured at 20° with respect to the beam axis. All coincident fragments and light charged particles were detected on a solid angle close to 4π sr by

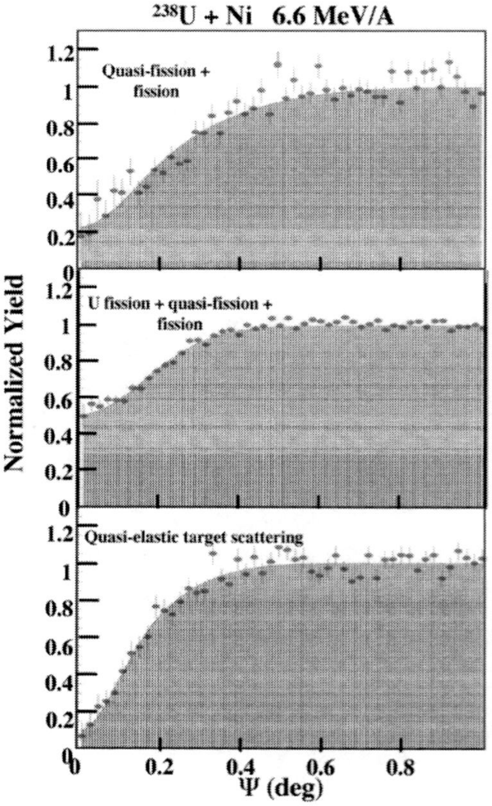

Figure 2. Blocking dips measured for three different selections of atomic numbers corresponding to different reaction mechanisms (see text).

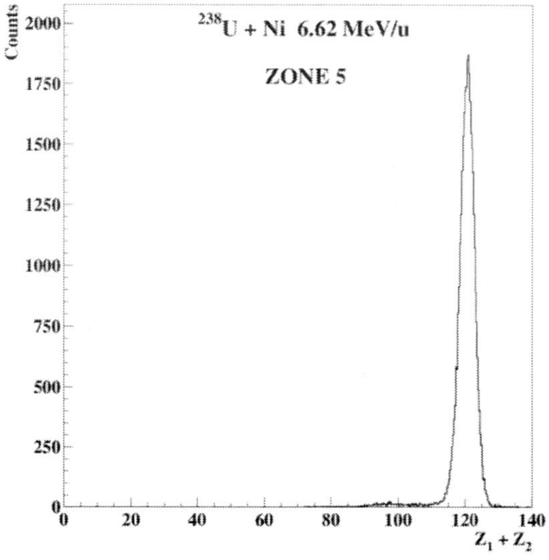

Figure 3. Sum of the atomic numbers of fission-like fragments when one of the fragments is detected at 20° with $65 < Z_1 < 85$.

the multidetector INDRA [12]. The correlation between the atomic number Z and the energy E of the fragments detected at 20° had rather similar behaviors [11], presented for example in Fig. 1 for the ^{238}U + Ni system. For Z close to the atomic number of the target, a quasi-elastic pic is observed, with an inelastic tail towards lower Z and E values. For Z $\approx$ 80, fission-like fragments are measured with very high yields. These fission-like events arise either from quasi-fission or from true fusion-fission reactions. The fragments with Z in between the two previous regions arise either from quasi-fission processes, or from fusion-fission or from a sequential fission of the projectile-like fragment (U or Pb). Finally, for atomic number close to the projectile, a more or less pronounced deep-inelastic branch of the projectile-like fragment is observed, depending on the system considered.

The blocking dips associated to three different mechanisms selected in the Z versus E maps are presented as an example in Fig. 2 for the ^{238}U + Ni system. For the selection presented in the lower bin of the figure, the dip associated with quasi-elastic target scattering presents a minimum value at $\psi = 0°$ (in the direction of the crystal axis) significantly lower than the one associated with the two other selections ($38 < Z < 58$ for the middle bin and $70 < Z < 85$ for the upper bin). As previously stressed, such an increase of the yield measured at $\psi = 0°$ indicates composite system lifetimes longer than 10^{-18}s. It has to be noted that the blocking dips might be smeared by specific experimental conditions, for instance the size of the beam spot or defects in the crystalline structure of the targets with respect to perfect crystals. Such smearing effects might result in an artificial increase of the measured yields in the $\psi = 0°$ direction, but such an increase would be the strongest for the narrowest dips, which are in our analyses the ones associated to the quasi-elastic target recoils. For the selection presented in the middle bin, the presence of sequential fission of uranium nuclei is responsible for the strong observed yield increase at $\psi = 0°$. However, no sequential fission events are present in the selection done in the upper bin. For this selection, only two heavy fragments are detected in the whole space. The sum Z_1+Z_2 of their atomic numbers is shown in Fig. 3 for the ^{238}U + Ni system, exhibiting a

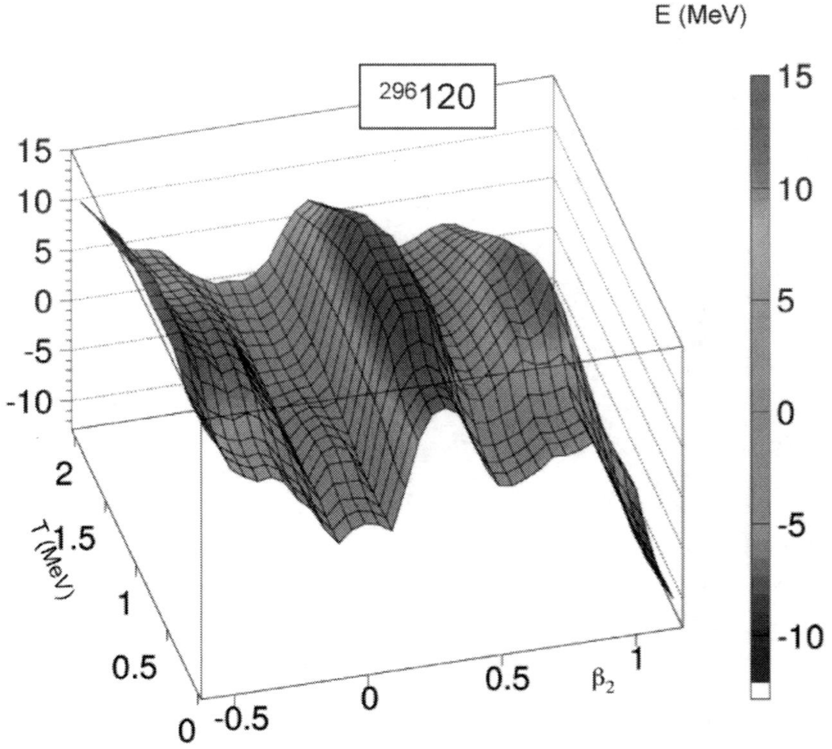

Figure 4. Hartree-Fock-Bogoliubov potential as a function of the temperature and of the deformation parameter β_2.

narrow distribution, peaked at $Z_1+Z_2 = 120$ with a width corresponding to the experimental Z resolution. In addition, the associated light charged particle and cluster multiplicities measured on 4π sr are very low, a few percents, providing thus an extra evidence that these fragments arise from composite systems formed by all the nucleons from the target and the projectiles (at least by all the protons; no information on possible neutron preequilibirum emission can be obtained in the present experiments). The sizable increase of the yield measured at $\psi = 0°$ in Fig.2 is thus a quite direct evidence for the formation of nuclei with Z = 120 that survive against fission more than 10^{-18}s. Similar conclusions are obtained for the formation of nuclei with Z = 124 in the ^{238}U + Ge system, but not for the formation of Z = 114 in the ^{208}Pb + Ge system [6, 11].

3. Dependence of fission barriers on temperature: effect on fission times

From the data presented in the previous Section, three main conclusions can be drawn: i) for the ^{238}U + Ni and ^{238}U + Ge systems at 6.6 and 6.1 MeV/A, respectively, fusion occurs with sizeable cross-sections; ii) the fission barriers of all the transient nuclei formed during the decay of the Z = 120 and 124 nuclei formed at high excitation energies are high enough to permit these nuclei to survive against fission more than 10^{-18}s; iii) the observed fission is essentially asymmetric, with one of the fragments with $Z \approx 80$, suggesting thus a fission process at rather low residual excitation energies, compatible with the long fission times evidenced. The two last conclusions seem hardly compatible with the temperature evolution of the fission barriers usually considered in statistical models applied to super-heavy fragment decay: considering that the fission barriers of super-heavy nuclei are essentially due to shell effects, an exponential

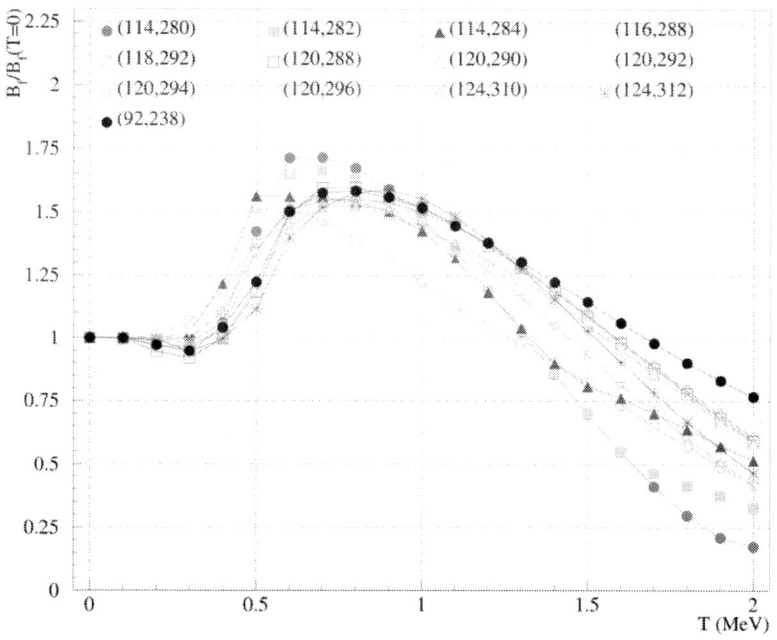

Figure 5. Evolution of the maxima of the fission barriers with temperature.

decrease of the fission barriers is usually assumed in order to take into account the shell effect damping with temperature [13]. However, the shell damping coefficient must often be considered as a free parameter in order to reproduce the data.

In order to have a better insight into the evolution of the super-heavy nucleus fission barriers with temperature, Hartree-Fock-Bogoliubov (HFB) calculations at finite temperature have been recently undertaken [14, 15], using the effective interaction D1S from D. Gogny [16, 17]. The HFB potentials have been calculated for a series of nuclei (super-heavy nuclei and ^{238}U) as a function of their temperature and deformation. A typical two-dimensional plot of these energies is presented in Fig. 4 for the nucleus (Z=120 , A=296). From such plots, the fission barriers can be inferred at a given temperature from the difference between the maximum potential obtained at a given deformation β_2 and the HFB ground state energies considered at T = 0 MeV. Surprisingly, the fission barrier maximum is not found at T = 0 MeV, but at a much higher temperature that can be better determined in Fig. 5 where the fission barrier maxima at temperature T have been normalized with respect to the one at T = 0 MeV for all the studied nuclei: between T $\approx$ 0.6 and T $\approx$ 0.8 MeV the maxima of the fission barriers can be larger than the one at T = 0 MeV by more than 50%.

The maximum observed in Fig 5 arises from the evolution of the pairing energies with deformation and temperature. Indeed, to determine the height of a fission barrier, the pairing effects have to be taken into account both at the deformation corresponding to the ground state and at the deformation corresponding to the top of the barrier. Fig. 6 presents for (Z = 120 , A=296), as a function of the deformation, the total HFB energy at T = 0 MeV in the top panel and, in the lower panels, the pairing energies at various temperatures. For this nucleus, the ground state corresponds to two minima, symmetric with respect to $\beta_2 = 0$, at $\beta_2 \approx -0.1$ and $\beta_2 \approx +0.1$, whereas the maximum of the fission barrier is at $\beta_2 \approx 0.3$. Fig. 6 shows that the contribution of the pairing energies to the potential energy at T = 0 is much larger for

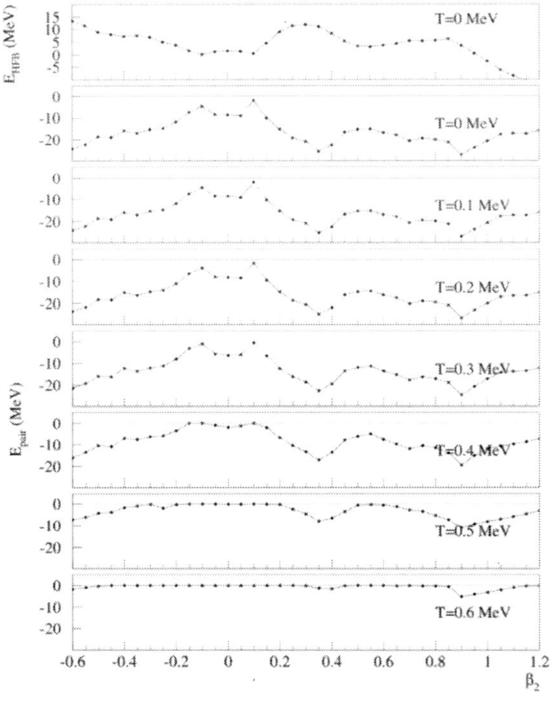

Figure 6. Hatree-Fock-Bogoliubov calculations for Z = 120, A = 296. Upper panel: Hartree-Fock-Bogoliubov energy at T = 0 MeV; lower panels: pairing energies at various temperatures.

deformations corresponding to the maximum of the barrier ($\beta_2 \approx 0.3$) than for deformations corresponding to the ground state configurations ($\beta_2 = \pm 0.1$). For increasing temperatures, the pairing energy at $\beta_2 \approx 0.3$ decreases in absolute value, but remains of sizable importance up to rather high temperatures, whereas the pairing energy for the ground state configurations becomes rapidly negligible. Such strong evolutions of the differences of the paring energies at the top of the fission barriers and close to the ground states are responsible for the increases of the fission barriers between 0 and about 800 keV. Let us mention that the Gogny effective interaction is known to provide an accurate description of nuclear pairing and is usually considered as a reference in this domain in the theoretical nuclear community.

This unexpected evolution of the fission barriers with temperature should imply quite dramatic changes in the statistical competition between particle (essentially neutrons) emission and fission for super-heavy nuclei. In the following, we describe a schematic scenario that can take into account long fission times together with tiny survival probabilities. It considers at each step of the decay the non-monotonic evolution of the fission barriers with temperature from Fig.5, coupled with the increasing neutron deficiency of the surviving nuclei. For instance, the nucleus with Z = 120 formed in ^{238}U + Ni reactions at T $\approx$ 1.6 MeV (see previous chapter) presents a fission barrier height similar to the one at T= 0 MeV. Due to nuclear matter viscosity, particle emission is strongly favored with respect to fission during a transient time [18] and the initial compound nucleus will cool down very fast towards a region of higher fission barriers by evaporation of a few neutrons. Once the statistical equilibrium for the fission width is reached [19, 20], a regular competition between evaporation and fission takes place, but involving nuclei with higher fission barriers and neutron bindings, implying thus longer lifetimes. The initially formed nucleus has thus, thanks to neutron emission, a sizable probability to survive against fission and to reach the region between 0.6 and 0.9 MeV corresponding to very high fission barriers. There, it will have very long lifetimes and a high probability to decay through neutron emission, leading thus to overall long fission times. However, after a few more neutron emissions these nuclei enter a domain of temperature ($T \lesssim 0.4$ MeV) where the fission barriers becomes lower due both to the temperature decrease and to higher and higher neutron deficiency. Considering the lower fission barriers and the higher involved neutron bindings, the nucleus will finally undergo fission with a very high probability, but at a very long overall fission time.

4. Conclusion

Direct evidence for the formation by fusion of nuclei with Z = 120 and 124 has been reached thanks to their very long measured fission times. A quite unexpected evolution of the fission barriers with the temperature is obtained by Hartree-Fock-Bogoliubov calculations at finite temperature. Considering such an evolution, a schematic scenario for the decay of these nuclei can be suggested that takes into account qualitatively the apparent discrepancy between the long measured fission times and the extremely difficult synthesis of super-heavy elements. It stresses the need for realistic parameters when statistical models are used in the super-heavy nucleus domain and the possible misleading conclusions that might be drawn from rough parameter extrapolation. In order to reach quantitative reproductions of the super-heavy nucleus decay through statistical models, more efforts are now clearly needed to fix the model parameters.

References

[1] Lebrun C *et al.* 1979 *Nucl. Phys. A* **321** 207
[2] Bock R *et al.* 1982 *Nucl. Phys. A* **388** 334
[3] Toke J *et al.* 1985 *Nucl. Phys. A* **440** 327
[4] Gregoire C *et al.* 1982 *Nucl. Phys. A* **387** 37c
[5] Aritomo Y 2006 *Nucl. Phys. A* **780** 222
[6] Morjean M *et al.* 2007 *Eur. Phys. J. D* **45** 27
[7] Jacquet D and Morjean M 2009 *Prog. in Part. and Nucl. Phys.* **63** 155
[8] Andersen J U *et al.* 1976 *Phys. Rev. Lett.* **36** 1539
[9] Goldenbaum F *et al.* 1999 *Phys. Rev. Lett.* **82** 5012
[10] Gemmel D S 1974 *Rev. Mod. Phys.* **46** 129
[11] Morjean M *et al.* 2008 *Phys. Rev. Lett.* **101** 072701
[12] Pouthas J *et al.* 1995 *Nucl. Instr. Methods A* **357** 418
[13] Ignatyuk A, Smirenkin G N and Tishin A S 1975 *Yad. Fiz.* **21** 485
[14] Laget M 2007 *PhD thesis* **IPNO-T-08-01**
[15] Berger J F, Hirata D and Girod M 2003 *Acta Phys. Pol. B* **34** 1909
[16] Dechargé J and Gogny D 1980 *Phys. Rev. C* **21** 1568
[17] Berger J F *et al.* 1991 *Comp. Phys. Com.* **63** 365
[18] Grangé P, Jun-Qing L and Weidenmüller H A 1983 *Phys. Rev. C* **27** 2063–2077
[19] Bohr N and Wheeler J A 1939 *Phys. Rev.* **56** 426
[20] Kramers H A 1940 *Physica VII* **4** 284

Journal of Physics: Conference Series **282** (2011) 012010 doi:10.1088/1742-6596/282/1/012010

Dynamics of dinuclear system formation and its decay in heavy ion collisions

Avazbek Nasirov[1,2], **Giorgio Giardina**[3], **Giuseppe Mandaglio**[3], **Marina Manganaro**[3], **Werner Scheid**[4]

[1]Joint Institute for Nuclear Research, 141980 Dubna, Russia
[2]Institute of Nuclear Physics, 100214, Tashkent, Uzbekistan
[3]Dipartimento di Fisica dell' Universitá di Messina, 98166 Messina, and Istituto Nazionale di Fisica Nucleare, Sezione di Catania, Italy
[4]Institute für Theoretische Physik der Justus-Liebig-Universität, Giessen, Germany

E-mail: `nasirov@jinr.ru`

Abstract. A variety of phenomena connected with the formation of a dinuclear complex is observed in the heavy ion collisions at low energies. The dinuclear system model allows us to analyze the experimental data and to interpret them by comparison of the partial capture, fusion and evaporation residue cross sections measured for the different reactions leading to the same compound nucleus. The comparison of theoretical and experimental values of the mass and angular distributions of the reaction products gives us a detailed information about reaction mechanism forming the observed yields. The observed very small cross sections of the evaporation residues may be explained by the strong fusion hindrance and/or instability of the heated and rotating compound nucleus and smallness its survival probability. The fusion hindrance arises due to competition between complete fusion and quasifission while the smallness of survival probability is connected with the decrease of the fission barrier at large excitation energy and angular momentum of compound nucleus.

1. Introduction

In the heavy ion collisions at low energies, a variety of phenomena is observed, which are connected with the formation of a dinuclear complex. Deeply inelastic collisions have been studied extensively over a wide range of energies and masses [1]. The multinucleon transfer and energy dissipation mechanisms of these reactions are very similar with the ones of the full momentum transfer (capture) reactions. The difference between deep inelastic collisions and quasifission reactions is determined by the lifetimes of dinuclear system (DNS) formed as intermediate system which consists of interacting nuclei with indestructible corns and excited nucleons in highlieing quantum states. Some of these excited nucleons form the neck or overlap region between constituents of DNS. The nucleon exchange mechanism at dissipation of the relative motion energy of the colliding nuclei was theoretically studied in Ref.[2, 3]. The basic points of those methods were developed to study deep inelastic and quasifission processes which are used to explain the hindrance to complete fusion [4].

The set of the successful experimental results in the synthesis of superheavy elements stimulated great efforts to experimental and theoretical investigations of the fusion-fission processes with massive nuclei. The very small cross sections σ_{ER} of the evaporation residues observed in the recent experiments [5, 6, 7] may be explained by the strong fusion hindrance and/or instability of the heated and rotating compound nucleus [8].

The experimental knowledge about fusion-fission reactions at sub- and near-barrier energies has grown considerably in the last twenty years. The theoretical models are able to reproduce the main features of such processes, even to make predictions of the cross sections for synthesis of superheavy elements which are more or less close to the experimental data. But properly understanding the fusion dynamics for heavy systems requires many more ingredients. The necessity of more experimental data to disentangle various concurrent effects, is clearly felt. A full understanding of all steps of the reaction dynamics is very important for the challenging issue of superheavy element production and new isotopes far from the valley of stability.

Dynamics of complete fusion and role of the entrance channel in formation of the reaction products in heavy ion collisions are questionable or they have different interpretation still nowadays. For example, what mechanism of fusion makes the main contribution to formation of compound nucleus: an increase of the neck between interacting nucleus or multinucleon transfer at a relatively restricted neck size? How is large the overlap between angular momentum distributions of dinuclear system and compound nucleus which determine the angular distribution of reaction products, cross sections of evaporation residue, fusion-fission and quasifission products? There is an ambiguity in separation of fusion-fission fragments from the quasifission and fast fission products. Still unclear surely a law of the distribution of the excitation energy of DNS between its constituent fragments. Therefore, it is interesting to study the mechanism of deep inelastic and quasifission reactions forming binary fragments accompanied by neutron and gamma quantum emissions. The analysis of a correlation angular, mass-energy distributions of the registered fragments and neutron and gamma quantum characteristics allows us to obtain useful knowledge about complete fusion mechanism.

2. Difference and similarity of quasifission and deep inelastic reactions

An investigation of dissipation dynamics of the deep inelastic and quasifission reactions is useful to establish relaxation times of different degrees of freedom in heavy ion collisions at low energies. From an analysis of the experimental data of the mass-angle distributions of the binary reaction products and gamma-quanta of giant dipole resonances we can conclude that a study of the DNS stage of these dissipative processes is perspective. Because a sufficient part of the reaction time of the dissipative processes in heavy ion collisions belongs to this stage, particularly for the massive system, as well as for the case of colliding nuclei with the nearly equal charge (mass) numbers. Qualitative difference between deep inelastic and quasifission reactions is that the full momentum transfer does not take place in the former mechanism while it occurs in the latter one. The classification of the experimental data of binary products as ones corresponding to the deep inelastic collision or capture reactions (quasifission and fusion-fission) is based on the mass distribution characteristics only: projectile-like and target-like products with the total kinetic energies (TKE) around the Viola systematics [9] are considered as products of deep inelastic collisions. The authors of Ref. [10] attributed the all events with energy losses larger than 8–15 MeV and with fragments of below 40 mass units to deep inelastic collision at the analysis of the experimental data of the $^{32}S+^{182}W$ reaction.

The capture, deep inelastic and complete fusion reaction cross sections are presented in Fig. 1. The difference between capture and complete fusion cross sections is the quasifission cross section:

$$\sigma_{\rm cap} = \sigma_{\rm fus} + \sigma_{\rm qfis}. \qquad (1)$$

Unfortunately, a possibility of the mixing mass-angle distributions of the quasifission and deep inelastic reactions is not studied well because the nature of quasifission has been not established yet.

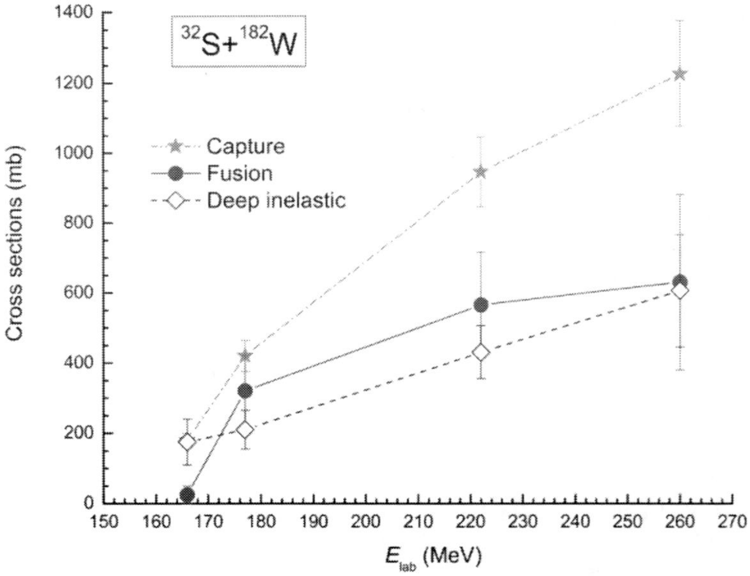

Figure 1. The experimental data for capture (dot-dashed line), complete fusion (solid line) and deep inelastic collision (dashed line) events for the ^{32}S+^{182}W reaction [10].

The difference between capture and deep inelastic reactions can be found theoretically from the analysis of the results of dynamical calculations by solving the equations of motion for the radial distance (Eq. 2), angular momentum (Eq. 6) and surface vibrations for quadrupole and octupole multipolarity (Eq. 4) (see Fig. 2) for the ^{48}Ca+^{208}Pb reaction.

The quasifission occurs at the full momentum transfer and it belongs to the capture reactions when the system traps into potential well as in Fig. 2(a) and the decay time of DNS, which is formed in this mechanism, is determined by the size of the potential well, by the values of excitation energy $E^*_{\rm DNS}$ and friction coefficients (radial and tangential).

In the deep inelastic reactions, DNS is not trapped into potential well because the DNS momentum decreases up to zero and relative distance between centers of mass of projectile and target nuclei reached the minimum value (see Fig. 2(b). Then the relative distance R increases due to repulsive forces. Although the kinetic energy of the relative motion goes on to be dissipated DNS can overcome the Coulomb barrier from intrinsic part to outside of the potential well and we observed two products after interaction and exchanging nucleons. Therefore, the interaction time of colliding nuclei in deep inelastic collision is small in comparison with the one in capture reactions. The friction coefficients and the size of the potential well plays crucial role in calculations of capture events by solving the following equations [4, 11]:

$$\mu(R,\alpha_1,\alpha_2)\ddot{R} + \gamma_R(R,\alpha_1,\alpha_2)\dot{R}(t) = F(R), \tag{2}$$

$$F(R,\alpha_1,\alpha_2) = -\frac{\partial V(R,\alpha_1,\alpha_2)}{\partial R} - \dot{R}^2\frac{\partial\mu(R)}{\partial R}, \tag{3}$$

$$D_{\beta_i}\ddot{\beta}_i(t) + \gamma_\beta(R)\dot{\beta}_i(\alpha_1,\alpha_2,t) = F_{\beta_i}(R) \tag{4}$$

$$F_\beta(R) = -\frac{\partial V(\beta_i,\alpha_1,\alpha_2)}{\partial\beta_i} \tag{5}$$

$$\frac{dL}{dt} = \gamma_\theta(R,\alpha_1,\alpha_2)R(t)\left(\dot{\theta}R(t) - \dot{\theta}_1 R_{1eff} - \dot{\theta}_2 R_{2eff}\right), \tag{6}$$

Journal of Physics: Conference Series **282** (2011) 012010 doi:10.1088/1742-6596/282/1/012010

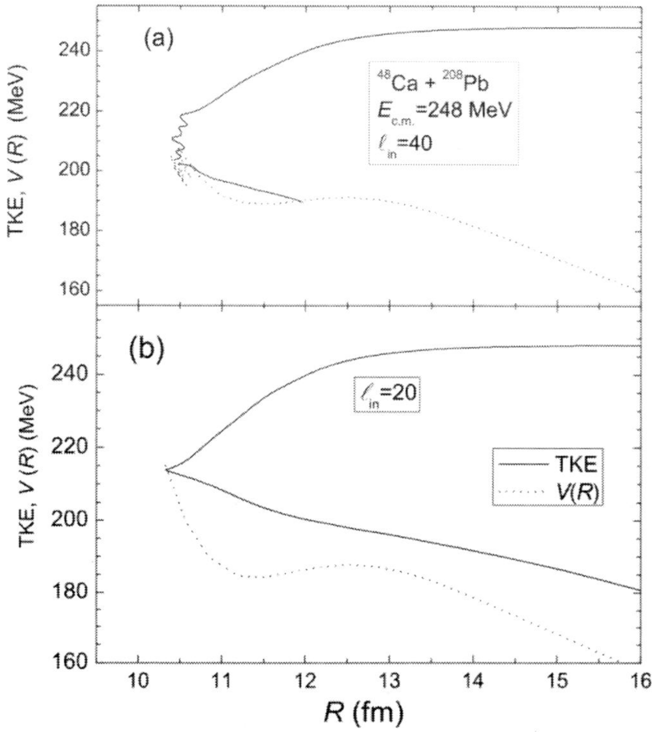

Figure 2. Capture (a) and deep inelastic collision (b) for the ^{48}Ca+^{208}Pb reaction.

$$L_0 = J_R(R,\alpha_1,\alpha_2)\dot{\theta} + J_1\dot{\theta}_1 + J_2\dot{\theta}_2\,, \tag{7}$$

$$E_{\rm rot} = \mu(R,\alpha_1,\alpha_2)\dot{R}^2/2 + \frac{J_R(R,\alpha_1,\alpha_2)\dot{\theta}^2}{2} + \frac{J_1\dot{\theta_1}^2}{2} + \frac{J_2\dot{\theta_2}^2}{2}\,, \tag{8}$$

where $R \equiv R(t)$ is the relative motion coordinate; $\dot{R}(t)$ is the corresponding velocity; α_1 and α_2 are the orientation angles between beam direction and axial symmetry axis of the projectile and target, respectively; L_0 ($L_0 = \ell_0\hbar$) and $E_{\rm rot}$ are defined by initial conditions; J_R and $\dot{\theta}$, J_1 and $\dot{\theta}_1$, J_2 and $\dot{\theta}_2$ are moment of inertia and angular velocities of the DNS and its fragments, respectively; γ_R and γ_θ are the friction coefficients for the relative motion along R and the tangential motion when two nuclei roll on each other's surfaces, respectively; $\omega_\lambda^{(i)}$, γ_λ and $D_\lambda^{(i)}$ are frequency, damping and mass coefficients for the surface vibration multipolarity λ, respectively; $V(R,\alpha_1,\alpha_2)$ is the nucleus-nucleus potential calculated by the double folding procedure [4, 12].

The capture includes complete fusion and quasifission events. The quasifission time is estimated by the formula

$$\tau_{\rm DNS}(T_Z) = \frac{\hbar}{\Gamma_{\rm qfiss}(T_Z)} \tag{9}$$

if we know the excitation energy $E^*_{\rm DNS}$ and quasifission barrier $B_{\rm qf}$ of the dinuclear system for its decay on fragments with charge numbers Z and $Z_{\rm tot} - Z$, by using the one-dimensional Kramers rate [13, 14, 15]

$$\begin{aligned}\Gamma_{\rm qfiss}(\Theta) &= K_{\rm rot}^{(\rm DNS)}/K_{\rm rot}^{(\rm CN)}\,\omega_m\left(\sqrt{\gamma^2/(2\mu_{\rm qf})^2 + \omega_{\rm qf}^2} - \gamma/(2\mu_{\rm qf})\right)\\ &\times\ \exp\left(-B_{\rm qf}/T_Z\right)\big)/(2\pi\omega_{\rm qf}).\end{aligned} \tag{10}$$

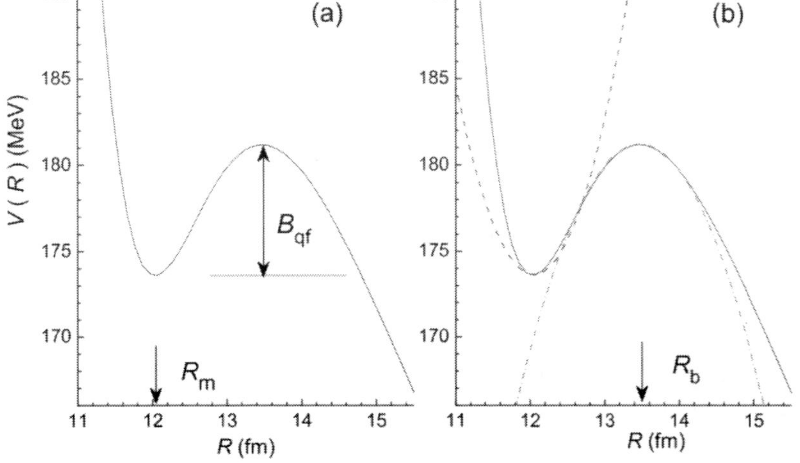

Figure 3. Quasifission barrier is a depth of the well in the nucleus-nucleus interaction potential $V(R)$ (a); potential well is replaced by the harmonic oscillator with frequency $\omega_{\rm m}$ and potential barrier is replaced by the inverted harmonic oscillator with frequency $\omega_{\rm b}$ which are used to calculate the decay time of DNS into two fragments.

Here the frequency $\omega_{\rm m}$ and $\omega_{\rm b}$ are found by the harmonic oscillator approximation to the shape of the nucleus-nucleus potential $V(R)$ for the given DNS configuration $(Z, Z_{\rm tot}-Z)$ on the bottom of its pocket placed at $R_{\rm m}$ and on the top (quasifission barrier) placed at $R_{\rm b}$ (see Fig. 3), respectively:

$$\omega_m^2 = \mu_{\rm qf}^{-1}\left|\frac{\partial^2 V(R)}{\partial R^2}\right|_{R=R_m}, \quad (11)$$

$$\omega_{\rm b}^2 = \mu_{\rm qf}^{-1}\left|\frac{\partial^2 V(R)}{\partial R^2}\right|_{R=R_{\rm b}}. \quad (12)$$

The nucleus-nucleus potential $V(Z, A, \ell, R)$ includes the Coulomb $V_{\rm Coul}(Z, A, R)$, nuclear $V_{\rm N}(Z, A, R)$ and rotational $V_{\rm rot}(Z, A, R, \ell)$ parts:

$$V(Z, A, \ell, R) = V_{\rm Coul}(Z, A, R) + V_{\rm N}(Z, A, R) + V_{\rm rot}(Z, A, R, \ell), \quad (13)$$

where R is the distance between the centers of the nuclei. Details of the calculation can be found in Refs. [16, 17].

The calculated values of $\hbar\omega_{\rm m}$ and $\hbar\omega_{\rm qf}$ were equal to 46.52 MeV and 22.37 MeV, respectively. The used value of the friction coefficient γ is equal to $8 \cdot 10^{-22}$ MeV fm^{-2}s which was found from our calculations; $\mu_{\rm qf} \approx \mu = A_1 \cdot A_2 / A_{\rm CN}$, where A_1 and A_2 are the mass numbers of the quasifission fragments.

The collective enhancement factor of the rotational motion $K_{\rm rot}$ to the level density should be included because the dinuclear system is a good rotator. It is calculated by the well known expression [18]:

$$K_{\rm rot}(E_{\rm DNS}) = \begin{cases} (\sigma_\perp^2 - 1) f(E_{\rm DNS}) + 1, & \text{if } \sigma_\perp > 1 \\ 1, & \text{if } \sigma_\perp \le 1, \end{cases}$$

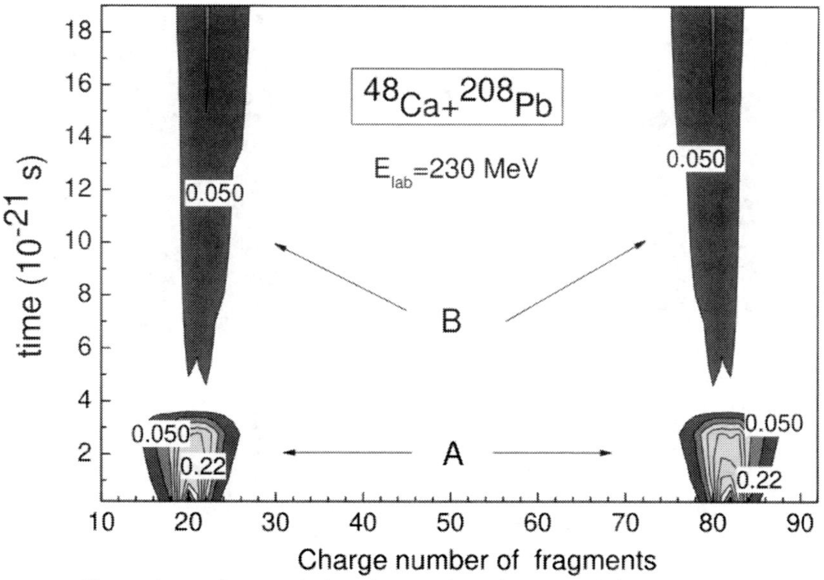

Figure 4. Time dependence of the charge distribution of deep inelastic collision (A) and quasifission (B) products for the ^{48}Ca + ^{208}Pb reaction.

where $\sigma_\perp = J_{\rm DNS}T/\hbar^2$; $f(E) = (1+\exp[(E-E_{\rm cr})/d_{\rm cr}])$; $E_{\rm cr} = 120\widetilde{\beta}_2^2 A^{1/3}$ MeV; $d_{\rm cr} = 1400\widetilde{\beta}_2^2 A^{2/3}$. $\widetilde{\beta}$ is the effective quadrupole deformation for DNS. We find it from the calculated $\mathcal{J}_\perp^{\rm DNS}$.

DNS can have long lifetime ($\tau_{\rm DNS}$) if it is formed at capture: the value of $\tau_{\rm DNS}$ depends on the depth of the potential well ($B_{\rm qf}$), DNS excitation energy ($E^*_{\rm DNS}$) at the given value of angular momentum ℓ and its moment of inertia $\mathcal{J}_\parallel^{DNS}$ [19].

$$\mathcal{J}_\parallel^{DNS} = \mathcal{J}_1 + \mathcal{J}_2 + M_1 d_\parallel^{(1)2} + M_2 d_\parallel^{(2)2}, \tag{14}$$

$$\mathcal{J}_\perp^{DNS} = \mathcal{J}_1 + \mathcal{J}_2 + M_1 d_\perp^{(1)2} + M_2 d_\perp^{(2)2} \tag{15}$$

where $d_\perp^{(i)}$ ($d_\parallel^{(i)}$) is the distance between the center of mass of the fragment i ($i = 1, 2$) and the axis corresponding to the largest (smallest) moment of inertia of DNS.

Due to the shell effects as the quantum states of the neutron and proton systems of the interacting nuclei we have DNS as a molecule of nuclei which don't fuse immediately. Therefore, the peculiarities of the DNS stage influence the formation of the reaction products with magic proton or neutron numbers. Shell effects are observed as cluster states in the large amplitude collective motions of nuclei. The observed cluster emission, mass-charge distribution of the quasifission fragments and spontaneous asymmetric fission of Th, U and Cf isotopes proved the strong role of shell structure. Reactions of heavy ion collisions and fission (spontaneous and induced) processes can be studied well using the DNS concept.

A characteristic feature of the deep inelastic processes is that the target and projectile identity is largely preserved. In Ref. [20], only for low TKE of the outgoing products (energies below the Coulomb repulsion, V_C, of touching spheres) a small mass drift is found. In these studies, it was difficult to establish the magnitude of such a drift unambiguously, because a relatively small initial mass asymmetry made it difficult to separate drifting projectile- and target-like distributions from the wings of a mass symmetric component.

We would like to stress the importance to distinguish the scattering processes, with no or negligible mass drift, i.e. quasi- and deep inelastic processes, from capture reactions that exhibit a small or large mass drift, usually but not always going to the full mass symmetry configuration and leading to decay into two almost equally large fragments.

The analysis of the mass (charge) distribution of the binary reaction products showed in Refs. [20] and [21, 22] showed that in collisions of nuclei with the magic proton or/and neutron

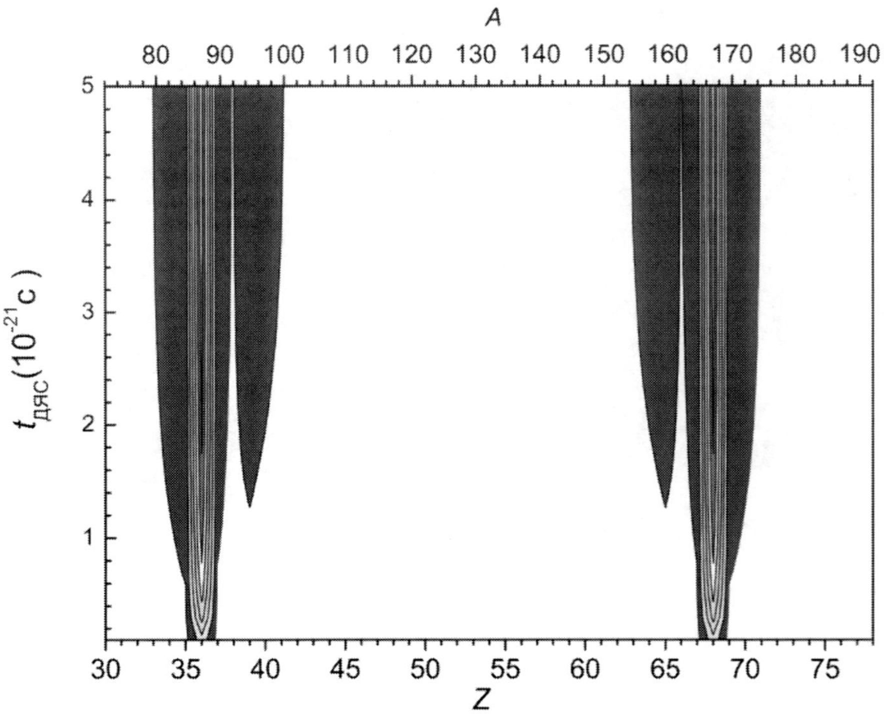

Figure 5. Time dependence of the charge distribution of deep inelastic collision and quasifission products for the ^{86}Kr + ^{166}Er reaction.

numbers at the around Coulomb barrier energies charge or mass drift may be small in the capture reactions too. In Ref. [20], an averaged charge number of the projectile-like products ^{86}Kr+^{166}Er was equal to the initial value $< Z_{\rm PLF} >= 36$ up to the total kinetic energy (TKE) loss of 180 MeV. Our theoretical results presented in Fig.5 show that deep inelastic and quasifission reactions products are mixed having the same charge number because the driving potential has a minimum at $Z = 36$ corresponding to projectile charge number. The similar phenomenon was observed in the ^{48}Ca+^{208}Pb reaction [22]. The small drift from the initial charge number of the projectile-like products does not mean that the deep inelastic reactions occured only. A maximum of the charge distribution can be concentrated at the charge value corresponding to the minimum of the potential energy surface. This phenomenon was reproduced by the DNS model and our results of the charge and mass distributions of the deep inelastic collisions (A) and quasifission (B) fragments are presented in Fig.4.

Understanding the mass drift is basic for investigation of the reaction mechanism. It is relevant to the deep inelastic process and to the capture processes. In capture reactions with heavy nuclei, the direction of the mass drift could influence the probability for the compound nucleus formation: a drift towards asymmetry could favour true compound nucleus formation (with subsequent evaporation of neutrons leading to the observed evaporation residues or fission fragments), whereas a drift towards symmetry could favour quasifission without compound nucleus formation.

3. Difference between capture and complete fusion for light and massive system

For light or medium-heavy systems, capture inside the Coulomb barrier leads to complete fusion, so that the capture (or barrier-passing) cross section may be equal to the complete fusion cross-

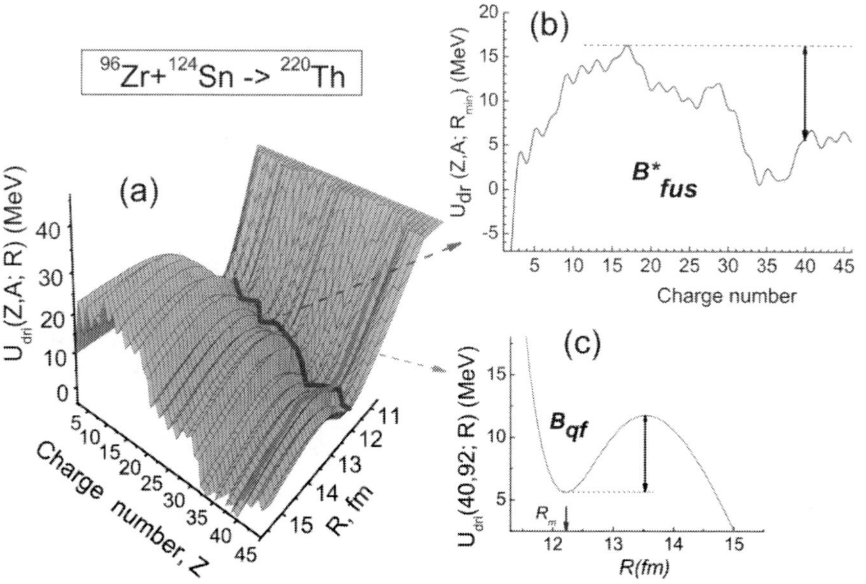

Figure 6. Potential energy surface calculated for the DNS formed in the ^{96}Zr+^{124}Sn reaction (a); driving potential built by connection of minima of the potential wells for all values of charge asymmetry of DNS (b): $B^*_{\rm fus}$ being considered as the intrinsic fusion barrier for the charge asymmetry $Z = 40$; potential well with depth $B_{\rm qf}$ which is used as a quasifission barrier (c). The minimum value of the potential well is at $R = R_m$.

section. According to the DNS concept the hindrance to complete fusion is connected with the presence of quasifission process as a competing channel which leads to formation of binary products without formation of compound nucleus [4, 12]. For the analysis and study of the hindrance to complete fusion we calculate the potential energy surface for DNS formed during interaction of nuclei as a sum of the reaction energy balance ($Q_{\rm gg} - value$) and nucleus-nucleus interaction $V(R)$:

$$U_{\rm driv}(Z_1, R, \ell) = Q_{\rm gg} + V(R), \tag{16}$$

where $Q_{\rm gg} = B_1 + B_2 - B_{\rm CN}$; B_1, B_2 and B_{CN} are the binding energies of the constituents of the DNS and compound nucleus, respectively. The values of the binding energies are obtained from [23, 24]. The evolution of DNS with given initial charge numbers of projectile- and target-nucleus (Z_P and Z_T) is determined by the potential energy surface. As an example, the potential energy surface calculated for the ^{96}Zr+^{124}Sn reaction is presented in Fig. 6(a). The approaching path of the projectile to the target nucleus occurs along the R-axis which is the distance between mass centers of colliding nuclei. Overcoming the Coulomb barrier the system starts to lose its kinetic energy of relative motion due to particle-hole excitation of nucleons and nucleon exchange as it is shown in Fig. 2. In this Section and further we consider capture reactions when the full momentum transfer in heavy ion collision occurred. The evolution of DNS along charge asymmetry takes place along valley on the potential energy surface. The driving potential $U_{\rm driv}$ is the curve lying on the bottom of this valley. A presence of a hindrance to complete fusion by multinucleon transfer can be found by estimation of the intrinsic fusion barrier $B^*_{\rm fus}$ as the difference between the values of $U_{\rm driv}$ corresponding to the initial charge asymmetry and

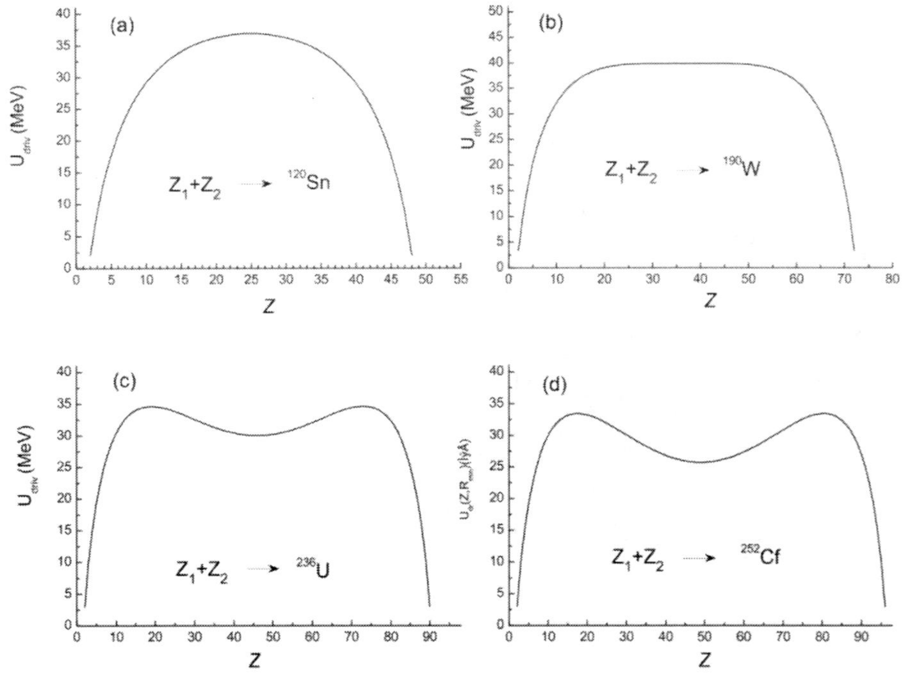

Figure 7. Driving potentials calculated for the DNS leading to formation of the compound nuclei ^{120}Sn (a), ^{190}W (b), ^{236}U (c) and ^{252}Cf (d).

its maximum value to the way of complete fusion $B^*_{\rm fus}(Z_P) = U_{\rm driv}(Z_{\rm max}) - U_{\rm driv}(Z_P)$. For example, in Fig. 6(b), $B^*_{\rm fus}(Z_P)$=10.2 MeV, because $U_{\rm driv}(17) = 16$ and $U_{\rm driv}(40) = 5.8$ MeV. The quasifission barrier $B_{\rm qf}$ for the charge asymmetry $Z_P = 40$ is equal to 5.68 MeV. The excitation energy of DNS $E^*_{\rm DNS} = E_{\rm c.m.} - V(R_m) + \Delta Q_{\rm gg}$, as well as $B^*_{\rm fus}$ and $B_{\rm qf}$ determine a hindrance to complete fusion.

Taking into account the dependence of the hindrance to complete fusion on the orbital angular momentum ℓ we calculate partial fusion cross section as a function of the collision energy $E_{\rm c.m.}$ [4, 12, 17]:

$$\sigma^{(\ell)}_{\rm fus}(E_{\rm c.m.}) = \sigma^{(\ell)}_{\rm cap}(E_{\rm c.m.}) P^{(\ell)}_{\rm CN}(E_{\rm c.m.}), \tag{17}$$

$P^{(\ell)}_{\rm CN}$ is the fusion probability in competition between complete fusion and quasifission. For light systems the capture and complete fusion cross sections are equal because $P^{(\ell)}_{\rm CN} = 1$ for the values of ℓ for which intrinsic fusion barrier $B^*_{\rm fus}$ is equal to zero. To see a dependence of the fusion probability on the total charge and mass numbers of the colliding nuclei we compare the driving potentials ($\ell = 0$) which were calculated by the use of the liquid-drop model to obtain the binding energies B_1, B_2, and $B_{\rm CN}$ for the four reactions leading to compound nuclei ^{120}Sn, ^{190}W, ^{236}U and ^{252}Cf (see, Fig. 7). One can say that for all reactions leading to the CN ^{120}Sn there is no hindrance to complete fusion because the maximum value of $U_{\rm driv}$ is in the middle part $Z_1 = Z_2$. Therefore, there is no intrinsic fusion barrier, $B^*_{\rm fus}$=0. For the reactions leading to the CN ^{190}W it is seen that approximately $B^*_{\rm fus}$=0, but there is the plateau at $20 < Z < 50$ which can be transformed to hollow at large values of angular momentum. As a result we can observe hindrance to complete fusion for the colliding energies $E_{\rm c.m.}$ at the near and above Coulomb barrier. The dependence of the driving potential on the DNS angular momentum for

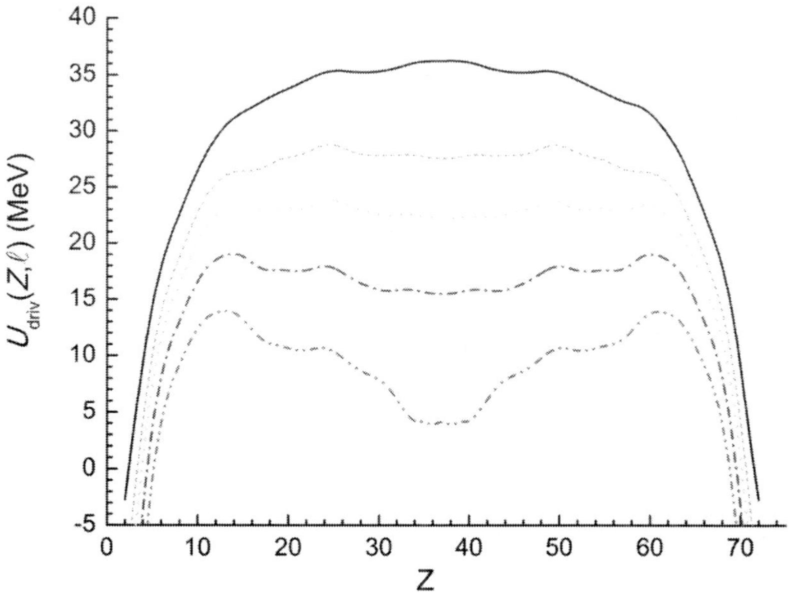

Figure 8. The dependence of the driving potential on the DNS angular momentum $\ell_{\rm DNS}$ for the reactions leading to the CN ^{190}W: $\ell = 0$ (solid line), 40 (dashed line), 60 (dotted line), 80 (dot-dashed line), and 100 (double dot-dashed line).

the reactions leading to the CN ^{190}W is demonstrated in Fig. 8. The increase of ℓ leads to the charge symmetric part of the driving potential. Consequently, the intrinsic fusion barrier, *i.e.* a hindrance to the complete fusion appears for the charge symmetric reactions leading to formation of ^{190}W at large angular momentum. The increase of the hindrance for the massive system was analyzed in Ref. [25].

4. Cold and hot fusion mechanisms in DNS model

Dynamics of complete fusion and role of the entrance channel in formation of heavy ion collision reactions are questionable or/and they have different interpretation still now. For example, what mechanism of fusion makes the main contribution to formation of compound nucleus:

– the increase of the radius of neck between interacting nuclei or multinucleon transfer at relatively restricted neck size?

– peculiarities of angular momentum distribution of DNS and compound nucleus which determine the angular distribution of reaction products, cross sections of evaporation residue, fusion-fission and quasifission products;

– separation of fusion-fission fragments from the quasifission and fast fission products;

– distribution of the excitation energy between different degrees of freedom, as well as between reaction products.

In the DNS concept [26], the evaporation residue cross section at collision energy $E_{\rm c.m.}$ is factorized as follows:

$$\sigma_{\rm ER}(E_{\rm c.m.}) = \sum_{\ell=0}^{\ell_{\rm f}} \sigma_{\rm fus}^{(\ell)}(E_{\rm c.m.}) W_{\rm sur}^{(\ell)}(E_{\rm c.m.}), \tag{18}$$

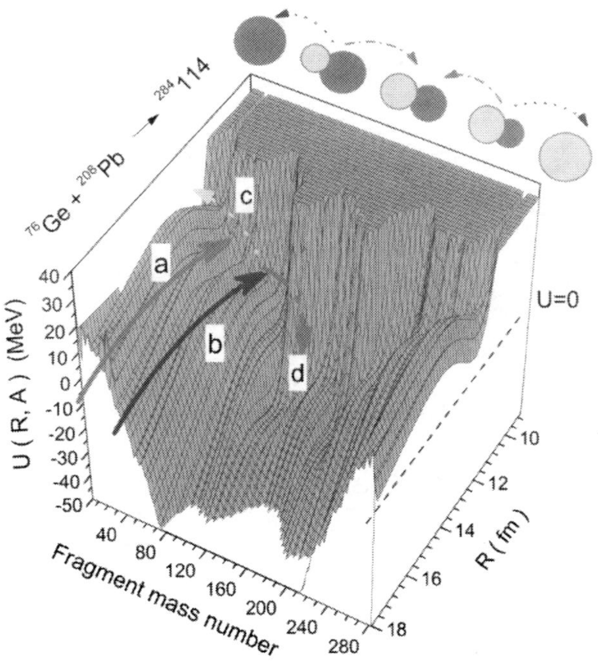

Figure 9. An explanation of the difference between "hot" (a) and "cold" (b) fusion on the potential energy surface of DNS for the ^{76}Ge+^{208}Pb reaction. The arrows (c) and (d) show the complete fusion and quasifission directions, respectively.

where $\sigma_{\rm fus}^{(\ell)}$ is the partial cross section of the complete fusion of the projectile and target nuclei; $W_{\rm sur}^{(\ell)}$ is the survival probability against fission of the heated and rotating nucleus at each step of the de-excitation cascade by evaporation of neutrons, protons and light particles up to formation of the evaporation residue; $\ell_{\rm f}$ is the value of CN angular momentum at which irs fission barrier disappears: $W_{\rm sur}^{(\ell)}(E) = 0$ for $\ell > \ell_{\rm f}$. The decrease of $W_{\rm sur}$ by increasing the excitation energy is determined by the increase of rate of competition between fission and emission of particles.

At synthesis of the superheavy elements $Z > 114$ the observed cross sections of the evaporation residues are near or less than 1 pb. The smallness of $\sigma_{\rm fus}$ or/and $W_{\rm sur}$ in formula (18) leads to small values of the measured cross section $\sigma_{\rm ER} \sim$ 1pb.

The experience of the synthesis of superheavy elements showed that the used reactions can be separated into "cold" and "hot" fusion reactions. In the "cold fusion" reactions the excitation energy the formed compound nucleus $E^*_{\rm CN}$ is less than 20 MeV [6, 7] while in "hot fusion" reactions $E^*_{\rm CN}$ is more 25 MeV [5]. In the "cold fusion" reactions the evaporation residue nuclei are formed after emission 1 or 2 neutrons from the heated and rotating compound nucleus. In the "hot fusion" reactions 3 or more neutron emission cascade preceded the evaporation residue formation. The charge asymmetry of the entrance channel determines the type of reaction: "cold fusion" reactions with the ^{208}Pb and ^{209}Bi targets used to synthesize superheavy elements Darmstadtium, Roentgenium and Copernicium were more mass symmetric than the "hot fusion" reactions with the ^{48}Ca projectile on the actinide targets used to synthesize more heaviest elements $Z = 114$–118 in the Flerov Laboratory of the Nuclear Reactions of Joint Institute for Nuclear Research. In Fig.9, the arrow "a" shows the path of the entrance channel leading to the "hot fusion" reactions while the arrow "b" shows the path leading to "cold fusion" reactions. Due to peculiarities of the landscape of the potential energy surface for the massive system, the excitation energies of DNS and compound nucleus for the symmetric charge numbers (large reaction Q_{gg} values) are smaller in the entrance channel with the ^{208}Pb target-nucleus (around arrow "b"): $E^*_{\rm CN} = E_{\rm c.m.} + Q_{\rm gg}$. It is seen from the shape of the potential energy surface that the "hot fusion" reaction can not be made "colder" because capture of projectile by target-nucleus becomes impossible by decreasing the beam energy. Note, for this system leading to formation of ^{284}114, the "cold fusion" reaction ^{76}Ge+^{208}Pb can not be made "warmer" because if we increase the beam energy significantly higher than the Coulomb barrier the system can not be captured due to small size of the potential well in the nucleus-nucleus interaction and restricted value of the friction coefficient for radial motion: we observe only deep inelastic collisions as in Fig. 2(b). One can say that, for the "cold fusion" reaction, the increase of the beam energy enough higher than the Coulomb barrier does not lead to increase the fusion cross section as it is expected from the extra-extra push model [27]. In the "cold fusion" reactions with ^{208}Pb and ^{209}Bi targets an increase of the projectile charge number leads to the drastic hindrance to complete fusion as a result of the increase of contributions of the quasifission and deep inelastic collisions. This phenomenon explained the observed smallest cross section in synthesis of superheavy element $Z = 113$ in the ^{64}Zn+^{209}Bi reaction [7].

Table 1. Comparison of the hindrance to formation of compound nucleus in the "cold" and "hot fusion" reactions used at synthesis of superheavy elements.

'Cold fusion" reactions	$\eta = \frac{A_2-A_1}{A_1+A_2}$	$P_{\rm CN}$	"Hot fusion" reactions	$\eta = \frac{A_2-A_1}{A_1+A_2}$	$P_{\rm CN}$
^{64}Ni+^{208}Pb†	0.529	$1.4 \cdot 10^{-7}$	^{48}Ca+^{244}Pu	0.671	$4.96 \cdot 10^{-2}$
^{64}Ni+^{209}Bi†	0.531	$7.0 \cdot 10^{-8}$	^{48}Ca+^{243}Am‡	0.670	$5.02 \cdot 10^{-2}$
^{70}Zn+^{208}Pb†	0.496	$2.5 \cdot 10^{-9}$	^{48}Ca+^{248}Cm‡	0.676	$1.13 \cdot 10^{-2}$
^{70}Zn+^{209}Bi†	0.498	$5.2 \cdot 10^{-10}$	^{48}Ca+^{249}Bk§	0.677	$5.06 \cdot 10^{-3}$
^{76}Ge+^{208}Pb†	0.465	$1.2 \cdot 10^{-10}$	^{48}Ca+^{249}Cf‡	0.677	$7.14 \cdot 10^{-3}$

†The estimations from Ref.[4]
‡The estimations from Ref. [16]
§The estimations from Ref. [28]

In Fig. 10 we presented comparison of the ER cross sections obtained by the DNS model (dot-dashed line) with the experimental data for the synthesis of Roentgenium and Copernicium in the ^{64}Ni+^{209}Bi and ^{70}Zn+^{208}Pb reactions, respectively. The dashed and solid lines in Fig. 10 show the theoretical results for the quasifission and complete fusion excitation functions. The ratio of the complete fusion to capture excitation function ($P_{\rm CN}$) shows how is strong the hindrance to complete fusion (see Eq. (17)) due to very small values of fusion cross section in comparison to capture cross section. The event at $E^*_{\rm CN} = 9$ MeV in the ^{70}Zn+^{208}Pb reactions (right panel) was described and published in [4] while the second event at $E^*_{\rm CN} = 12$ MeV was measured after the appearance of the cited paper. One can say that the second event was predicted in [4]. So, we concluded that smallest cross section at synthesis of Copernicium is caused mainly by the huge contribution of the quasifission events that is inherent for the "cold fusion" reactions. This conclusion was deduced from the theoretical analysis which included the realistic nuclear shell effects in calculation of the potential energy surface. The dotted lines in Fig. 10 shows the fission barrier of compound nucleus $B_{\rm fis}$ which is a function of its excitation energy and angular momentum.

The "hot fusion" reactions were favorable for the synthesis of the superheavy elements $Z > 112$: new superheavy elements $Z = 114$–118 have been obtained at the Flerov Laboratory of

Nuclear Reactions of JINR (Dubna, Russia) [5] during the last decade. The ER cross sections in synthesis of superheavy elements $Z = 114$ and $Z = 116$ were confirmed in the recent experiments performed in the Lawrence Berkeley Laboratory [29] and GSI (Darmstadt) [30], respectively. The main reason allowing the experimentalists to succeed in the "hot fusion" reactions or the synthesis of the superheavy elements $Z = 114$–118 is smallness of the intrinsic fusion barrier $B^*_{\rm fus}$ which decreased the hindrance to complete fusion. But large values of the excitation energy and angular momentum of the compound nucleus formed in the "hot fusion" reactions decrease its survival probability against fission. Therefore, the cross sections of evaporation residues in synthesis of superheavy elements decreases from 3–4 picobarns for $Z = 114$ up to 0.5 pb for $Z = 118$. In Table 1 we compare the hindrance to formation of compound nucleus in the "cold" and "hot fusion" reactions which were used at synthesis of superheavy elements. The difference in the hindrances presented by $P_{\rm CN} = \sigma_{\rm fus}/(\sigma_{\rm fus} + \sigma_{\rm qfis})$ in both types of reactions is evidently seen: i) the hindrance to complete fusion is stronger for the "cold fusion" reactions in comparison with "hot fusion" reactions; ii) the hindrance to "cold fusion" reactions increases drastically by increasing of the projectile charge number while it changes slowly in "hot fusion" reactions by increasing the target charge number. This is explained by strong change of the potential energy surface around the charge asymmetry $Z = 82$ than in the region $Z = 20$ for the considered systems.

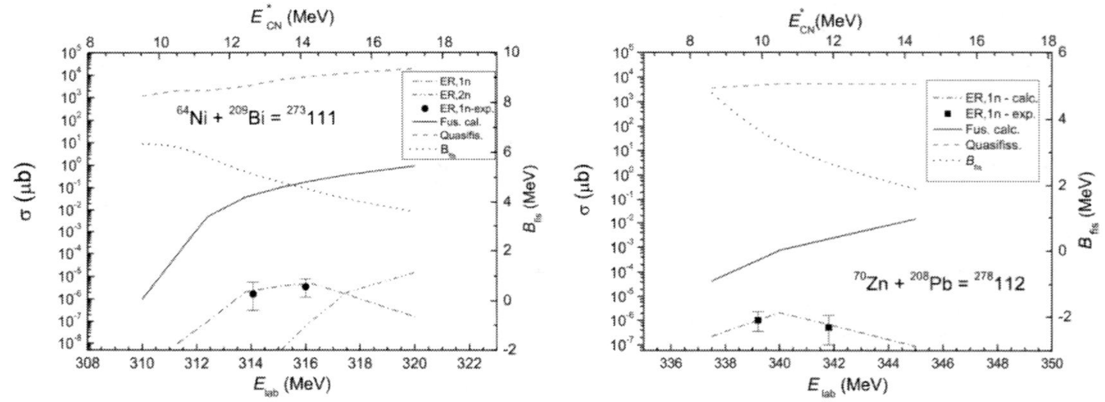

Figure 10. Comparison of the ER results obtained by the DNS model (dot-dashed line) with the experimental data for the synthesis of Roentgenium and Copernicium in the ^{64}Ni+^{209}Bi (left panel) and ^{70}Zn+^{208}Pb (right panel) reactions, respectively. Quasifission and fusion excitation functions are shown by dashed and solid lines, respectively. The dotted line is the fission barrier.

5. Mixing of mass-angle distributions of the quasifission and fusion-fission products

The mass (charge) and angular distributions of the reaction products in the heavy ion collisions are the main indicator used to make conclusions about a reaction mechanism. According to the established opinion the products with masses close to the ones of projectile and target nuclei are considered as products formed in deep inelastic reactions; the products with the symmetric mass distributions are considered as products of the fusion-fission reactions, *i.e.* fission products of the heated and rotating compound nucleus formed at complete fusion; the products having intermediate masses between projectilelike and fusion-fission products are considered as quasifission products. The last kind of products appear when proton or neutron numbers in the heavy or light fragment are close to the magic numbers 28, 50, 82 or 126 [31, 32]. So, the symmetric mass distributions from capture reactions can either be the result of compound

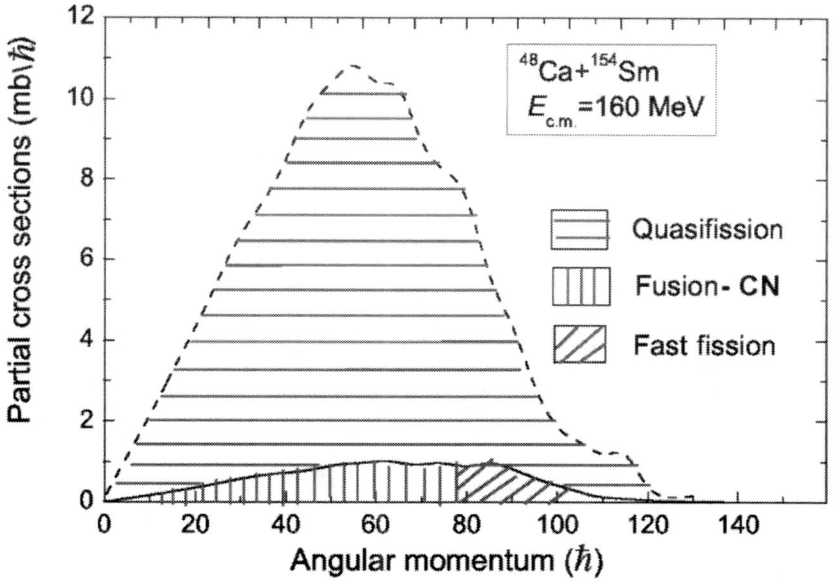

Figure 11. Partial cross sections (angular momentum distributions) of quasifission (area filled by the horizontal lines), fast fission (area filled by the skew-eyed lines) and compound nucleus (area filled by the vertical lines) events calculated by the DNS model for the ^{48}Ca+^{154}Sm reaction.

nucleus formation followed by fission, or - if the criteria for compound decay are violated - they are ascribed to the quasifission and fast fission channels.

We should remind that quasifission products are formed at decay of DNS without reaching the stage of compound nucleus or one can say without reaching saddle point to arrive to compact shape. The quasifission means that TKE of binary fragments is close to that of fusion-fission products or Viola systematics.

The experimental methods used to estimate the fusion probability depend on the unambiguity of identification of the complete fusion reaction products among the quasifission products. The difficulties arise when the mass (charge) and angular distributions of the quasifission and fusion-fission fragments strongly overlap depending on the reaction dynamics. As a result, the complete fusion cross sections may be overestimated:

$$\sigma_{\rm fus} = \sigma_{\rm ffis} + \sigma_{\rm ER}, \tag{19}$$

where $\sigma_{\rm ER}$ is measured by good accuracy while $\sigma_{\rm ffis}$ may include contribution of the quasifission and fast fission products which are formed at decay of DNS and at decay of the deformed mononucleus-no compound nucleus. We remind the difference between quasifission and fast fission processes:

– quasifission is the decay of DNS which is formed in the capture–full momentum transfer reactions. The angular momentum distribution for the quasifission events can extend from $\ell = 0$ up to $\ell = \ell_{\rm d}$ where $\ell = \ell_{\rm d}$ the maximum value of the DNS angular momentum (see Fig. 11, area filled by the horizontal lines). The mass (charge) distribution of quasifission products may be in the wide range from masses (charges) of projectile- and target-like fragments up to symmetric masses mixing with the fusion-fission products;

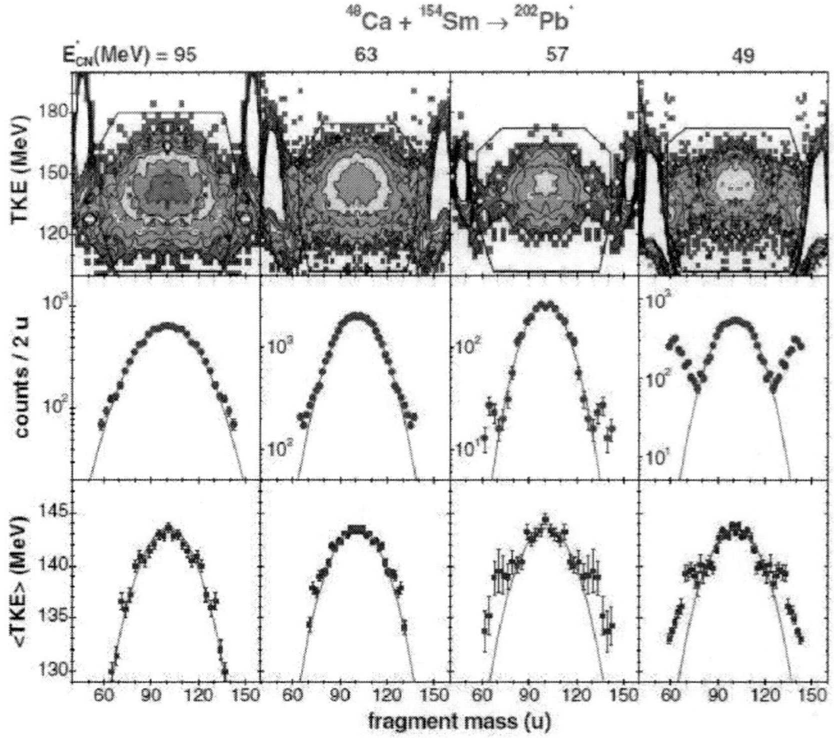

Figure 12. Two-dimensional TKEmass matrices (upper panels), yields of fragments and their TKE as a function of the fragment mass (middle and bottom panels, respectively) in the ^{48}Ca+^{154}Sm reaction at the different $E^{*}_{\rm CN}$ excitation energies (designated above upper panels). Solid lines in the middle and bottom panels are Gaussian and parabola fits to the mass and TKE distributions, respectively. This figure was copied from Ref. [32].

– fast fission is the decay of the mononucleus which is a non-equilibrated system survived against quasifission but unstable to be formed as compound nucleus due to absence of the fission barrier caused by its fast rotational velocity [33]. Therefore, the partial cross section of fast fission is populated only at large values of the angular momentum ℓ (see Fig.11, the area filled by skew-eyed lines).

We tried to analyze reasons for the lack or disappearance of the quasifission feature in the experimental data for the ^{48}Ca+^{144}Sm and ^{48}Ca+^{154}Sm reactions presented in the paper [32] as the main conclusion of study. The authors established the fusion suppression and the presence of quasifission for the reactions with the deformed ^{154}Sm target at the beam energies near and below the Coulomb barrier. The contribution of the quasifission fragments with masses in the range $55 < A < 145$ to the total mass distribution of fission fragments decreases, with respect to the contribution of the symmetric compound nucleus-fission, as the ^{48}Ca projectile energy

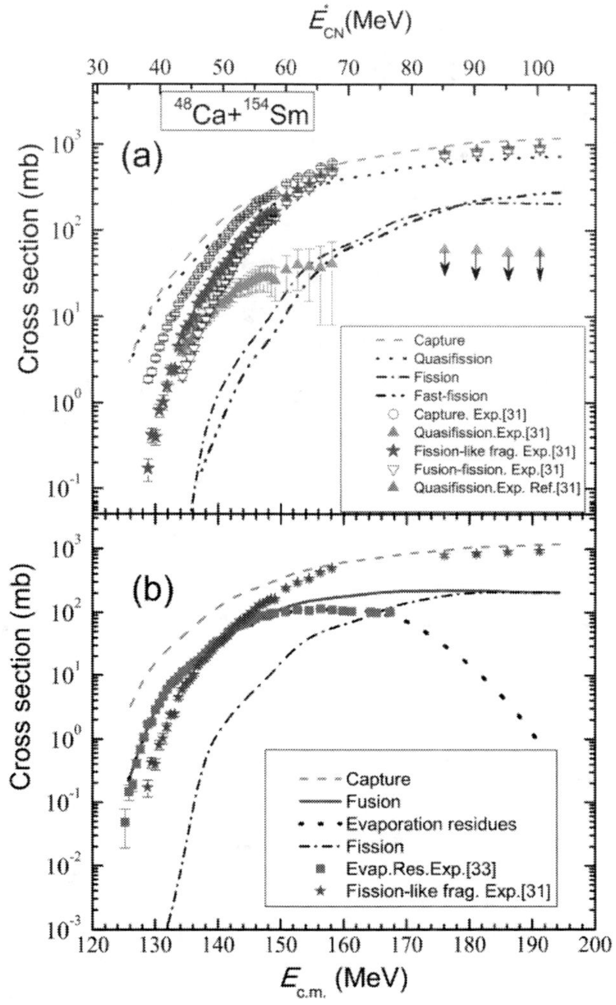

Figure 13. Comparison of the results of this work by the DNS model for the capture, complete fusion, quasifission, fast fission and evaporation residue cross sections with the measured data of the fusion-fission and quasifission given in Ref. [32] (panel (a)) and with data of the evaporation residues obtained from Ref. [34] (panel (b)) for the ^{48}Ca+^{154}Sm reaction.

increases (see Fig. 12). The authors did not consider a possibility of the quasifission product yield with masses outside the range $55 < A < 145$. The quasifission products are formed in the ranges $A < 55$ and $A > 145$ too. But those can be mixed with the products of deep inelastic reactions as in the ^{48}Ca+^{208}Pb reaction which was discussed in Section 2. This is a reason why our theoretical results of the capture cross section (long dashed line) overestimated the experimental data (open circles) presented in Ref. [32] at the low energies (see Fig. 13(a)).

The origination of the measured fission-like fragments (stars) at the large bombarding energies is explained by the sum of the quasifission (short dashed line), fusion-fission (dash dotted line) and fast fission (dash-double dotted line) fragments. A cross section of the fast fission channel increases by increasing the bombarding energy due to the increase of the angular momentum of mononucleus. At low energies the contribution of the fusion-fission to the yield of binary fragments is small in comparison with the quasifission contribution. The small calculated fusion-

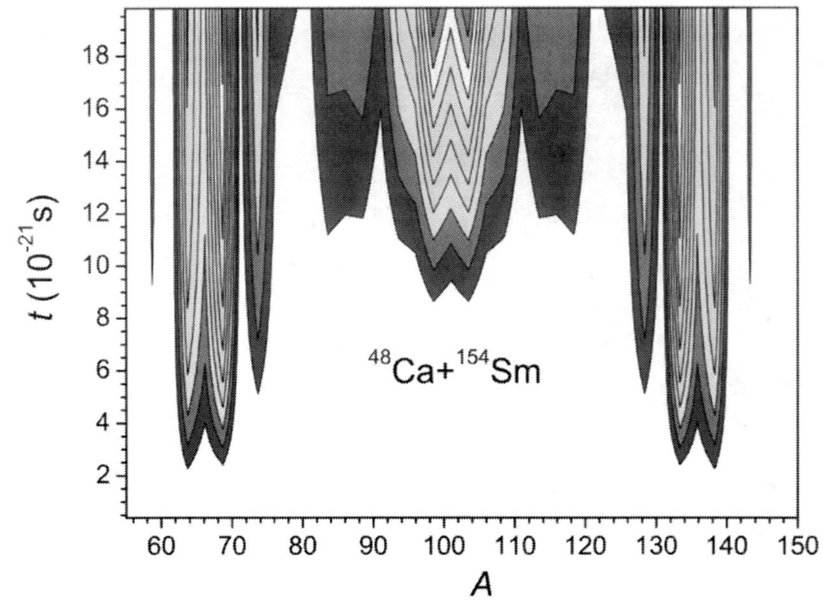

Figure 14. Mass distribution of quasifission products formed in the ^{48}Ca+^{154}Sm reaction as a function of time.

fission cross section is explained by the large fission barrier (B_f=12.33 MeV) for the ^{202}Pb nucleus according to the rotating finite range model by A. J. Sierk [33] and by the additional barrier $B_f^{(micr)} = -\delta W = -(\delta W_{saddle-point} - \delta W_{gs}) \cong 8.22$ MeV caused by the nuclear shell structure. We conclude that the experimental fusion-fission data obtained at low energy collisions contain a huge contribution of quasifission fragments with masses $A > 83$ which show an isotropic distribution as presented in Ref. [32]. This is not a new phenomenon and it was discussed as a result of theoretical studies, for example, in our previous papers [11, 35] and in Ref. [36]. The experimental results confirming this conclusion appeared recently in Ref. [37, 38]. The contribution of the mass symmetric products mixed with the fusion-fission products with similar masses increases ambiguity at estimation of fusion cross section by formula (19). This is a reason of increasing the difference our theoretical results and extracted from the experimental data for quasifission.

At the large energy $E_{c.m.}$=154 MeV (E^*_{CN}=63 MeV) the experimental values of the quasifission cross section are much lower than that of the fusion-fission cross section [32]. According to our theoretical result a sufficient part of the quasifission fragments shows the behaviour of the fusion-fission fragments: the mass distribution can reach the mass symmetric region and their angular distribution can be isotropic due to the possibility that the dinuclear system can rotate by large angles for large values of its angular momentum [39]. The authors of Ref. [32] did not exclude such a behaviour of the quasifission fragments. It is difficult to separate the quasifission fragments from the fusion-fission fragments when both, their mass and angle distributions, overlap in the region of symmetric masses. So, ignoring the quasifission products mixed with deep inelastic collisions and consideration of the all mass symmetric products as fusion-fission products are reasons of difference between the theoretical (dotted line in Fig. 13(a)) and experimental (up filled triangle in Fig. 13(a)) values of quasifission cross sections.

At low energies the projectile-like quasifission fragments with $A < 70$ give a large contribution to the cross section for the considered ^{48}Ca+^{154}Sm reaction since the excitation energy of the

DNS is too small to shift the maximum of the mass distribution to more mass symmetric configurations of DNS. The observed quasifission feature at low energies is connected with the peculiarities of the shell structure of the interacting nuclei. The increase in the beam energy leads to a decrease of the shell effects and the yield of the quasifission fragments near the asymmetric shoulders decreases because the main contribution of quasifission moves to the mass symmetric range. As it is seen from Fig. 14 that the yield of products with the masses in the range $85 < A < 125$ appears at times $t > 8.5 \cdot 10^{-21}$s. The evaporation residue and fusion-fission excitation function were calculated by the advanced statistical model [40, 41]. In this model, the partial fusion cross sections obtained by the DNS model were used as input data.

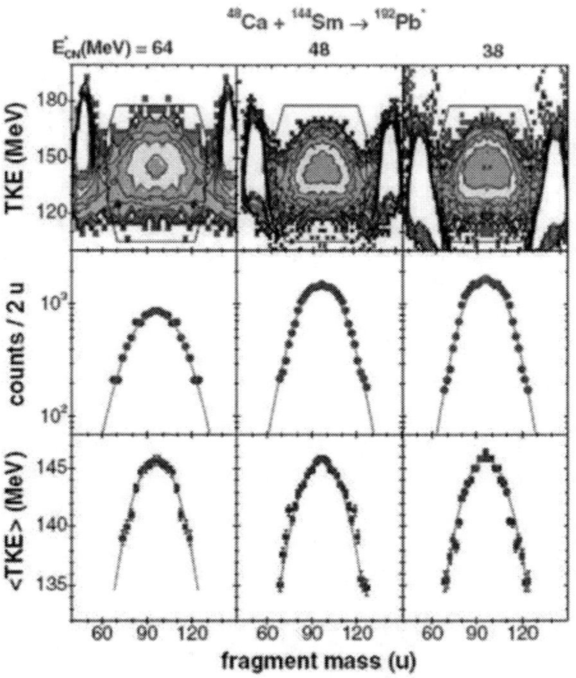

Figure 15. Two-dimensional TKEmass matrices (upper panels), yields of fragments and their TKE as a function of the fragment mass (middle and bottom panels, respectively) in the ^{48}Ca+^{144}Sm reaction at the different $E^*_{\rm CN}$ excitation energies (designated above upper panels). Solid lines in the middle and bottom panels are Gaussian and parabola fits to the mass and TKE distributions, respectively. This figure was copied from Ref. [32].

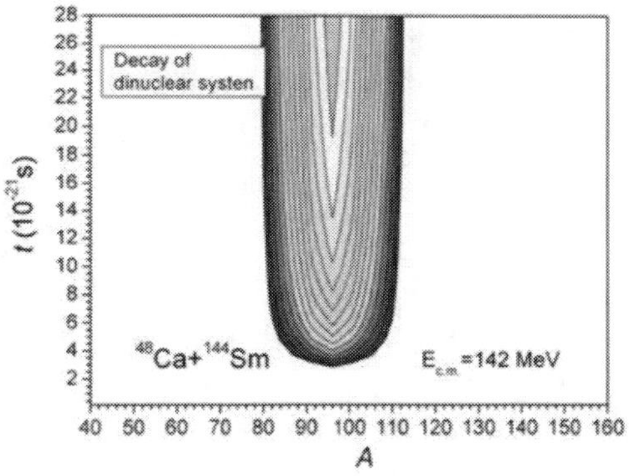

Figure 16. The mass distribution the quasifission products formed in the ^{48}Ca+^{144}Sm reaction as a function of time.

Due to large fission barrier of the compound nucleus ^{202}Pb fission probability is very small and the yield of the evaporation residues after neutron emission cascade dominates. Therefore, our theoretical fission excitation function (dot-dashed line) presented in Fig. 14(a) is sufficiently lower than the experimental data (down open triangles). As we have discussed above the latter data included a contribution of the quasifission events. It is important to note that the evaporation residue excitation function (thick dotted line in Fig. 14(b)) is in good agreement with the experimental data (solid squares in Fig. 14(b)) which is the unambiguous physical quantity obtained from Ref. [34]. This fact confirms the correctness of our partial fusion cross sections.

The difficulties in identification of the quasifission products are connected with absence of the appropriate knowledge about reaction mechanism and complete analysis of the yield of quasifission and accompanying particles. The assumption about possibility to observe the yield of quasifission products only in the mass range corresponding to the middle between projectile-like and fusion-fission products is not completely true: if there is no sufficient yield of reaction products in this intermediate region it does not mean that there is no yield of the quasifission products. For example, the authors of Ref. [32] came to the same conclusion about the ^{48}Ca+^{144}Sm reaction where they did not observe the characteristic peak of the quasifission products in the mass distribution (see Fig. 15) Our theoretical results of the charge and mass distributions showed that there are two reasons of a lack of the usual quasifission peak: (i) one part of the mass distribution of the quasifission fragments is in the mass range of projectile-like products $48 < A < 60$; (ii) another part of the quasifission fragments is mixed with the fusion-fission fragments and has similar isotropic distributions. The isotope ^{144}Sm is a magic nucleus with the neutron number N=82. Therefore, the concentration of the asymmetric mode of the quasifission fragments in the mass range $48 < A < 60$ is explained by the effect of the shell structure of the double magic projectile-nucleus ^{48}Ca and magic target-nucleus ^{144}Sm on the mass distribution of the reaction fragments. As a result, the mass distributions of the products of deep-inelastic collisions and asymmetric quasifission overlap in this mass range. This case is similar to the ^{48}Ca+^{208}Pb reaction where the presence of the quasifission feature was doubtful. But our investigation showed that due to the collision of the double magic ^{48}Ca and ^{208}Pb nuclei the mass distribution of the quasifission fragments is concentrated around the initial masses (see Fig. 2) because the potential energy surface has a local minimum in this region. In Fig. 16, we present the time dependence of the mass distribution of quasifission products of the ^{48}Ca+^{144}Sm reaction. One can see that the mass numbers of the quasifission products are concentrated in the mass range $80 < A < 110$ which overlaps completely with the mass range of fusion-fission products. In different from the ^{48}Ca+^{154}Sm reaction, in the ^{48}Ca+^{144}Sm reaction no yield of the quasifission products were found in the intermediate mass range where quasifission was used to be observed.

The lack of quasifission events in the experimental studies of the ^{48}Ca+^{144}Sm reaction or disappearance of quasifission events by increasing the beam energy are connected with the measurement and analysis of the experimental data. More advanced experiments at different beam energy and reaction charge asymmetry must be performed by registration in coincidence of the fissionlike products, neutrons and gamma-quanta. The results of such experiments should be analysed by study the different correlation functions of the observed quantities to distinguish quasifission feature in the case that the mass-angle distributions of the quasifission and fusion-fission fragments strongly overlap in the mass symmetric region. We should know the information about the angular momentum and excitation energy of the system going to fission (compound nucleus or DNS) additionally to the mass, angle and kinetic energy distributions of the fission fragments and accompanying them neutrons. The theoretical models must be enough perfect and they should have as possible a small amount of free parameters to make unambiguous conclusions about reaction mechanism after the good description of the observed experimental data which are interpreted without additional assumptions.

References

[1] Wilpert M, Gebauer B, Wilpert Th, von Oertzen W, Bohlen H G, and Speer J 1995 *Phys. Rev.* C **51** 680
[2] Adamian G G, Jolos R V, Nasirov A K, Muminov A I 1996 *Phys. Rev.* C **53** 871
[3] Jolos R V, Nasirov A K, Adamian G G, and Muminov A I 2000 *Eur. Phys. J.* A **8** 115
[4] Giardina G, Hofmann S, Muminov A I, and Nasirov A K 2000 *Eur. Phys. J.* A **8** 205
[5] Oganessian Yu Ts, *et al.* 2004 *Phys. Rev.* C **70** 064609; Oganessian Yu Ts, *et al.* 2009 *Phys. Rev.* C **79** 024603

[6] Hofmann S 1997 *Z. Phys.* A **358** 125; Hofmann S 1998 *Rep. Progr. Phys.* **61** 639; Hofmann S and Munzenberg G 2000 *Rev. Mod. Phys.* **72** 733
[7] Morita K 2004 *J. Phys. Soc. Jpn.* **73** 2593; Morita K 2007 *J. Phys. Soc. Jpn.* **76** 043201
[8] Fazio G *et al.* 2005 *Phys. Rev.* C **72** 064614
[9] Viola V E, Kwiatkowski K, Walker M 1985 *Phys. Rev.* C **31** 1550
[10] Keller J G, Back B B, Glagola B G, Henderson D, Kaufman S B, Sanders S J, Siemssen R H, Videback F, Wilkins B D, Worsham A 1987 *Phys. Rev.* C **36** 1364
[11] Nasirov A K, Giardina G, Muminov A I, Scheid W and Yakhshiev U T 2003 *Proc. Symposium Nuclear Clusters, Rauischholzhausen, Germany*, 5-9 August 2002 (ed. R. V. Jolos and W. Scheid, EP Systema, Debrecen) p.415; 2004 *Acta Physica Hungarica* A **19** 109
[12] Fazio G *et al.* 2003 *J. Phys. Soc. Jpn* **72** 2509
[13] Kramers H A 1940 Physica **7** 284; Strutinsky V M 1973 *Phys. Lett.* B**47** 121
[14] Grangé P *et al.* 1983 *Phys. Rev.* C **27** 2063; Grangé P 1984 *Nucl. Phys.* A **428** 37c
[15] Fröbrich P and Tillack G R 1992 *Nucl. Phys.* A**540** 353
[16] Fazio G, Giardina G, Lamberto A, Ruggeri R, Saccà C, Palamara R, Muminov A I, Nasirov A K, Yakhshiev U T, Hanappe F 2004 *Eur. Phys. J.* A **19** 89
[17] Nasirov A, Fukushima A, Toyoshima Y, Aritomo Y, Muminov A, Kalandarov Sh, Utamuratov R 2005 *Nucl. Phys.* A **759** 342
[18] Junghans A R, de Jong M, Clerc H -G, Ignatyuk A V, Kudyaev G A, Schmidt K -H 1998 *Nucl. Phys.* A**629** 635
[19] Nasirov A K, Muminov A I, Utamuratov R K, Fazio G, Giardina G, Hanappe F, Mandaglio G, Manganaro M, and Scheid W 2007 *Eur. Phys. J.* A **34** 325
[20] Rudolf G, Gobbi A, Stelzer H, Lynen U, Olmi A, Sann H, Stockstad R G and Pelte D 1979 *Nucl. Phys.* A **330** 243
[21] Fazio G, Giardina G, Mandaglio G, Hanappe F, Muminov A I, Nasirov A K, Scheid W, Stittge L 2005 *Mod. Phys. Lett.* A **20** 391
[22] Prokhorova E V *et al.* 2008 *Nucl. Phys.* A **802** 45
[23] Audi G, Wapstra A H 1995 *Nucl. Phys.* A**595** 509
[24] Möller P, Nix J R 1988 *Atom. Data and Nucl. Data Tabl.* **39** 213
[25] Fazio G *et al.* 2005 *Phys. Rev C* **72** 064614
[26] Antonenko N A, Cherepanov E A, Nasirov A K, Permjakov V P, Volkov V V, *Phys. Lett. B* **319** (1993) 425; 1995 *Phys. Rev.* C **51** 2635
[27] Blocki J P, Feldmeier H, Swiatecki W J 1986 *Nucl. Phys.* A **459** 145
[28] Giardina G, *et al.* Contributions of this Symposium
[29] Düllmann Ch E 2010 *Phys. Rev. Lett.* **104** 252701
[30] Heinz S, *et al.* 2010 Contributions of this Symposium
[31] Itkis M G *et al.* 2004 *Nucl. Phys.* A **734** 136
[32] Knyazheva G N, Kozulin E M, Sagaidak R N, Chizhov A Yu, Itkis M G, Kondratiev N A, Voskressensky V M, Stefanini M, Behera B R, Corradi L. 2007 *Phys. Rev.* C **75** 064602
[33] Sierk A J 1986 *Phys. Rev.* C **33** 2039
[34] Stefanini A M, *et al.* 2005 *Eur. Phys. J.* A **23**) 473
[35] Fazio G, Giardina G, Lamberto A, Ruggeri R, Hanappe F, Materna T, Jovliev U Yu, Khugaev A V, Muminov A I, Nasirov A K 2003 *Proc. Int. Symposium "New Prospects and Lines of Research in Nuclear Physics", Messina, Italy*, 24-26 October 2002, ed. G. Fazio and F. Hanappe, World Scientific, Singapore p.258
[36] Aritomo Y, Ohta M 2004 *Nucl. Phys.* A **744** 3
[37] Hinde D J, du Rietz R, Dasgupta M, Thomas R G, and Gasques L R 2008 *Phys. Rev. Lett.* **101** 092701
[38] Knyazheva G N, Itkis M G, Kozulin E M, Lyapin V G, Rubchenya V A, Trsacka V, Khlebnikov S V 2008 *Particles and Nuclei, Letters* **5** 40 (in Russian)
[39] Nasirov A K, Muminov A I, Giardina G, Mandaglio G, Manganaro M 2010 *Journal of Physics: Conference Series* **205** 012018
[40] D'Arrigo A, Giardina G, Herman M and Taccone A 1992 *Phys. Rev.* C **46** 1437
[41] D'Arrigo A, Giardina G, Herman M, Ignatyuk A V and Taccone A 1994 *J. Phys.* G **20** 365

Journal of Physics: Conference Series **282** (2011) 012011 doi:10.1088/1742-6596/282/1/012011

Study for fusion-fission and quasifission in reactions using ^{238}U target nucleus

Katsuhisa Nishio

Advanced Science Reserch Center, Japan Atomic Energy Agency,
Tokai, Ibraki 319-1195, Japan

E-mail: `nishio.katsuhisa@jaea.go.jp`

Abstract. Fragment mass distributions for fission after full momentum transfer (FMT) were measured for the ^{30}Si,^{31}P,^{34,36}S,^{40}Ar + ^{238}U reactions at bombarding energies around the Coulomb barrier. The experiment was carried out at the JAEA tandem accelerator facility. We observed strong variation of the mass distribution on beam energy and projectile nucleus. In the reaction of ^{36}S +^{238}U we observed a transition from symmetry to asymmetry mass distributions when the beam energies were decreased from the above-barrier to sub-barrier values. The mass asymmetry was $A_L\ /A_H = 74/200$, which corresponds to the fission valley leading to the nuclei close to the doubly closed-shell nuclei ^{78}Ni /^{208}Pb. The fission channel is populated by quasifission, which is the disintegration without forming the compound nucleus. The incident-energy dependence is strongly correlated with the prolate deformation of ^{238}U and the orientation at the initial impact. The results indicate that the reaction starting at the polar collisions on ^{238}U has larger quasifission probability. The mass distributions are nicely reproduced by a model calculation using Langevin equation. The ^{40}Ar + ^{238}U reaction also has the similar mass asymmetry ($A_L/A_H \approx 78/200$) in quasifission. In the reactions of ^{31}P + ^{238}U and ^{30}Si + ^{238}U, mass asymmetries are $A_L/A_H \approx 81/188$ and 90/178, respectively. The results suggest that the system produced by the reaction using lighter projectile approaches closer to the shape of the compound nucleus even when quasifission occurs. In the reactions of ^{30}Si + ^{238}U and ^{34}S + ^{238}U, we also measured the evaporation residue (ER) cross sections to obtain information on the fusion probability. From the ER cross sections for the ^{34}S + ^{238}U reaction it was suggested that the symmetric fission is also dominated by quasifission.

1. Introduction

Experimental challenges to produce superheavy nuclei (SHN) have been carried out by using heavy ion fusion reactions [1, 2, 3]. Development of a theoretical model to predict cross sections for nuclei located at the extreme end of heavy elements is important for the proper selection of target and projectile as well as the bombarding energy to produce these nuclei. The reaction is considered to proceed in three steps; (1) penetration of the Coulomb barrier between two colliding nuclei, (2) formation of a compound nucleus after the system is captured inside the Coulomb barrier and (3) survival of the exited compound nucleus to produce evaporation residue (ER) against fission (fusion-fission).

The first step, penetrating the Coulomb barrier, is relatively well understood. Enhancing of the capture cross section (σ_{cap}) relative to the one-dimensional barrier penetration model has been observed at the sub-barrier energy, which is explained by the distribution of the Coulomb barrier, as a result of coupling the incident flux to the collective states in the interacting nuclei.

Deformation of the target and projectile is another reason for enhancing the capture cross sections. The second process, forming a compound nucleus (fusion probability), is not well understood. A theoretical model must treat the dynamic evolution of a system from the initial touching configuration up to the so called compound nucleus state. In a reaction using a heavy target and projectile, which is the case for the production of the heaviest element, quasifission competes against fusion. Once the fusion cross section is calculated, it is multiplied by the survival probability, which can be determined by a statistical model, to calculate the cross section to produce SHN.

Measurement of the ER cross sections gives information on the fusion probability. However, because of the low production rate for SHN, available data with high statistical accuracy are limited. When a model can treat fusion-fission and quasifission in a consistent framework, such as the unified theory [4], the measurement of fission properties can be another benchmark for testing the model, as fusion-fission and quasifission would have different decay properties.

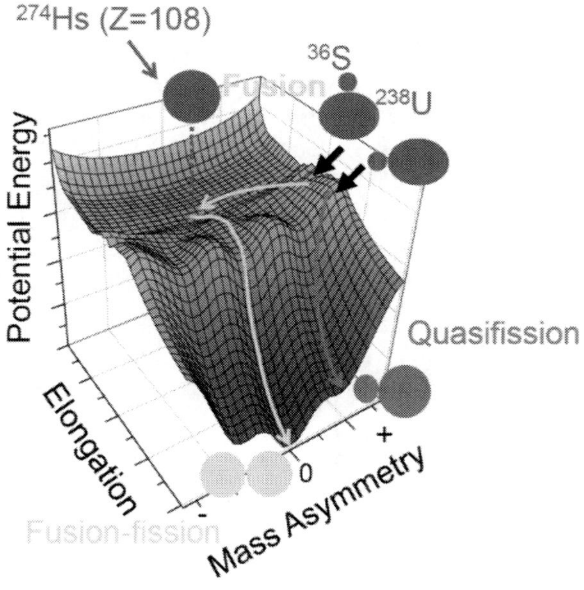

Figure 1. Potential energy for ^{274}Hs produced in the fusion reaction of $^{36}S + ^{238}U$.

Fusion reactions using actinide target nuclei are extensively used to investigate SHN. The reasons are ; (1) a relatively neutron rich SHN than the cold fusion reactions are produced, thus the decay properties of these nuclei have information on the structure in the vicinity of the spherically closed-shell at N=184, (2) nuclei having a relatively long half-lives allows a study for the chemical property, and (3) the cross sections maintain values of a few picobarn even for the production of the heaviest elements [1, 5]. The relatively large cross sections for SHN produced using a ^{48}Ca beam are explained by a high survival probability of the compound nuclei in competition with fission owing to large fission barriers of nuclei in the vicinity of the $N = 184$ shell closure. Another reason could be a higher fusion probability in reactions using actinide targets. Since actinide nuclei are prolately deformed, there exists a configuration at which the projectiles hit the equatorial region of the deformed target nuclei. In this case a compact configuration is achieved and the system may have a larger fusion probability than the reactions using spherical target nuclei such as lead or bismuth.

We have investigated the effects of nuclear orientation on fusion and/or quasifission by measuring the fission fragment mass distributions and the ER cross sections in the reactions using ^{238}U target nuclei to obtain information on fusion probability. The evolution of the nuclear shape from the contact point is largely influenced by the structure of the potential energy and the initial touching point as shown in Fig. 1. At the equatorial collisions, the system has a larger probability to form the compound nucleus, whereas the polar collisions would give higher probability to disintegrate as quasifission. The quasifission is expected to fission through the mass-asymmetric channel.

We also made a theoretical calculation based on a dynamical model using Langevin equation to interpret the measured mass distributions, where fusion-fission and quasifission fragments are separately determined.

2. Experimental details

2.1. In-beam fission measurement

The mass distributions and cross sections of fission fragments in the reactions of $^{30}Si, ^{34,36}S, ^{31}P, ^{40}Ar + ^{238}U$ were measured using beams supplied by the tandem accelerator of the Japan Atomic Energy Agency (JAEA) in Tokai, Japan. The experimental set-up and the analysis method were described in [6].

Beam energies were changed from above-barrier to sub-barrier values. The beam intensities were typically from 0.1 to 1.0 p-nA. The ^{238}U target was prepared by electrodeposition of UO_2 on a 90-μg/cm^2 thick nickel backing. Both fission fragments (FFs) were detected in coincidence by position-sensitive multiwire proportional counters (MWPCs). The detectors were located on both sides of the target each at a distance of 211 mm and at angles of $\theta_1 = -61.0°$ for MWPC1 and $\theta_2 = +90.0°$ for MWPC2 in the reactions of $^{30}Si, ^{34,36}S, ^{31}P + ^{238}U$. The angles of $\theta_1 = -72.0°$ and $\theta_2 = +72.0°$ were chosen in the $^{40}Ar + ^{238}U$ reaction. The MWPCs covered the emission angles of $\pm 25.0°$ around the detector center. We also determined the our-of-plane angles of both fragments ϕ_1 and ϕ_2.

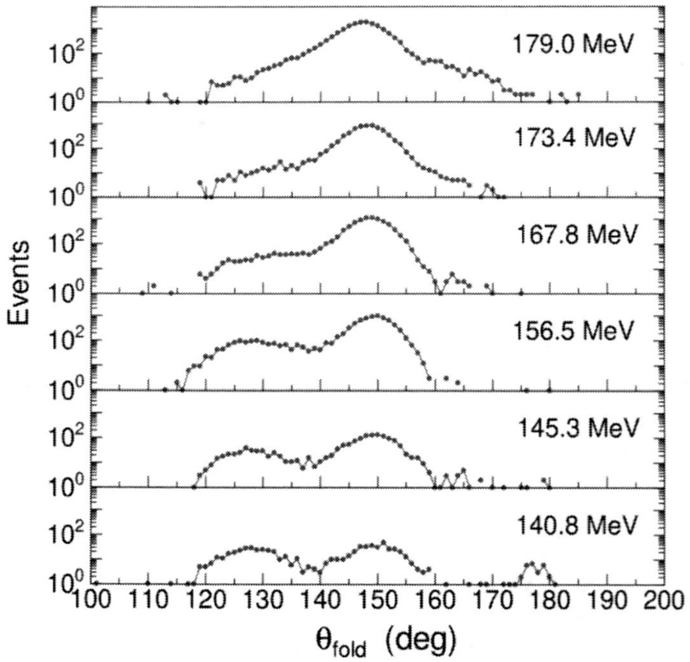

Figure 2. Distribution for folding angle of fission fragments produced in the $^{30}Si + ^{238}U$ reaction. The beam energy in the laboratory frame is indicated.

The time difference, ΔT, between the signals from MWPC1 and MWPC2 was measured. The charges induced in both MWPCs contain information on the energy deposition ΔE_1 and ΔE_2

of particles traversing the detectors and were recorded. In the two dimensional spectrum of ΔT versus $\Delta E_1 + \Delta E_2$, fission events were well separated from elastically scattered projectile-target events.

Fission events occurring after complete transfer of the projectile momentum to the composite system (full momentum transfer (FMT) fission) are separated from those fission events following nuclear transfer. These latter events occur when fissile targets like ^{238}U are irradiated. Figure. 2 shows the folding angle distributions for fission fragments produced in the ^{30}Si + ^{238}U reaction. The FMT fissions are located around $\theta_{\rm fold} = 148°$ at the incident energy of $E_{\rm lab} = 179.3$ MeV and around $\theta_{\rm fold} = 150°$ at the lowest energy 140.8 MeV. Fission following nucleon transfer has a larger folding angle than the FMT fission in the high energy region of $E_{\rm lab}$= 179.0 MeV, as seen in the tail on the spectra which extends over $\theta_{\rm fold} = 160°$. In the low energy region, nucleon transfer fission is observed with a smaller folding angle of $\theta_{\rm fold}$=120°−140° than the FMT fission. These trends are explained by the angular dependence of the transfer reactions, which preferentially occur at grazing angles. The FMT fission fragments were separated from the nuclear transfer fission on the ($\theta_{\rm fold}$, $\phi_{\rm sum}$) plane, and used for the analysis.

2.2. Evaporation residue measurement
Measurements of ER cross sections were carried out in the reactions of ^{30}Si + ^{238}U and ^{34}S + ^{238}U. The experiments were performed at the linear accelerator UNILAC and the velocity filter SHIP at GSI in Darmstadt. The SHIP set-up was essentially the same as described in [2]. The ^{30}Si beam was extracted from a 14 GHz ECR ion source using isotopically enriched material, ^{30}SiO (99.5% isotopic enrichment). A $^{34}SO_2$ gas with a 99 % isotopic enrichment was used to extract the ^{34}S beam. Average beam intensities at the target position were typically 0.7−1.0 $p\mu$A for ^{30}Si and 2.0−2.5 $p\mu$A for ^{34}S. The beam had a pulse structure of 5.0 ms width at 50 Hz repetition frequency. Details of the experiments are described in [7, 8].

For the ^{30}Si + ^{238}U run, the uranium targets were prepared by evaporation of isotopically depleted $^{238}UF_4$ and condensation on a carbon backing. In the experiment of ^{34}S + ^{238}U, the targets were prepared by sputtering the depleted ^{238}U metal on a carbon backing.

The efficiency of SHIP was determined using a Monte Carlo calculation. We obtained a value of 11 % and 15 % for ^{30}Si + ^{238}U and ^{34}S + ^{238}U, respectively.

In the focal plane of the SHIP, ERs and their subsequent α decay and/or spontaneous fission (sf) were detected by a position sensitive 16-strip Si PIPS detector (stop detector) with an active area of 80 mm × 35 mm. The energy resolution for fully stopped α's was typically 25∼26 keV (FWHM). Escaping α particles or a complementary fission fragments were detected by a 'box detector' which covered 85% of the area of the backward hemisphere. Timing detectors were located in front of the silicon detector array to distinguish signals from implanted ERs or background particles from radioactive decays in the stop detector. The correlated events is identified primarily based on a coincidence of the positions of implanted ER, subsequent α decays and/or sf.

3. Experimental results
3.1. In-beam fission measurement
The cross sections for the FMT fissions ($\sigma_{\rm fiss}$) of ^{30}Si + ^{238}U are shown in Fig.3(a) as a function of the center-of-mass energy, $E_{\rm c.m.}$, as well as an excitation energy of the compound nucleus, E^*. The cross section was obtained by fitting the fragment angular distribution in the center-of-mass angle to a function described in [9] and integrating the fitted curve over the solid angle. The fission cross section is almost equal to the capture cross section ($\sigma_{\rm cap}$).

In order to see the effects of nuclear properties on the capture cross sections, we show in Fig. 3(a) the calculation using the coupled-channels code, CCDEGEN [10]. The dotted curve is the result without considering any collective properties or deformation of the target and

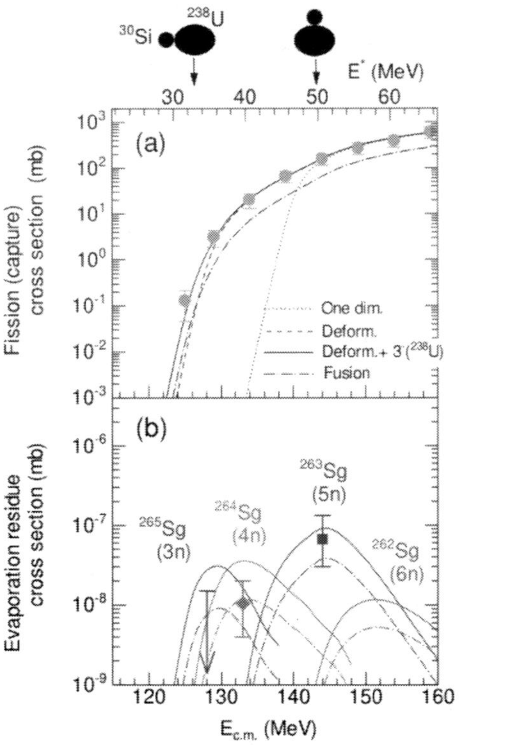

Figure 3. (a) Cross sections for FMT fission and (b) evaporation residue cross sections for $^{30}Si + ^{238}U$. Curves are the model calculations (see text).

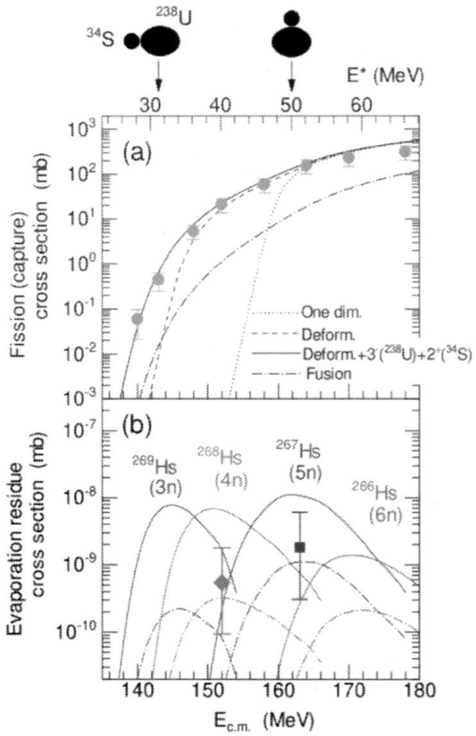

Figure 4. (a) Cross sections for FMT fission and (b) evaporation residue cross sections for $^{34}S + ^{238}U$. Curves are the model calculations (see text).

projectile (one-dimensional barrier penetration model). The dashed curve is the result which takes into account the prolate deformation of ^{238}U with $(\beta_2, \beta_4) = (0.275, 0.050)$ [11]. The calculation reproduces within the error the experimental data down to $E_{\mathrm{c.m.}}$ =129.0 MeV. The solid curve is the result additionally taking into account the coupling to the 3^- state at 0.73 MeV in ^{238}U [12] (β_3 =0.086 [13]), which reproduces the data at the lowest incident energy of 125.0 MeV. Low Coulomb barrier for the polar collisions, as marked in the upper part of Fig. 3, is the main reason for the enhancement of the capture cross section at the sub-barrier energies relative to the one-dimensional barrier penetration model.

The fission cross sections for $^{34}S + ^{238}U$ are shown in Fig. 4(a). The data are compared with the coupled-channels calculations. The assumptions given to draw the dotted and dashed curves are the same as in Fig.3(a). To reproduce the lowest data point at $E_{\mathrm{c.m.}}$ =140.0 MeV, couplings to 3^- state (0.73 MeV) in ^{238}U and to the 2^+ state (2.13 MeV [12]) in ^{34}S (β_2 =0.25 [14]) are taken into account (solid curve).

Figure 5(a) shows the mass distributions for the FMT fissions in the $^{30}Si + ^{238}U$ reaction [15]. To draw the mass distribution, we assumed that the mass distributions do not depend on $\theta_{\mathrm{c.m.}}$. The distributions are Gaussian with mass symmetry in the energy range from $E_{\mathrm{c.m.}} = 139.0$ MeV to 154.0 MeV. The shape of the distributions, however, are different for the sub-barrier energies at $E_{\mathrm{c.m.}} = 134.0$ and 129.0 MeV. They have asymmetric component around $A_{\mathrm{L}}/A_{\mathrm{H}} \approx 90/178$. The difference of the mass distribution at the lowest energy data is characterized also by the standard deviation σ_{m} as indicated in each panel of Fig. 5(a). The value increases from 28.1±1.0 u

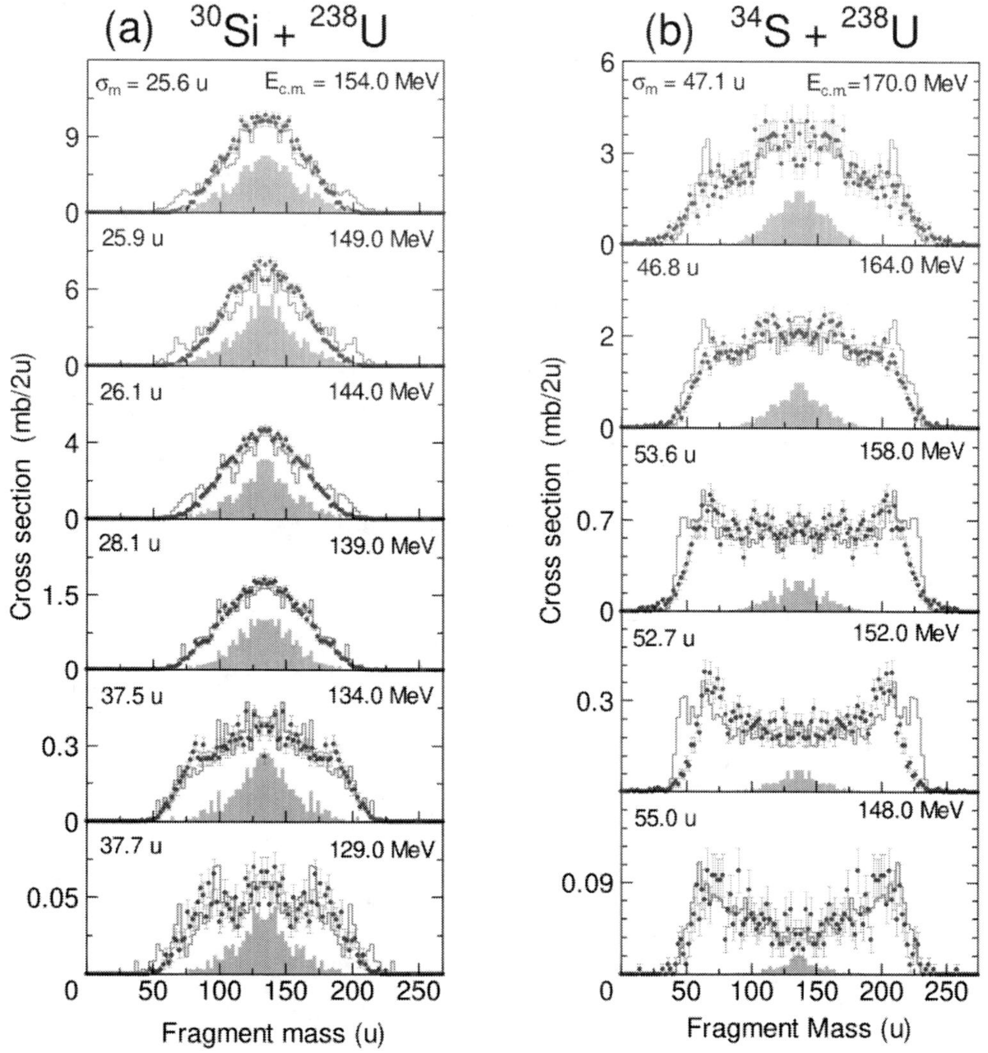

Figure 5. Fragment mass distributions for FMT fissions of the reactions (a) $^{30}\mathrm{Si} + {}^{238}\mathrm{U}$ and (b) $^{34}\mathrm{S} + {}^{238}\mathrm{U}$. Histograms shows a model calculation. The calculated fusion-fission spectrum is shown by the filled area. Reaction energy in the center-of-mass system and the standard deviation σ_{m} of the measured spectrum are shown.

(139.0 MeV) to 37.5±1.0 u (134.0 MeV). Considering the measured ER cross sections (Fig. 3(b)), as discussed in the following, we conclude that the asymmetric fission channel originates from quasifission.

The mass distributions in the $^{34}\mathrm{S} + {}^{238}\mathrm{U}$ reaction are shown in Fig. 5(b) [8]. At the highest energy of 170.0 MeV, the distribution is Gaussian with mass symmetry. The σ_{m} value of the spectrum, however, is far larger than the one for $^{30}\mathrm{Si} + {}^{238}\mathrm{U}$. Toward the low incident energy, the asymmetric fission yield increases sharply around the mass asymmetry $A_{\mathrm{L}} = 68$ and $A_{\mathrm{H}} = 204$. The mass asymmetry corresponds to the fragments near the double-closed shell nuclei, $^{78}\mathrm{Ni}$ and $^{132}\mathrm{Sn}$ (see Fig. 1). A similar observation was made in the study of the reaction $^{36}\mathrm{S} + {}^{238}\mathrm{U} \to {}^{274}\mathrm{Hs}^{*}$ [6]. The phenomenon was interpreted by the effects of nuclear orientation on fusion

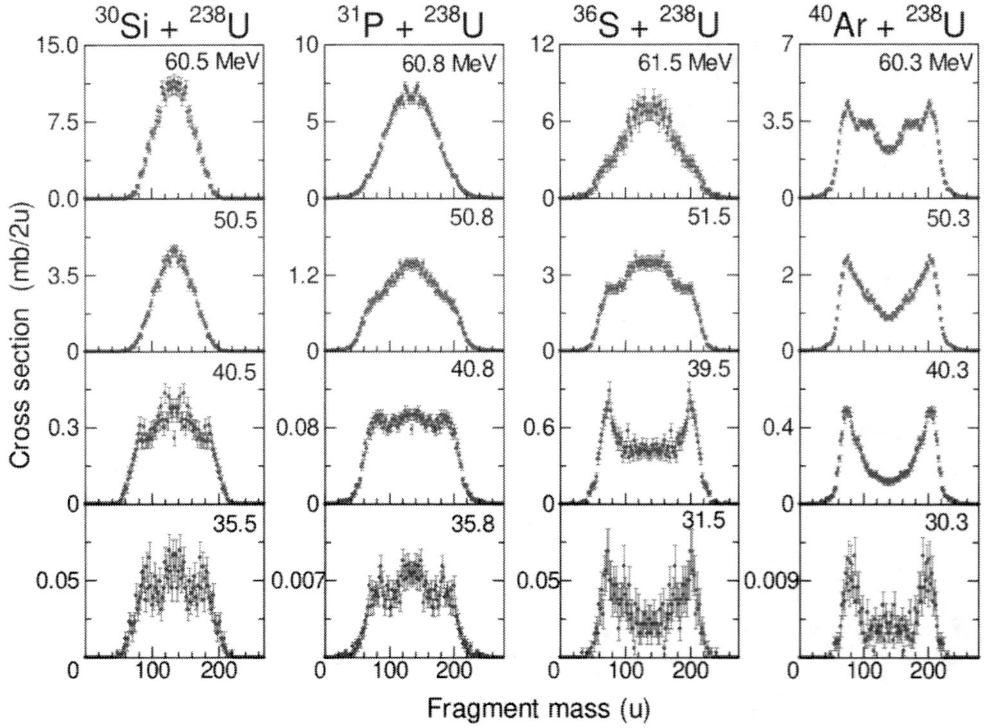

Figure 6. Fission fragment mass distributions in the reactions of ^{30}Si,^{31}P,^{36}S,^{40}Ar + ^{238}U. The excitation energy E^* of a compound nucleus is shown in each spectrum.

and/or quasifission. At the sub-barrier energy, projectiles collide on the polar sides of the ^{238}U nucleus. In this case the reaction starts from a distant contact point with a large charge-center distance, which results in a larger quasifission probability than the reactions starting from the equatorial collisions (see Fig. 1).

Figure 6 shows the mass distributions in the reactions using different projectiles of ^{30}Si,^{31}P,^{36}S and ^{40}Ar. The excitation energy of a compound nucleus is shown in each section of the figure. For a fixed excitation energy, a yield for an asymmetric fission channel increases with projectile charge, showing that quasifission probability increases due to the larger Coulomb repulsive force in the reaction process. The enhanced quasifission yield toward the low incident energy are observed for all the reactions owing to the orientation effects.

In the reactions of ^{34}S+^{238}U (Fig.5(b)), ^{36}S+^{238}U (Fig.6) and ^{40}Ar+^{238}U (Fig.6), the quasifission mass asymmetries are nearly the same, and centered around $A_{\rm L}/A_{\rm H} \approx 68/204$, 74/200 and 76/202, respectively. According to the mass asymmetry parameter $\alpha = (A_{\rm H} - A_{\rm L})/(A_{\rm H} + A_{\rm L})$ defined by the light (L) and heavy (H) fragment masses, they corresponds to 0.50, 0.46 and 0.45. These mass asymmetry are closely correlated to the fission channel formed by the shells near the double-closed shell nuclei, ^{78}Ni and ^{208}Pb (see Fig.1). This channel is the same as observed in reactions using heavier projectiles bombarded to actinide target nuclei [16], where the reactions of ^{48}Ca + ^{238}U, ^{244}Pu, ^{248}Cm produce asymmetric fission centered at $A_{\rm H} \approx 210$ ($\alpha = 0.42 \sim 0.46$).

The observed mass asymmetry in the quasifission for ^{30}Si + ^{238}U is $\alpha = 0.33$ ($A_{\rm L}/A_{\rm H} \approx 90/178$). The value is significantly smaller than those obtained for ^{34}S ,^{36}S ,^{40}Ar +^{238}U. The asymmetric fission channel in the ^{31}P + ^{238}U reaction, $\alpha = 0.40$ ($A_{\rm L}/A_{\rm H} \approx 81/188$), also does not fit the shells of ^{78}Ni/^{208}Pb. The potential energy landscape of ^{268}Sg produced by ^{30}Si +^{238}U

Journal of Physics: Conference Series **282** (2011) 012011 doi:10.1088/1742-6596/282/1/012011

has almost the same structure as ^{274}Hs produced by ^{36}S + ^{238}U (see Fig. 4 in [6]). The observation of the different mass asymmetry indicates the difference in the evolution of nuclear shape in quasifission.

3.2. Evaporation residue measurement

The measurement of ER cross sections in the fusion of ^{30}Si + ^{238}U was carried out from April 8 to May 1, 2006. We used three different beam energies of $E_{\rm c.m.} = 144.0$, 133.0 and 128.0 MeV, which correspond to the anticipated maximum cross sections of $5n$-, $4n$-, and $3n$-evaporation channels. Details of the experimental results are described in [7].

At 144.0 MeV, a beam dose of 1.8×10^{18} was accumulated. We observed three α-decay chains starting from ^{263}Sg($5n$). We obtained the cross section of 67^{+67}_{-37} pb. At the sub-barrier energy of $E_{\rm c.m.} = 133.0$ MeV, we observed four spontaneous fission events with a beam dose of 4.0×10^{18}. The lifetime of three sf events is similar and relatively short. The half-life was determined to be 120^{+126}_{-44} ms from three sf events. The nucleus was assigned to the spontaneously fissioning isotope ^{264}Sg ($4n$). The cross section for the three ^{264}Sg sf events was 10^{+10}_{-6} pb. At the lowest beam energy of 145.5 MeV, no decay events were measured with a beam dose of 1.7×10^{18}. An upper cross section limit was determined to be 15 pb at 68 % confidence level (one event would have had a cross section of 8.2 pb). Measured cross sections and/or the cross section limit are shown in Fig. 3(b).

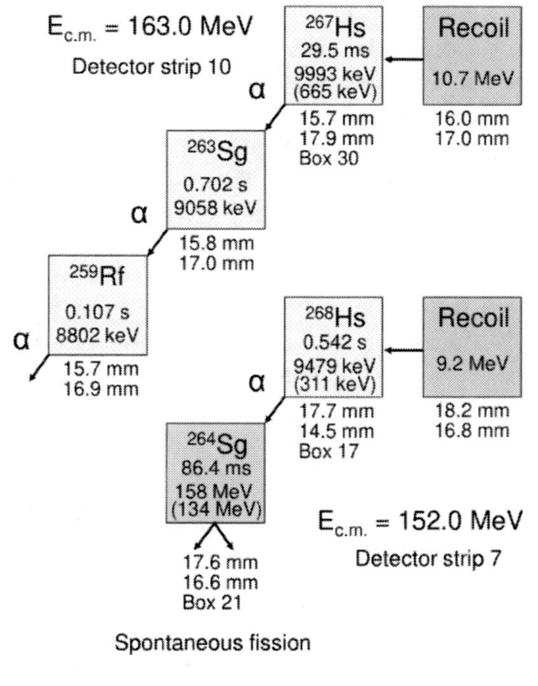

Figure 7. Decay chains observed in the ^{34}S + ^{238}U reaction. For the details, see [8]

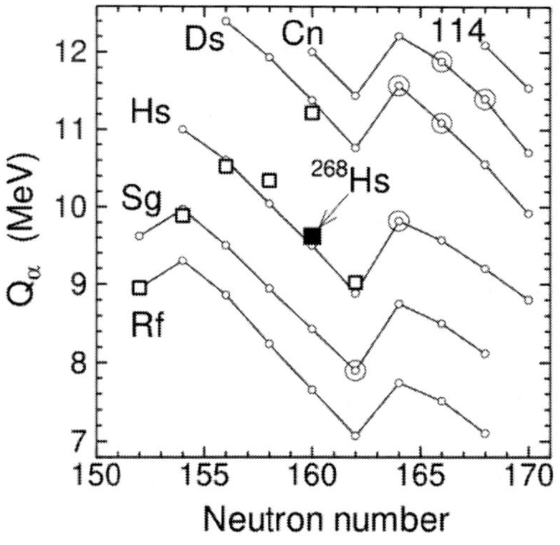

Figure 8. Alpha-decay Q-values of even-even nuclei in the region of N=162. Present new isotope ^{268}Hs is shown. Lines represent the theory [18]. Double circles represent nuclei and their daughter products which could be studied in the reactions ^{34,36}S + ^{248}Cm,^{244}Pu and $4n$-evaporation channels.

Production of hassium isotopes in the fusion of ^{34}S + ^{238}U was carried out from January 1 to February 8, 2009. Two different beam energies of $E_{\rm c.m.} = 163.0$ and 152.0 MeV were chosen, corresponding to the maximum $5n$- and $4n$-evaporation cross sections. Details of the experimental results are described in [8].

Journal of Physics: Conference Series **282** (2011) 012011 doi:10.1088/1742-6596/282/1/012011

At $E_{c.m.} = 163.0$ MeV, a beam dose of 4.8×10^{18} was accumulated. We observed one decay chain, which we assign to the production of ^{267}Hs as shown in Fig. 7. The cross section for ^{267}Hs was determined to be $1.8^{+4.2}_{-1.5}$ pb as plotted in Fig. 3(b). The value agrees with 2.5 pb measured in [17]. At the sub-barrier energy of $E_{c.m.} = 152.0$ MeV, a beam dose of 1.2×10^{19} was accumulated. In this irradiation, one decay chain was identified, as shown in Fig.7. We assigned the decay chain starting from the new isotope ^{268}Hs. The 9749-keV α-decay was followed by the sf of ^{264}Sg which was produced directly in the $4n$-evaporation in fusion of $^{30}\mathrm{Si} + {}^{238}\mathrm{U}$ [7]. The obtained cross section $0.54^{+1.3}_{-0.45}$ pb is shown in Fig. 4(b).

The α-decay Q value, Q_α, of ^{268}Hs is shown in Fig. 8 together with those for the other isotopes. The theoretical data [18] are also shown, which predicts the decreasing Q_α value toward the deformed neutron-shell at $N = 162$. The present data for ^{268}Hs fits perfectly into the systematics as well as theoretical value.

4. Discussions

For a quantitative analysis of the mass distributions, we performed a model calculation combining the coupled channels method and a dynamical description of the reaction based on the three-dimensional Langevin equation [19]. The dynamical calculation based on the Monte Carlo method was used for describing the reaction paths in the potential energy landscape. The two-center shell model was used to calculated the potential energy of a nucleus whose shape is defined by z (charge center distance), δ (deformation) and α (mass asymmetry). The deformation of the reaction partners and their orientation in the reaction plane was considered. The coupled channels method was first used to compute the penetration probability of the Coulomb barrier for a fixed orientation angle. The dynamical calculation was then started from the shape at contact configuration for each orientation. Fusion is defined as the case when the trajectory enters inside the local energy minimum corresponding to the compound nucleus, whereas quasifission is defined as disintegration without reaching the minimum.

The calculated distributions of FMT fragments for $^{30}\mathrm{Si} + {}^{238}\mathrm{U}$ are represented by the histograms in Fig. 5(a). The model reproduces the general shape of the distribution as well as the feature of the appearance of the asymmetric fission channel and increase of the standard deviation σ_m at the sub-barrier energies. The calculated fusion-fission events in this model are shown by the filled areas in Fig. 5(a). It is Gaussian with mass symmetry, and the standard deviation of the spectrum σ_m is nearly constant with 21~25 u in the entire energy rage. The value is significantly smaller than 37 ~ 38 u for the measured distributions at the sub-barrier energies of 129.0 and 134.0 MeV. Furthermore the calculated fusion-fission spectrum does not show any asymmetric fission channels. The results support the observed asymmetric fission to be quasifission. In the calculation, we can determine the fusion probability P_{fus} as the ratio of fusion-fission events to the FMT fission events. By multiplying the P_{fus} to capture cross section (σ_{cap}), a fusion cross section σ_{fus} is obtained. The results for $^{30}\mathrm{Si} + {}^{238}\mathrm{U}$ are shown by the dash-dotted curve in Fig. 3(a).

The same calculation was made in the $^{34}\mathrm{S} + {}^{238}\mathrm{U}$ reaction as shown in Fig. 5(b). The results reproduce the measured distribution, especially transition from symmetric to asymmetric mass distributions with different incident energy is well reproduced. The σ_m value in the calculated fusion-fission spectrum is 16–20 u, which is nearly the same as the $^{30}\mathrm{Si} + {}^{238}\mathrm{U}$ reaction. The value is about factor two less than those for the measured distribution. A remarkable difference between $^{30}\mathrm{Si} + {}^{238}\mathrm{U}$ and $^{34}\mathrm{S} + {}^{238}\mathrm{U}$ reactions is the fusion-fission yield among the FMT fissions. Even in the symmetric mass region, the yield drops significantly in the case of ^{34}S projectile. The fusion cross sections σ_{fus} for $^{34}\mathrm{S} + {}^{238}\mathrm{U}$ are shown in Fig. 4(a) by the dash-dotted curve.

To see if the above model is appropriate to estimate the fusion probability, P_{fus}, a statistical model code HIVAP [20] was used to calculate the ER cross sections by inputting the fusion cross sections for $^{30}\mathrm{Si} + {}^{238}\mathrm{U}$ (dash-dotted curve in Fig. 3(a)), and the results are compared with the experimental data. The calculation reproduces the cross sections for ^{263,264}Sg within errors as well as the cross section limit for ^{265}Sg. Similarly, the model can account for the cross sections for ^{267,268}Hs produced in the fusion of ^{34}S + ^{238}U as shown in Fig. 4(b). If we assume all the fragments were arising from the compound nucleus fission, the ER cross sections would have values shown by the solid curves in Fig. 3(b) or Fig. 4(b). Apparently, this assumption contradicts the cross sections of ^{264,265}Sg and ^{267,268}Hs.

It is shown in the calculated ER cross sections for ^{34}S + ^{238}U in Fig. 4(b) that the probability for quasifission after the capture process becomes larger toward the low bombarding energy, demonstrating the orientation effects. It is consistent with the ER cross sections for ^{268}Hs and ^{267}Hs.

In spite of the reduced fusion probability to produce SHN using heavier projectiles, ^{34}S, our cross section calculations reveal a considerable yield for producing relatively neutron rich nuclei at energies below the Bass barrier, when targets of deformed actinide nuclei are used. Figure 4(b) shows the cross sections for the $3n$ evaporation channel, ^{269}Hs. The cross section is expected to be comparable to the one for the $4n$ evaporation residue, ^{268}Hs. By applying fusion reaction at the energy below the Bass barrier, the new isotopes $^{274-277}$Ds (N=164–167) could be produced in the reactions ^{34,36}S + ^{244}Pu in $3n$–$4n$ evaporation channels or $^{278-281}$Cn (N=166–169) in ^{34,36}S + ^{248}Cm reactions also in $3n$–$4n$ channels. Alpha decay from these nuclei would populate isotopes of hassium and seaborgium, from which the isotopes ^{272,273}Hs (N=164 and 165) and ^{268,269}Sg (N=162 and 163) are not yet known. Similarly, isotopes of odd elements could be produced in reactions with ^{34,36}S beams and a ^{243}Am target. The ERs and their daughter nuclei includes new isotopes of ^{273,275,276}Rg (N=162, 164 and 165) and ^{269,271,272}Mt (N=160, 162 and 163).

The knowledge of synthesis and decay properties of the nuclei discussed above is especially important. First, they build a bridge between the well established, but relatively neutron deficient nuclei produced in cold fusion reactions and those with higher neutron number produced in hot fusion, which are not connected to known nuclei, because their α-decay chains are terminated by sf. Second, the α-decay Q-values of these new nuclei together with those for the known nuclei provide information on the increased binding energy at the deformed shells with Z =108 and N =162. Due to the large gaps in the single particle energies, increased Q_α values are expected beyond these neutron and proton numbers. In Fig. 8 the predicted Q_α values not observed so far are marked by double circles.

5. Conclusions

The mass distributions for FMT fission fragments in the reactions of ^{30}Si,^{34,36}S,^{31}P,^{40}Ar + ^{238}U were measured at bombarding energies around the Coulomb barrier. The probability of asymmetric fission increases at the sub-barrier energy. The phenomenon is interpreted as the enhanced quasifission probability, which represents orientation effects on fusion and/or quasifission. The ER cross sections were measured in the reactions of ^{30}Si + ^{238}U and ^{34}S + ^{238}U to obtain information on fusion probability $P_{\rm fus}$. The $P_{\rm fus}$ decreased significantly for the latter reaction, thus showing the enhanced quasifission probability. The conclusion was consistent with the theoretical model calculation which is based on the Langevin equation and takes into account the orientation of the deformed nucleus ^{238}U, in which fusion-fission and quasifission are separately determined. In spite of the significantly low fusion probability at the sub-barrier energy, there are fusion probabilities large enough to produce $3n$- and $4n$-evaporation residues. In the combination between ^{34,36}S projectile and several actinide target nuclei, totally 18 new isotopes could be produced as ER or α-decay descendants. They are located in the region of elements from seaborgium to copernicium with neutron numbers ≥162 so that the gap between the known nuclei produced in cold and hot fusion reactions could be filled.

The investigations were accomplished with the collaborators of H. Ikezoe, S. Mitsuoka, Y. Nagame, I. Nishinaka, K. Tsukada, K. Tsuruta (JAEA), S. Hofmann, D. Ackermann, V.F. Comas, Ch.E. Düllmann, S. Heinz, J.A. Heredia, F.P. Heßberger, J. Khuyagbaatar, B. Kindler, I. Kojouharov, P. Kuusiniemi, B. Lommel, R. Mann, M. Mazzocco, M. Schädel, H.J. Schött, B. Sulignano (GSI), Y. Aritomo (JAEA,FLNR), K. Hagino, K. Hirose, T. Ohtsuki (Tohoku University), A.G. Popeko, A. Svirikhin, A.V. Yeremin(FLNR), S. Antalic, S. Saro (Comenius University), A. Gorshkov, R. Graeger, A. Türler, A. Yakushev (TU München), Y. Watanabe (KEK), Z. Gan (IMP). We are grateful to W. Hartmann, J. Steiner, H.G. Burkhard, and J. Maurer of GSI.

We would like to thank the UNILAC staff and the crew of the JAEA tandem accelerator facility for preparation and operation of the beams. This work was supported by a Grant-in-Aid for Scientific Research of the Japan Society for the Promotion of Science.

References

[1] Yu.Ts. Oganessian, J. Phys. G **34**, R165 (2007).
[2] S. Hofmann and G. Münzenberg, Rev. Mod. Phys. **72**, 733 (2000).
[3] K. Morita *et al.*, J. Phys. Soc. Jpn. **73**, 1738 (2004).
[4] V. Zagrebaev and W. Greiner, J. Phys. G, **31** 825 (2005).
[5] S. Hofmann *et al.*, Eur. Phys. J. A **32**, 251 (2007).
[6] K. Nishio *et al.*, Phys. Rev. C **77**, 064607 (2008).
[7] K. Nishio *et al.*, Eur. Phys. J. A **29**, 281 (2006).
[8] K. Nishio *et al*,. Phys. Rev. C, **82** 024611 (2010).
[9] R. Vandenbosch and J.R. Huizenga, *Nuclear Fisson* (Academic Press, New York, 1973).
[10] modified version of the CCFULL code, K. Hagino *et al.*, Computer Phys. Comm. **123**, 143 (1999).
[11] K. Nishio *et al.*, Phys. Rev. Lett. **93**, 162701 (2004).
[12] R.B. Firestone *et al*,. *Table of Isotopes Eighth Edition* (JohnWiley & Sons, Inc.)
[13] R.H. Spear *et al.*, At. Data Nucl. Data Tables, **42**, 55 (1989).
[14] S. Raman *et al.*, At. Data Nucl. Data Tables, **36**, 1 (1987).
[15] K. Nishio *et al.*, K. Nishio *et al*,. Phys. Rev. C, **82**, 044604 (2010).
[16] M.G. Itkis *et al.*, Nucl. Phys. **A787**, 150c (2007).
[17] Yu.A. Lazarev *et al.*, Phys. Rev. Lett. **75**, 1903 (1995).
[18] I. Muntian *et al*,. Acta. Phys. Pol. B **34** 2073 (2003).
[19] Calculation is based on a model in Y. Aritomo, Phys. Rev. C, **80**, 064604 (2009), but effects of nuclear orientaion was taken into account (Y. Aritomo *et al.*, to be submitted).
[20] W. Reisdorf and M. Schädel, Z. Phys. A **343**, 47 (1992).

Statistics vs. dynamics: hints from systems of intermediate fissility

E Vardaci[1,2], A. Di Nitto[1,2], P. Nadtochy[2], A. Brondi[1,2], G. La Rana[1,2], R. Moro[1,2], A. Ordine[2], A. Boiano[2], M. Cinausero[3], G. Prete[3], V. Rizzi[3], N. Gelli[4], F. Lucarelli[4], G.N. Knyazheva[5], E.M. Kozulin[5], T.A. Loktev[5], S. Smirnov[5]

[1] Dipartimento di Scienze Fisiche, Universitá di Napoli "Federico II", 80126 Napoli, Italy
[2] Istituto Nazionale di Fisica Nucleare, 80126 Napoli, Italy
[3] Laboratori Nazionali di Legnaro, Istituto Nazionale di Fisica Nucleare, Legnaro (Padova), Italy
[4] Dipartimento di Fisica and Istituto Nazionale di Fisica Nucleare, Firenze, Italy
[5] Flerov Laboratory of Nuclear Reactions, JINR,141980, Dubna

E-mail: Emanuele.Vardaci@na.infn.it

Abstract. Systems of intermediate fissility are characterized by an evaporation residues cross section comparable or larger than the fission cross section, and by a relatively higher probability for charged particle emission in the pre-scission channel. In a theoretical framework in which time scale estimates of the fission process rely on statistical model calculations, the analysis of particle emission in the evaporation residues channel is the source of additional constraints on statistical and dynamical models. This contribution will focus on our statistical and dynamical analysis of a more complete set of data from the system ^{32}S + ^{100}Mo at E_{Lab} = 200 MeV. Statistical model fails in reproducing the whole set of data and no convincing estimate is possible of the fission time scale. In particular, while pre-scission multiplicities can be reproduced without delay, the model strongly overestimates proton and alpha particle multiplicities in the evaporation residues channel irrespective of the statistical model input parameters and prescriptions used for the level density and the transmission coefficients. The analysis of the same set of data with a dynamical model produces a very good agreement with the full set of data and indicates that one-body dissipation plays a dominant role in the fission process, implying a fission delay of 23-25x10^{-21}s.

1. Introduction

A large variety of experimental studies of induced fission in heavy ion reactions [1] have shown that pre-scission multiplicities of light particles (neutrons, protons and alpha particles) increase monotonically with the bombarding energy, in contrast with the calculations of the standard statistical model (SM) which start from the complete thermalization of the compound system. Since the fission process is considered to be affected by nuclear dissipation [1, 2], this result is considered as the evidence that fission is a slow process with respect to the lifetime for the evaporation of light particles. With increasing excitation energy, the particle decay lifetime decreases and becomes smaller than the time necessary for the build-up of the collective motion of the nuclear matter toward the saddle point.

Consequently, fission does not compete as effectively as predicted by the SM in the early stage of the decay, and light particles, and possibly GDR γ-ray, emissions can occur with higher probability.
The overall cause of this transient effect is believed to be associated with the nuclear matter viscosity which slows down the collective flow of mass from the equilibrium to the scission point and does not allow the fission decay lifetime to be downscaled with the excitation energy as in the case of light particles evaporation. This is equivalent to consider that fission is delayed with respect to the picture of the SM in which the fission width has its full Bohr-Wheeler value already at the beginning of the decay. An energy domain has further been identified [3] above which the SM predictions begin to deviate from the data.
Several variants of the SM have been proposed in the literature to take explicitly into account time scales as well as nuclear viscosity. In the simplest fashion, known as the "neutron clock" [1], the SM is modified so to include another free parameter, the fission delay τ_d: at the beginning of the evaporative cascade the fission decay width is kept to zero for a time τ_d. After τ_d the fission width is set to the full Bohr-Wheeler value. The time scale is defined by the light particle life time. This means that at the beginning of the decay cascade fission does not compete with light particle evaporation until a time τ_d has passed. Estimates of τ_d are obtained by the fit of the experimental multiplicities to the ones predicted by the SM which includes this new parameter.
Other refinements of the neutron clock approach have been proposed to distinguish between different time steps during the fission process [4]. The common approach is to split the path from the equilibrium-to-scission into two regions, the pre- and the post-saddle. The total fission delay is defined as the sum of τ_d and τ_{ssc}, where τ_{ssc} is the time necessary to travel the path from saddle to scission. The relevant observables are computed using τ_d and τ_{ssc} as free parameters, along with the other input parameters relative to the specific ingredients of the model, and fit to the experimental data. In spite of the extensive work, estimates of the fission time scales are however quite controversial. The reported values range from 0 [5] to 500 x 10^{-21} s [6], depending on the system and on the experimental probe. Furthermore, such estimates are weakened by the fact that different sets of input parameters can result in equally good fits within the same model [7, 8].
It must be pointed out that only neutron multiplicities have been measured in most of the studies and mostly for heavy systems (A > 200), and the lack of a sufficient number of constraints to the models could, in several cases, be the source of discrepancies. In order to withdraw a more consistent picture of nuclear dissipation it is crucial, in our opinion, to take into account simultaneously a larger number of observables and probes which can be expected to be sensitive to the nuclear dissipation and to try to reproduce the variety of observables with a unique set of input parameters.

2. Dissipation in systems of intermediate fissility

Systems of intermediate fissility (χ = 0.5-0.6) are very little studied although they offer several advantages. They are characterized by an evaporation residue (ER) cross section comparable or larger than the fusion-fission (FF) cross section, and by a shorter path in the deformation space from the saddle-to-scission point [9]. Consequently: 1) the input parameters of the models can be further constrained by the energy spectra and multiplicities of the light particles in the ER channel; 2) the effect of the fission delay over the fission and ER cross sections is much more pronounced with respect to heavier systems because the emission of a charged particle in the pre-saddle region strongly enhances the probability of producing an evaporation residue as consequence of both a reduction of the fissility and the large value of the angular momentum necessary to ignite fission. The use of the light particle multiplicities in the ER channel as further constraint grounds, however, also on the reliability of the statistical model to reproduce such multiplicities when all the necessary experimental constraints are given and this is not demonstrated yet.
We expect that the measurements of neutron and charged particle multiplicities and energy spectra in the two channels as well as the measurements of the cross sections of the channels themselves will

allow more severe constraints onto the models. This should provide more reliable values of fission delay and of the friction parameter, and contribute to a better comprehension of the origin of nuclear viscosity. In this framework, the 8πLP collaboration has started a research program at the Laboratori Nazionali di Legnaro (Padova, Italy) aimed at studying the fission dynamics in systems of intermediate fissility.

In this presentation we will report on our analysis of the reaction 200MeV ^{32}S + ^{100}Mo leading to the composite system ^{132}Ce at E_x = 122 MeV and fusion angular momentum $L_{fus} = 72\hbar$, derived from the measured fusion cross section in the sharp cut-off approximation. We will show the inability of the SM to provide an estimate of the fission time scale when the evaporation residue channel is included as a further constraint in the procedure used to estimate the fission delay time. Afterwards, our study with an advanced realistic dynamical approach based on a three-dimensional (3D) Langevin approach [10] will be discussed. It will prove to be a method that better reproduces the overall multitude of data.

3. Experimental procedure and data analysis

The experiment was performed at the XTU Tandem - ALPI Superconducting LINAC accelerator complex of the Laboratori Nazionali di Legnaro. A 200 MeV pulsed beam of ^{32}S of about 1 pnA intensity was used to bombard a self supporting ^{100}Mo target 300 μg/cm thick. A beam burst with frequency of about 1.25 MHz and duration of about 2 ns was used.

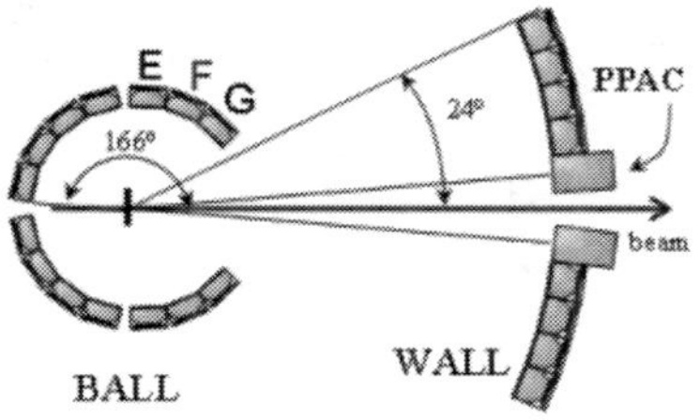

Figure 1. Schematic layout of the 8πLp apparatus

We used the BALL and the WALL sections of the 8πLP apparatus [11], shown schematically in figure 1, to detect light charged particles (LCP). The WALL consists of 116 telescopes placed at 60 cm from the target and covers an angular range from 2° to 24°. The BALL consists of 7 rings placed coaxially around the beam axis each with 18 telescopes which amount to a total number of 126 telescopes covering an angular range from 34° to 166°. The telescopes of the BALL are made out of a 300μm Si detector mounted in the flipped configuration (particle entering from the ohmic side) backed by a 3 mm CsI(Tl) crystal. The rings are labelled from A to G going from backward to forward angles. Each ring covers an angular opening of about 17°. The experimental method consists in measuring LCP in coincidence with both fission fragments and with the ER.

The fission fragments were detected in the telescopes of the ring F and G of the BALL. The Pulse Shape Discrimination technique allows the separation between heavy fragments and LCP stopping in the same detector. Evaporation residues were detected through 4 Parallel Plate Avalanche Counter modules (figure 1). Each one covers a forward angle between 2.5° and 7.5°, and subtends a solid angle of about 0.3 msr. A module consists of two coaxial PPACs mounted and operating in the same gas volume at a distance of 15 cm from each other. By adjusting the gas pressure, it is possible to stop the ER between the two PPACs, and let the lighter ions to impinge on the second PPAC. Consequently, ERs are sorted out from the first PPAC signals using the signals from the second PPAC as a veto.

In a separate experiment the ER cross section was measured by means of the electrostatic deflector of LNL and the FF cross section was measured with the double-arm time-of-flight spectrometer CORSET [12] at LNL as well.

For the system ^{32}S + ^{100}Mo at E_{lab} = 200 MeV we have measured most of the relevant quantities in the ER and FF channels: proton and alpha particle energy spectra and multiplicities, ER and FF cross sections as well as mass and total kinetic energy distributions of fission fragments. To extract the pre- and post-scission integrated multiplicities, particle energy spectra have been analyzed considering three evaporative sources: the composite nucleus prior to scission (CE) and the two fully accelerated fission fragments (F1 and F2). We used a well established procedure which employs the Monte Carlo statistical code GANES [13,14].

The full set of data is shown in Table 1 along with the results of the SM calculations performed with the code PACE2_N97 [15] and with a 3D Langevin dynamical code [10, 16] which implements the one-body and two-body dissipation models. The dynamical model is coupled with the statistical model Lilita_N97 to simulate the emission of LCP from ER and from the composite system before scission (pre-scission emission). The symbols are as follows: the multiplicities of the protons and alpha particles in the ER channel are, respectively, M_p and M_α (ER); M_p and M_α (PRE) are the prescission multiplicities. σ_{ER} and σ_{FF} are, respectively, the ER cross section and the FF cross section; <Mass> and σ_{Mass} are the mean and the standard deviation of the measured mass distribution, respectively; <TKE> and σ_{TKE} are the mean and the standard deviation of the measured TKE distribution, respectively.

Table 1. Comparison of measured observables (Exp.) and the best predictions of the code. One-body is obtained with full one-body dissipation, two-body with viscosity parameter μ = 0.46 TP.

	Mp (ER)	$M\alpha$ (ER)	Mp (PRE)	$M\alpha$ (PRE)	σ_{ER} (mb)	σ_{FF} (mb)	<Mass> (u)	σ_{Mass} (u)	<TKE> (MeV)	σ_{TKE} (MeV)
Exp.	0.9 (0.14)	0.56 (0.09)	0.055 (0.007)	0.038 (0.005)	828 (50)	130 (13)	66	15.4	90.9	11.4
SM	1.44	1.64	0.058	0.034	813	143				
OneBody	1.198	0.556	0.064	0.0399	786	150	65	14.6	82.1	7.2
TwoBody	1.18	0.57	0.059	0.031	758	178	65.5	14.9	79.4	7.3

4. Statistical model analysis

We analyzed the measured quantities in Table 1 with the SM implemented in the code PACE2. The original code has indeed been extended with the inclusion of new options for the level density and the transmission coefficients as well as the possibility to account for a fission delay according to the prescription widely used in the literature [1]. The fission delay parameter τ_d is used in such a way that the fission probability is zero up to the time τ_d and has the full Bohr-Wheeler value subsequently.

If we limit our analysis to the FF channel only, namely, if we try to reproduce only the multiplicities in the FF channel as usually done [1], the data shown in Table 1 can be reasonably well reproduced assuming $a_v = A/9$, $a_f/a_v = 1.04$, liquid drop model (LDM) yrast line and optical model (OM) transmission coefficients [17-19], without any delay. The parameter a_v is the Fermi gas level density parameter for particle evaporation and a_f is the level density parameter for fission.

From this result one could conclude that no transient effect takes place in this decay, in contrast with the systematics [3], although a different combination of input parameters does not exclude the presence of a relatively small fission delay. On the other hand, with the same parameters, the model strongly overestimates the ER particle multiplicities even though it reproduces the ER cross section. This is an evident contradiction: if the model is not able to reproduce the LCP multiplicities in the ER channel, once the ER cross section is well accounted for, the same model can not be supposed as a reliable tool to estimate the fission time scale through the pre-scission light particle multiplicities.

Table 2. Summary of the SM parameters used in the calculations for the level density (**a**$_v$),yrast line (YR) and the transmission coefficients (TC);

Prescription	a_v	YR	TC
a)	A/6	RS	OM
b)	A/12	LDM	OM
c)	A/6	RS	FS
d)	A/6	LDM	OM

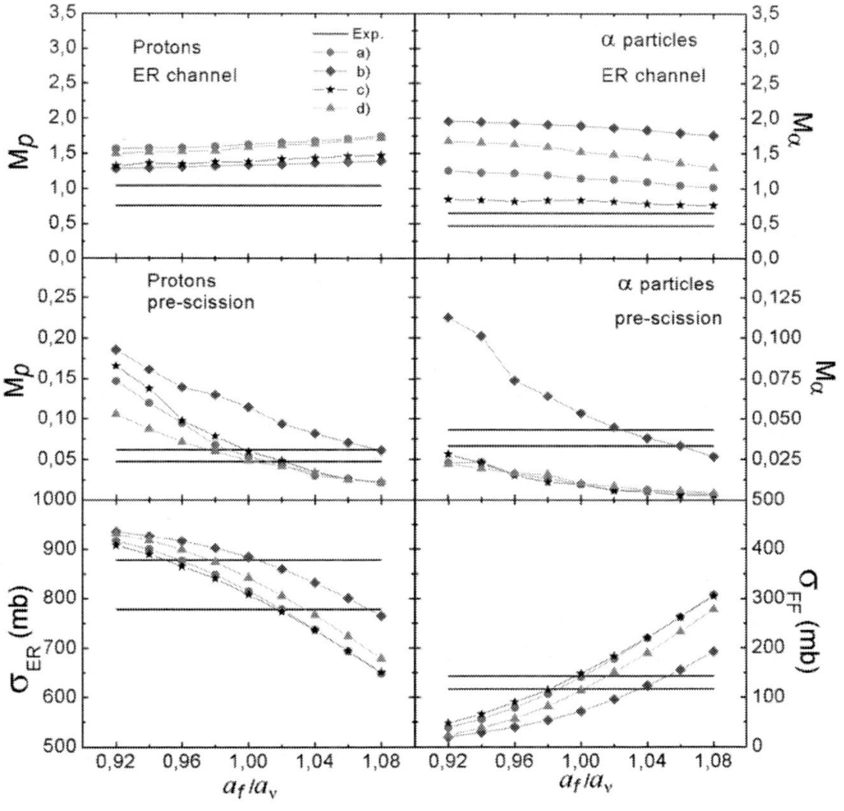

Figure 3. Measured evaporative (ER) and pre-scission (PRE) charged particle multiplicities together with the FF and ER cross sections (full lines indicating lower and upper limits of the uncertainty), compared to the predictions of the statistical model (symbols connected by lines to guide the eye) using prescriptions: a), b), c) and d) reported in Table 2. See text for details.

In order to explore the possibility to reproduce the data in both channels with a unique set of input parameters we performed an extensive analysis with different prescriptions of the level density parameter and transmission coefficients. Calculations have been carried out adopting three different and well known prescriptions for the yrast line: 1) Gilbert Cameron [20], 2) LDM and 3) sharp rigid sphere (RS) with radius parameter $r_0 = 1.2$ fm. Different prescriptions have also been used for the level density parameter a_v: 1) a constant value ranging from A/6 to A/12, 2) inclusion of shell effects [21] with a damping term [22] as a function of the excitation energy and 3) a temperature dependent prescription [23]. Transmission coefficients derived from: 1) optical model and 2) fusion systematics (FS) [24] have been used. Different values of fission delay and a_f/a_v have been adopted to modulate particle-fission competition. Calculations have been constrained by the sum of the measured evaporation residue and fission cross section $\sigma_{fus} = \sigma_{ER} + \sigma_{FF} = 958\, mb$.

In figure 3 we show the multiplicities for protons and alpha particles, in the ER and FF channels, as well as the measured channel cross sections, compared to the calculated values, as a function of the ratio a_f/a_v . We report in the figure the results corresponding to the prescriptions labeled as a), b), c) and d), whose peculiarities are reported in Table 2. The prescriptions a), b), c) and d) presented here have been chosen among the many combinations for which calculations have been performed as they allow to explore the full range of variability of the calculated values of the observables under examination. No fission delay has been included in the calculations. From figure 3 we infer that the SM strongly overestimates proton and alpha particle multiplicities in the ER channel for this system, irrespective of the input parameters and the prescriptions used for the level density and transmission coefficients. Same result is confirmed by the calculations performed with the well known codes Lilita_N97 [25] and Gemini [26]. Furthermore, the inclusion of a time delay to further suppress the fission does not change the overall pattern of the calculated data with respect to the experimental data. At the same time, the influence of nuclear deformation would further enhance the statistical model particle multiplicities predictions, resulting in a larger overestimation. On the other hand, the comparison of the measured proton and alpha particle energy spectra with the SM shows no evidence of nuclear deformation.

It should be pointed out that the overestimate in the ER channel found for the present compound system was found also in other systems of similar mass. We have in fact compared experimental data taken from the literature with the predictions of our code PACE2_N97. Indeed in the literature there are only few systems for which the ER channel LCP multiplicities have been measured. From calculations performed by us, once again we find that the SM overestimates protons and alpha particle multiplicities in the ER channel which makes us to suspect that the SM is behaving surprisingly at variance with what expected.

5. Dynamical model analysis

These contradictory results outline the necessity of considering dynamical models. Recently we have coupled the Lilita_N97 code with a dynamical model [10, 27] which describes the fission process by using a 3D Langevin stochastic approach. This coupling was necessary in order to allow the evaporation of light particles from the composite system during the evolution along trajectories in the phase space. At the moment we have performed several sets of calculations for the system $^{32}S + ^{100}Mo$ at E_{Lab} = 200 MeV assuming different prescriptions of transmission coefficients and level densities for particle evaporation, and by modulating the values of the strength of the one-body and two-body dissipation schemes. From Table 1 we see that the model is able to reproduce most of the measured quantities, including the ones in the ER channel, assuming full one-body dissipation, and with a viscosity parameter μ independent upon the temperature but dependent on the deformation of the fissioning system. In order to obtain a similar agreement with two-body dissipation, the unrealistic value of viscosity parameter μ = 0.24 TP had to be used, as already found in Ref. [28]. The full one-body dissipation implies a transient times for fission in the range of 23-25 x 10^{-21}s.

In figure 4 we show how the reduced friction coefficient varies with the deformation of the nucleus en route toward fission in the one-body dissipation model. The case that is able to give the best agreement with the full set of data is represented by the red line (K_s =1), namely, full one-body dissipation. The two-body dissipation case is represented by the black line. It is clear that one-body dissipation shows a stronger dependence on the deformation. Furthermore, the viscosity grows at the beginning of the

deformation until a maximum is reached; afterwards, it decreases monotonically for increasing deformation. This means that viscosity shows the maximum strength only at the beginning of the collective motion and when the shape is still fairly compact. No dependence on the temperature is assumed so far.
From the model and the computational method it is also possible to build the time distribution of all fission events. This is shown in figure 5 for the case of ^{132}Ce. The distribution has a maximum at 30zs but it extends up to 4000zs. This makes the average time for fission to be 1250zs. This figure is hence quite informative because it shows that fission can take place in quite a large interval of time. What is normally used in the statistical approach does not correspond to any of the characteristic times of the distribution above and this confirms the inadequacy of the statistical model approach to nuclear dissipation. The extension of the time distribution may also explain why different time scales are extracted with the statistical model approach when different probes are used.

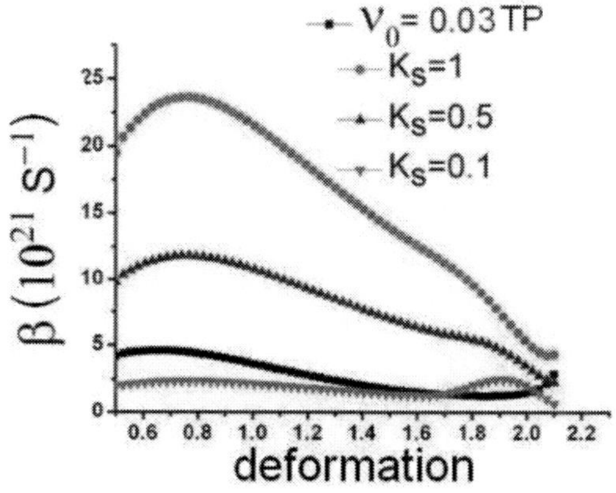

Figure 4. Friction coefficient as a function of the deformation of the fissioning nucleus in the case of the one-body dissipation mechanism. The black solid line is the functional dependency expected in the case of the two-body

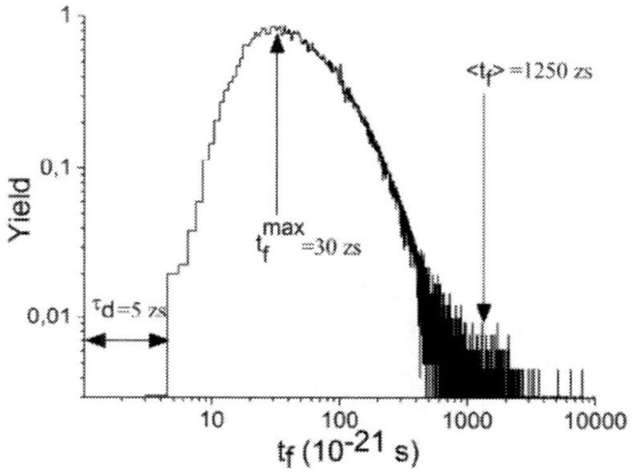

Figure 5. Time distribution of all the fission events for the nucleus ^{132}Ce.

6. Conclusions and perspectives

Our study on the system 200MeV ^{32}S + ^{100}Mo highlights the inadequacy of the SM in describing the LCP particle multiplicities in the ER channel. Same analysis performed on data from literature in the region of mass number A ≈ 150 and excitation energy E_x ≈ 100-200 MeV, for the ER channel, provides similar conclusions. This result pours some shade on the application of the SM in studies designed to investigate on the presence of transient effects. These findings also repropose the problem of the reliability of the SM in describing the compound nucleus decay and have a relevant impact on the extraction of the fission delay time through the use of the SM. The dynamical approach to fission decay is instead very promising in describing both fission and evaporation residues channel within the same model. Furthermore, a dynamical model allows to penetrate more intimate details of the fission process. For instance, the time distribution of the fission events provides hints to interpret the large variety of fission time scale found in the literature. At the same time, the model can be more and more refined if more observables are measured for the same system. Of course, the model should be oriented on the calculations of quantities that are directly linked to measured observables. In this respect, we have enlarged the computational capabilities of our code to include the calculation of energy spectra and angular distribution of the pre-scission particles. This is a novel feature that constraints even more the model parameters and in turn allows to disentangle more characteristic features of the fission process. One observable which we also consider important is the isospin degree of freedom. In [27] it is remarked the importance of selecting the proper probe for testing a dissipation model according to the isospin of the compound nucleus. One part missing in our computational model is the evaporation from the fission fragments. This is an important feature since post-scission light particle multiplicities are measured. The comparison of these observables with the predictions of a model that follows the full decay chain, from equilibrium to fragment decays, would probe more in detail models for the share of excitation energy and angular momentum, and would provide a more direct link to the features of a nucleus at the scission point. One of these is the temperature of the nucleus at the scission point. Such an extension of the model should also consider the possible dependence of the nuclear viscosity on the temperature. Consequently, experiments should be designed to explore this particular aspect. Both these developments are currently in progress.

7. References

[1] Paul P and Thoennesen M, 1994 *Ann. Rev. Nucl. Part. Sci.* **44** 65
[2] Kramer A 1940 *Physica (Amsterdam)* **7** 284
[3] Thoennessen M and Bertsch G F 1993 *Phys. Rev. Lett.* **71** 4303
[4] Lestone J P et al 1993 *Nucl. Phys.* **A 559** 277
[5] Lestone J P et al 1999 *Phys. Rev.* **C59** 1540
[6] Hofman D J, Back B B and Paul P 1996 *Nucl. Phys.* **A599** 23c
[7] Chatterjee A et al 1995 *Phys. Rev.* **C52** 3167
[8] Diószegi I et al 2000 *Phys. Rev.* **C61** 24613
[9] Itkis M G and Rusanov A Ya 1998 *Phys. Part. Nucl.* **29** 160
[10] Nadtochy P N et al 2003 *Phys. Rev.* **C65** 064615
[11] Fiorettto E. et al 1997 *IEEE Trans. Nucl. Scie.* **44** 1017
[12] Kozulin E M et al 2008 *Instrum. Exp. Tech* **51** 44
[13] Lacey R et al 1988 *Phys. Rev.* **C37** 2540
[14] La Rana G et al 2003 *Eur. Phys. J.* **A16** 199
[15] PACE2_N97 is a modified version of the original PACE2 by A. Gavron 1980 *Phys. Rev.* **C21** 230
[16] Di Nitto A 2009 PhD. Thesis, University of Naples, Italy, *www.infn.it*
[17] Huizenga J R and Igo G 1961 *Nucl. Phys.* **29** 462
[18] Perey F G 1963 *Phys. Rev.* **131** 745
[19] Willmore D and Hudson P E 1964 *Nucl. Phys.* **55** 673
[20] Gilbert A and Cameron A G W 1965 *Can. J. Phys.* **43** 1446
[21] Kataria S K, Ramamurthy V S and Kapoor S S 1978 *Phys. Rev.* **C18** 549
[22] Ignatyuk A V et al 1975 *Yad. Fiz.* **21** 485, [1975 *Sov. J. Nucl. Phys.* **21** 255]
[23] Lestone J P, *Phys. Rev.* **C52** 1118
[24] Vaz L C et al 1984 *Z. Phys.* **A318** 231
[25] Lilita_N97 is an extensively modified version of the original Lilita program made by J. Gomez del Campo and R. G. Stockstad, ORNL Report No. TM7295, 1981 (unpublished)
[26] R. J. Charity, Computer code GEMINI (unpublished)
[27] Nadtochy P , Vardaci E, Di Nitto A, Brondi A, La Rana G, Moro R, Cinausero M, Prete G, Gelli N, Lucarelli F 2010 Phys. Lett. **B685** 258
[28] Wada et al 1993 *Phys. Rev. Lett.* **70** 3539

Fusion-fission and quasifission competition in the $^{32}S+^{184}W$ reaction

H. Q. Zhang*, C. L. Zhang, and C. J. Lin
China Institute of Atomic Energy, P. O. Box 275, Beijing 102413, China
E-mail: huan@ciae.ac.cn

A. K. Nasirov
Institute of Nuclear Physics,Tashkent, Uzbekistan

G. Mandaglio, M. Manganaro, and G. Giardina
Dipartimento di Fisica dell' Università di Messina, 98166 Messina, and Istituto Nazionale di Fisica Nucleare, Sezione di Catania, Italy

Abstract. The angular distribution of fission fragments for the $^{32}S+^{184}W$ reaction at center-of-mass energies of 118.8, 123.1, 127.3, 131.5, 135.8, 141.1 and 144.4 MeV were measured. The experimental fission excitation function is obtained. The fragment angular anisotropy is found by extrapolating the fission angular distributions. The measured fission cross sections are decomposed into fusion-fission, quasifission and fast-fission contributions by the dinuclear system model. The total evaporation residue and fusion-fission excitation functions are calculated in the framework of the advanced statistical model. The hindrance to complete fusion at small collision energies increases due to the increase of quasifission events and it is explained by the elongated shape of the dinuclear system which is formed in collisions with small orientation angles to the beam direction. An increase of the hindrance to complete fusion at large beam energies is explained by the dependence of the quasifission and intrinsic fusion barriers of dinuclear system on its angular momentum: at large angular momentum the quasifission barrier decreases and the intrinsic fusion barrier increases. In this reaction the contributions of fusion-fission and quasifission fragments are comparable.

1. Introduction

Studies of fusion-fission reactions between heavy ion projectile and heavy target nuclei have demonstrated to be very useful in developing an understanding of the nuclear reaction dynamics.

The study of dynamics of processes in heavy ion collisions at the near Coulomb barrier energies showed that complete fusion does not occurs immediately in case of the massive nuclei collision [1, 2, 3, 4, 5]. The quasifission process competes with formation of compound nucleus (CN). This process occurs when the dinuclear system prefers to break down into fragments instead of to be transformed into fully equilibriated CN. The number of events going to quasifission increases drastically by increasing the sum of the Coulomb interaction and rotational energy in the entrance channel [6, 7]. Another reason decreasing yield of ER is the fission of a heated and rotating CN which is formed in competition with quasifission. The stability of massive CN decreases due to the decrease of the fission barrier by increasing its excitation energy E^*_{CN} and angular momentum ℓ [8, 9]. Because the stability of the transfermium nuclei are connected with the availability of shell correction in their binding energy [11, 12] which are sensitive to E^*_{CN} and values of the angular momentum. To find favorable reactions (projectile and target pair) and the optimal beam energy range leading to larger cross sections of synthesis of superheavy elements, we should establish conditions to increase as possible the events of ER formation.

The presence of quasifission fragments in the measured yield of fission-like fragments is determined by the large values of anisotropy in their angular distribution (see Refs.[1, 14, 13], and references therein) and by the increased yield of fragments with masses near proton magic numbers 28, 50, 82 and neutron magic numbers 50, 82, 126 [15, 16].

In this contribution we have analyzed the angular distribution of fission fragments of the $^{32}S+^{184}W$ reaction and we obtained fission excitation function at the center-of-mass energies of 118.8, 123.1, 127.3, 131.5, 135.8, 141.1 and 144.4 MeV. The fragment angular anisotropy $A_{\rm exp}$ is found by extrapolating the fission angular distributions to angles 0° and 90° by the method used in Ref. [17]. Then, the mean square angular momentum $\langle L^2\rangle$ values were obtained. Hereafter we use for simplicity ℓ from the definition $L = \ell\hbar$. We assumed the calculated capture cross sections to be equal to the experimental data of fission-like fragments in order to decompose the measured fission cross section into fusion-fission, quasifission and fast fission contributions by the dinuclear system model [2, 3, 7, 18]. The quasifission is a break of the DNS into two fragments bypassing the stage of the CN formation. The fast fission process is the inevitable decay of the fast rotating mononucleus into two fragments without reaching the equilibrium compact shape of CN. Such mononucleus is formed from the dinuclear system survived against quasifission. At large values of the angular momentum $\ell > \ell_f$, where ℓ_f is a value of ℓ at which the fission barrier of the corresponding CN disappears, mononucleus immediately decays into two fragments [19]. As distinct from fast fission, the quasifission can occur at all values of ℓ at which capture occurs. The total ER and fusion-fission excitation functions are calculated in the framework of the advanced statistical model [8, 9, 10]. The ER cross sections for the $^{32}S+^{184}W$ reaction are small in comparison with the other contributions forming capture cross section.

To determinate the ER cross section $\sigma_{\rm ER}(E)$ we have to know at first the partial capture cross sections $\sigma^l_{\rm capture}(E)$, the two factors $P_{\rm CN}$ and $W_{\rm sur}$ corresponding to competition between complete fusion and quasifission at the second stage, and to competition between cooling of the heated and rotating CN by evaporation of light particles and its fission into two fragments at the third stage of reaction. This way of calculation was suggested in the dinuclear system (DNS) concept [2]and the estimation of $P_{\rm CN}$ and $W_{\rm sur}$ is the key point for the research of the fusion reaction products, especially for the synthesis of superheavy elements.

2. EXPERIMENTAL PROCEDURE

The experiment was performed at HI-13 tandem accelerator at China Institute of Atomic Energy, Beijing. A collimated ^{32}S beam with incident energies $E_{\rm lab} = 140, 145, 150, 155, 160, 165$ and 170 MeV bombarded on a target of ^{184}W which was mounted at center of the scattering chamber. The ^{184}W target with thickness about 200 μg/cm^2 was evaporated on an about 20 μg/cm^2 carbon foil backing. Typical ^{32}S beam current range was 800-1000 enA monitored by a shielded suppressed Faraday cup at periphery of the chamber, because of the variety according to the bombarding energy and scattering angle. The beam energy loss in traveling half the target were calculated and was about 0.5 MeV.

At the forward angles, an array of five Si detectors with depletion depth ranging from 200 to 300 μm, which covered the angular range of θ_L=14°-35°, 35°-55° and 55°-75° were mounted on the movable arm in the chamber and five masks were placed in the front of each detector for assuring the angular resolution. The detectors to the target distance was 27 cm. The Rutherford scattering was monitored at forward angle of θ_L=15° by four Si(Au) surface barrier detectors

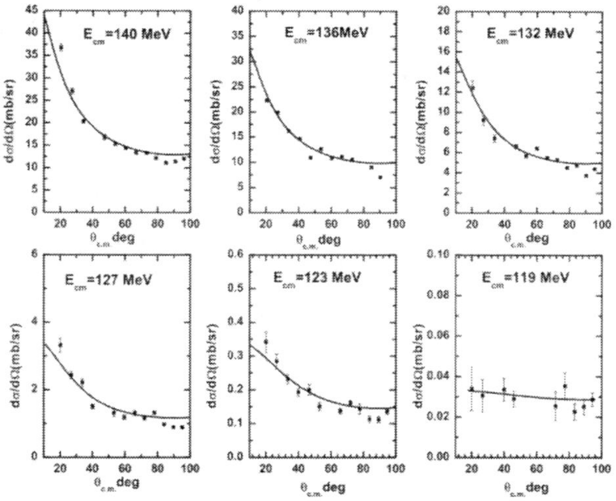

Figure 1. Fission fragment angular distribution for the ^{32}S+^{184}W reaction. Incident energies are shown in the figure. The experimental data are shown with the fitting curve, which is used to determine the anisotropy A_{exp} of the fragment angular distribution and mean square values of angular momentum from these events.

for the normalization of the cross section measurements.

In addition to these individual Si detectors, two groups of Si strip detectors were mounted on opposite sides of the beam. These Si strip detectors were 48×50 mm^2 in area and each detector consisted of 24 strips. Due to a lack of readout electronics, the strips were tied together in groups of eight for readout. Data from these strip detectors were recorded in the coincidence mode with the requirement that each detector was struck by a fission fragment and the folding angle between the hits corresponded to a full momentum transfer event. To calculate the kinematics it was assumed all processes observed can be treated as binary reactions. This assumption was tested by examining the folding angle distribution of coincident fragments in the Si detector and the Si strip detectors. The average folding angle agrees with the expectations based on total fission kinetic energies taken from the Viola systematics [20].

To obtain the absolute cross sections, we measured the solid angle by using the α-particles from the ^{241}Am source and the elastic scattering products. Centering of the beam on the target was ensured by the four monitor detectors. A gate was set on the fission event and data were collected by the coincident mode.

3. EXPERIMENTAL RESULTS

3.1. Fission fragment angular distributions

The fission fragment angular distributions were measured using the coincident detectors and are shown in Fig. 1. In fitting the angular distribution of the fission fragments we used the familiar expression as in Ref. [17]:

$$W(\theta) = \sum_{J=0}^{J_{\text{max}}} \frac{(2J+1)^2 \exp[-(J+\frac{1}{2})^2 \sin^2\theta/4K_0^2] J_0[i(J+\frac{1}{2})^2 \sin^2\theta/4K_0^2]}{\text{erf}[(J+\frac{1}{2})/(2K_0^2)^{1/2}]} \quad (1)$$

assuming M=0, *i.e.* assuming the spins of the target and projectile were zero, where J_0 is the zero order Bessel function with imaginary argument and error function erf(x) is defined as

$$\text{erf}(x) = (2/\pi^{1/2}) \int_0^x \exp(-t^2)dt. \quad (2)$$

The measured values of σ_{capture} and the deduced values of A_{exp} and K_0^2 for the ^{32}S+^{184}W reaction are presented in Table 1. J_{max} is obtained by reproducing the capture cross section.

Table 1. The measured capture cross sections and the deduced values of $A_{\rm exp}$, K_0^2 and $\langle l^2 \rangle$ for the ^{32}S+^{184}W reaction. The $E_{\rm c.m.}$ are the energies calculated as corresponding to the beam energies in the center of the target.

$E_{\rm c.m.}$ (MeV)	$E^*_{\rm CN}$ (MeV)	$\sigma_{\rm capture}$ (mb)	$A_{\rm exp}$	K_0^2 ($\hbar^2$)	$\langle l^2 \rangle$ ($\hbar^2$)
118.8	37.2	0.04	1.51	114.71	234
123.1	41.5	2.35	2.16	124.35	577
127.3	45.8	22.97	2.27	132.09	671
131.5	50.0	81.01	2.74	140.01	975
135.8	54.3	132.27	3.06	148.67	1225
141.1	58.5	189.33	3.28	157.35	1435
144.4	61.8	237.06	3.8	155.09	1737

The K_0^2 value is found by fitting the angular distribution at known $J_{\rm max}$ from the total fission cross section. It is seen from Fig. 1 that the anisotropy of angular distribution increases by increasing collision energy $E_{\rm c.m.}$.

3.2. Capture cross section

In order to deduce the capture cross section from the data, the Si strip detectors operating in the coincidence mode were used. Each fission event was selected on the basis of the correct value of energy and of the folding angle corresponding to complete momentum transfer using the forward Si detectors as a "trigger" detector. After correction for the efficiency of the Si strip detectors, a differential cross section $d\sigma/d\Omega(\theta)$ was obtained. The total cross section was deduced from the integration of the differential cross sections.

The resulting experimental values of the capture cross sections are shown in Table 1 and Fig. 3 where they are compared with the theoretical results. The total fission cross section is assumed to be equal to the theoretical capture cross section and it was decomposed into fusion-fission and quasifission parts in the framework of the DNS model mentioned in Section 1. The comparison of the measured capture cross section, anisotropy, mean square values of angular momentum, and variance K_0^2 with the corresponding experimental data is discussed in Section 4.

3.3. Anisotropy of fission-fragments and mean square angular momentum values

The experimental values of the anisotropy $A_{\rm exp}$ are found by extrapolating the fission angular distributions to angles 0° and 90° by the method used in Ref. [17]. The anisotropies as a function of center-of-mass energies are shown in Table 1 and Fig. 7 where they are compared with the theoretical results. Using the approximate relation between the anisotropy and the mean square angular momentum (see Section 4 for details), the mean square angular momentum values $\langle l^2 \rangle$ are deduced from the experimental anisotropies. The data are shown in Table 1 and Fig. 8.

4. Theoretical description and comparison with measured data

The experimental data for the excitation function of fission-like products in the ^{32}S+^{184}W reaction were analyzed in the framework of the dinuclear system model (DNS) [6, 7, 18, 21, 22]. The capture, fusion, quasifission and fast fission excitation functions have been calculated for this reaction.

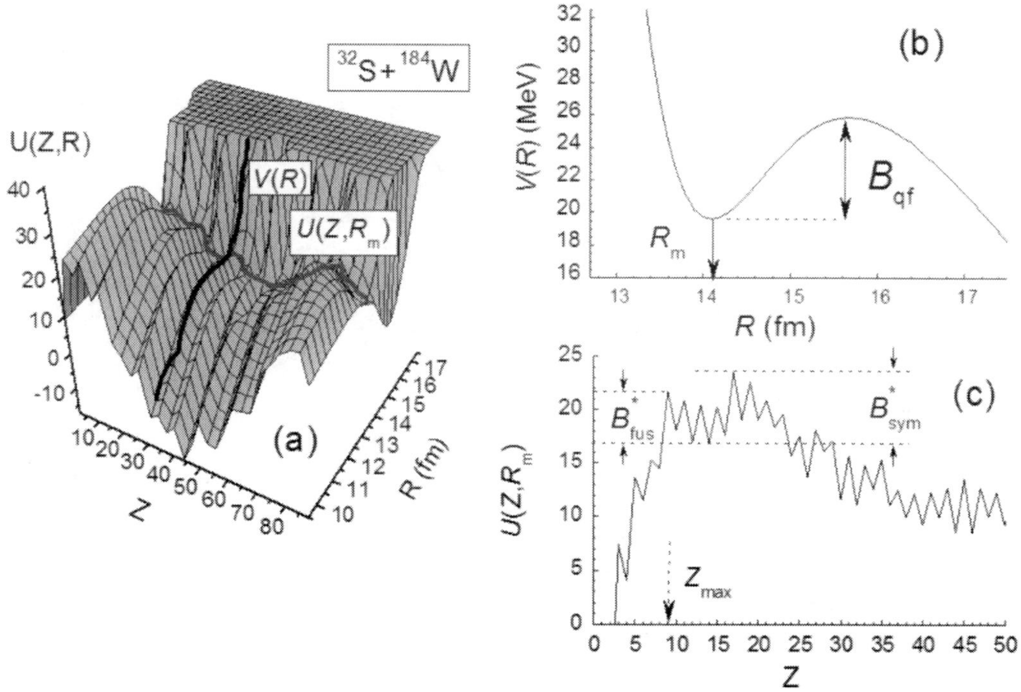

Figure 2. The potential energy surface for a dinuclear system leading to the formation of the ^{216}Th* CN as a function of the relative distance R between centers of interacting nuclei and their charge numbers Z, panel (a); the nucleus-nucleus interaction potential $V(R)$ shifted on the Q_{gg}-value for the ^{32}S + ^{184}W reaction, panel (b); the driving potential, $U_{\rm dr}(Z, R_m)$, which is a curve linking minimums corresponding to each charge asymmetry in the valley of the potential energy surface as a function of Z, panel (c).

According to the DNS model a capture event is the trapping of the collision path into the potential well after dissipation of the sufficient part of the relative kinetic energy of a projectile nucleus in the center-of-mass coordinate system. At capture the full momentum transfer from the relative motion of nuclei into excitation energy of dinuclear system takes place. Certainly the presence of a potential pocket and adequacy of the collision energy $E_{\rm c.m.}$ to overcome the interaction barrier of the entrance channel $V_{\rm B}$ are necessary conditions to occur capture. Thus capture leads to forming dinuclear system which characterized by mass (charge) asymmetry of its nuclei, rotational energy $V_{\rm rot}$ and excitation energy E^*_{DNS}. The relative energy of nuclei is relaxed, therefore, the total kinetic energy of fragments formed at its decay are close to the Viola systematics [20].

The nucleus-nucleus potential $V(Z, A, R)$ is a sum of the Coulomb $V_{\rm C}(Z, A, R)$ and nuclear interaction $V_{\rm N}(Z, A, R)$, as well as the rotational energy $V_{\rm rot}(Z, A, R, \ell)$:

$$V(Z, A, R, \ell) = V_{\rm C}(Z, A, R) + V_{\rm N}(Z, A, R) + V_{\rm rot}(Z, A, R, \ell) \quad (3)$$

where $Z = Z_1$ and $A = A_1$ are the charge and mass numbers of one of fragments forming DNS, respectively, while the charge and mass numbers of another fragment are equal to $Z_2 = Z_{\rm CN} - Z$ and $A_2 = A_{\rm CN} - A$, respectively, where $Z_{\rm CN}$ and $A_{\rm CN}$ the charge and mass numbers of being formed CN; R is the relative distance between the centers of nuclei forming the dinuclear system.

The partial capture cross section is found by solution of kinetic equation for the relative motion and orbital angular momentum ℓ for the different orientation angle α_T of the target nucleus as it was performed in Refs. [6, 18]. The fusion cross section is calculated from the branching ratio $P_{\rm CN}(Z)$ of the decay rates of overflowing the border of the potential well $(B_{\rm qf}^{(Z)})$ along R at a given mass asymmetry (decay of DNS–quasifission) over the barriers on mass asymmetry axis $B^*_{\rm fus}$ for the complete fusion or $B^{(Z)}_{\rm sym}$ in opposite direction to the symmetric configuration of DNS (see Fig. 2):

$$P^{(Z)}_{\rm CN}(E^*_{DNS}) \approx \frac{\Gamma^{(Z)}_{\rm fus}(B^*_{\rm fus}, E^*_{DNS})}{\Gamma^{(Z)}_{\rm (qf)}(B_{\rm qf}, E^*_{DNS}) + \Gamma^{(Z)}_{\rm (fus)}(B^*_{\rm fus}, E^*_{DNS}) + \Gamma^{(Z)}_{\rm sym}(B_{\rm sym}, E^*_{DNS})}, \quad (4)$$

where $\Gamma_{\rm fus}$, $\Gamma_{\rm qf}$ and $\Gamma_{\rm sym}$ are corresponding widths determined by the level densities on the barriers $B^*_{\rm fus}$, $B^*_{\rm sym}$ and $B_{\rm qf}$ involved in the calculation of $P_{\rm CN}$ are used in the model [18, 7, 4] based on the dinuclear system concept [23]. Here $E^*_{\rm DNS}(Z_P, A_P, \ell) = E_{\rm c.m.} - V(Z_P, A_P, \ell, R_{\rm m})$ is the excitation energy of dinuclear system in the entrance channel, where Z_P and A_P are charge and mass numbers of the projectile nucleus. $V(Z, A, R_{\rm m}, \ell)$ is the minimum value of the nucleus-nucleus potential well (for the DNS with charge asymmetry Z) and its position on the relative distance between the centers of nuclei is marked as $R = R_{\rm m}$ in Fig. 2b. The value of $B_{\rm qf}$ for the decay of DNS with the given charge asymmetry of fragments is equal to the depth of the potential well in the nuclear-nuclear interaction (see Fig. 2b). The intrinsic fusion barrier $B^*_{\rm fus}$ is connected with mass (charge) asymmetry degree of freedom of the dinuclear system and it is determined from the potential energy surface (Fig. 2a):

$$U(Z; R, \ell) = U(Z, \ell, \beta_1, \alpha_1; \beta_2, \alpha_2) = B_1 + B_2 + V(Z, \ell, \beta_1, \alpha_1; \beta_2, \alpha_2; R) - (B_{CN} + V_{CN}(\ell)). \quad (5)$$

Here, B_1, B_2 and $B_{\rm CN}$ are the binding energies of the nuclei in DNS and the CN, respectively, which were obtained from [24]; the fragment deformation parameters β_i are taken from the tables in [25, 26, 24] and α_i are the orientation angles of the reacting nuclei relative to the beam direction; $V_{\rm CN}(\ell)$ is the rotational energy of the CN. The distribution of neutrons between two fragments for the given proton numbers Z and Z_2 or ratios A/Z and A_2/Z_2 for both fragments were determined by minimizing the potential $U(Z; R)$ as a function of A for each Z.

The driving potential $U_{\rm dr}(Z) \equiv U(Z, R_m)$ is a curve linking minimums corresponding to each charge asymmetry Z in the valley of the potential energy surface from $Z = 0$ up to $Z = Z_{\rm CN}$ (see Fig. 2a and 2c). We define the intrinsic fusion barrier for the charge asymmetry Z of dinuclear system as $B^*_{\rm fus}(Z, \ell) = U(Z_{\rm max}, R_{\rm m}(Z_{\rm max}), \ell) - U(Z, R_{\rm m}(Z), \ell)$, where $U(Z_{\rm max}, \ell)$ is a maximum value of potential energy in the valley along the way of complete fusion from the given Z configuration.

The masses and charges of the projectile and target nuclei are not constant during capture and after formation of the dinuclear system. The intense proton and neutron exchange between constituents of DNS is taken into account by calculation of the complete fusion probability $P_{\rm CN}$ as fusion from all populated DNS configurations according to the formula

$$P_{CN}(E^*_{\rm DNS}(Z, A, \ell); \{\alpha_i\}) = \sum_{Z_{\rm sym}}^{Z_{\rm max}} Y_Z(E^*_{\rm DNS}(Z, A, \ell)) P^{(Z)}_{\rm CN}(E^*_{\rm DNS}(Z, A, \ell); \{\alpha_i\}) \quad (6)$$

where $E^*_{\rm DNS}(Z, A, \ell) = E^*_{\rm DNS}(Z_P, A_P, \ell) + \Delta Q_{\rm gg}(Z)$ is the excitation energy of DNS with angular momentum ℓ for a given value of its charge-asymmetry configuration Z and $Z_{\rm CN} - Z$; $\Delta Q_{\rm gg}(Z)$ is the change of $Q_{\rm gg}$-value by changing the charge (mass) asymmetry of DNS; $Y_Z(E^{*(Z)}_{\rm DNS})$ is the probability of population of the $(Z, Z_{\rm CN}\text{-}Z)$ configuration at $E^{*(Z)}_{\rm DNS}$ and given orientation angles (α_1, α_2). It was obtained by solving the master equation for the evolution of the dinuclear system charge asymmetry (for details see Refs. [18, 13]). $Z_{\rm sym} = (Z_1 + Z_2)/2$ while $Z_{\rm max}$ corresponds to the point where the driving potential reaches its maximum $(B^*_{\rm fus}(Z_{\rm max}) = 0)$ (see Refs. [7, 27]).

The calculations were performed for the $E_{\rm c.m.}$=119.5–149.5 MeV energy range and initial values of the orbital angular momentum ℓ =0–120$\hbar$. Due to the deformed shape of ^{184}W ($\beta_2 = 0.24$ and $\beta_4 = -0.095$) in the ground state we included in our calculations a dependence of the excitation function of capture, complete fusion and quasifission on the orientation angle α_T of its axial symmetry axis. The ground state shape of ^{32}S is spherical but the quadrupole (2^+) and octupole (3^-) collective excitations in spherical nuclei are taken into account as amplitudes of the zero-point motion mode of surface vibration. The deformation parameters of the first excited quadrupole state $\beta_2^{(2+)} = 0.312$ (taken from Ref. [25]) and the ones of the first excited octupole state $\beta_3^{(3-)}$=0.41 (taken from Ref. [26]). The final results of the capture and complete fusion are obtained by averaging the contributions calculated for the different orientation angles (α_T=0, 15, 30, ..., 90°) of the symmetry axis of the target nucleus:

$$\langle\sigma_{\rm fus}(E_{\rm c.m.}, l)\rangle = \int_0^{\pi/2} \sin\alpha_T \sigma_{\rm fus}(E_{\rm c.m.}, l; \alpha_T) d\alpha_T. \tag{7}$$

This methods were developed and used in the Refs. [18, 21, 22]. The results of calculation of the excitation functions are presented in Fig. 3. In this figure we compare theoretical results with the experimental data for the capture cross section of this work. The agreement between the experimental and theoretical capture cross sections was reached by changing the radius parameter C_R entered to rescale the nuclear radius:

$$R_1 = C_R\sqrt{(R_{\rm p}^2\, Z_1 + R_{\rm n}^2\, (A_1 - Z_1))/A_1}\,, \tag{8}$$

where $R_{\rm p}$ and $R_{\rm n}$ are the proton and neutron radii, respectively, obtained from Ref. [28]:

$$R_{\rm p} = 1.237(1 - 0.157(A - 2Z)/A - 0.646/A)A^{1/3}, \tag{9}$$
$$R_{\rm n} = 1.176(1 + 0.25(A - 2Z)/A + 2.806/A)A^{1/3}. \tag{10}$$

The presented results are obtained at $C_R = 0.925$ for all values of $E_{\rm c.m.}$. Using Eq.(7) we calculated the partial fusion cross sections which were used to estimate the cross sections of ER and fusion-fission by the advanced statistical model [8, 9]. Taking into account the dependence of the fission barrier ($B_{\rm f}$) of the rotating CN on its angular momentum we found a value of ℓ at which $B_{\rm f}$ disappears using the rotating finite range model by A. J. Sierk [29]: $\ell_{\rm B}$=68. Then we calculate the fast fission contribution for $\ell > \ell_{\rm B}$

$$\sigma_{\rm fast\,fission}(E_{\rm c.m.}) = \sum_{\ell=\ell_{\rm B}}^{\ell=\ell_{\rm max}} (2\ell + 1)\sigma_{\rm fus}(E_{\rm c.m.}, \ell) \tag{11}$$

where $\ell_{\rm max}$ is the maximum value of angular momentum of the dinuclear system for the given value of $E_{\rm c.m.}$. The value of $\ell_{\rm max}$ is found by solving the equations of motion for the radial distance and orbital angular momentum with the given values of $E_{\rm c.m.}$, ℓ_0 and $R_{\rm max} = 20$ fm.

In Fig. 3 are also presented the total fusion-fission and ER cross sections obtained by the advanced statistical model (see Refs. [8, 9, 10], and references therein) describing the full deexcitation cascade of the ^{216}Th* CN formed in the ^{32}S+^{184}W reaction. The code takes into account the competition between evaporation of light particles (n, p, α, and γ) and fission processes along each step of the deexcitation cascade of CN. The effective fission barrier for CN and intermediate excited nuclei along the cascade are obtained taking into account the macroscopic fission barrier, predicted by the rotating droplet model as parameterized by Sierk [29], together with the microscopic corrections allowing for the angular momentum and

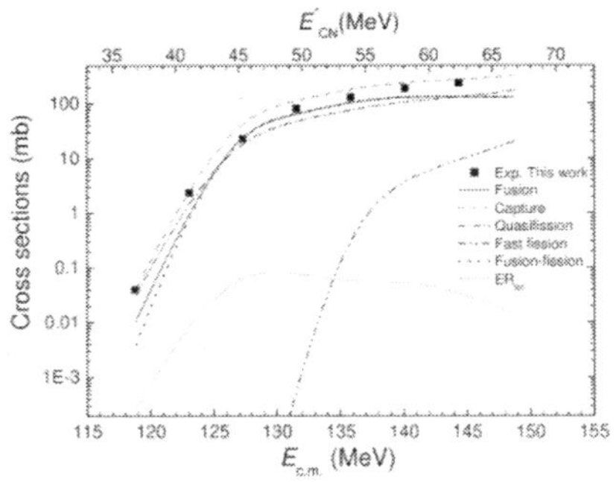

Figure 3. Comparison of the experimental data with theoretical results.

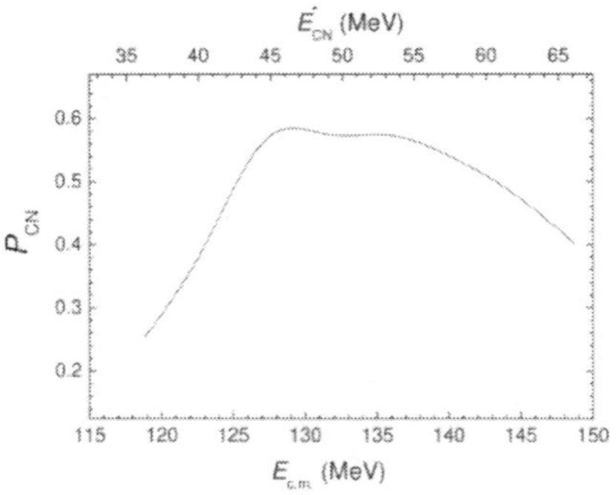

Figure 4. Theoretical value of $P_{\rm CN}(Z)$ for ^{32}S+^{184}W system.

temperature fade-out of shell corrections [24] to the fission barrier (see Refs. [3, 27, 30], and reference therein).

The cross section of ER formed at each step x of the deexcitation cascade after the emission of $\nu(x)$n+$y(x)$p+$k(x)\alpha+s(x)\gamma$ particles (ν, y, k, s are numbers of neutrons, protons, α-particles and γ-quanta) from the hot CN is calculated by the formula[7, 27, 3]:

$$\sigma_{\rm ER}(E^*_x) = \sum_{Z_{l=0}}^{l_d} \sigma^l_{(x-1)}(E^*_x) W_{{\rm sur}(x-1)}(E^*_x, l), \tag{12}$$

where $\sigma^l_{(x-1)}(E^*_x)$ is the partial cross section of the intermediate nucleus formation at the $(x-1)$th step and $W_{{\rm sur}(x-1)}(E^*_x, \ell)$ is the survival probability of the $(x-1)$th intermediate nucleus against fission along the deexcitation cascade of CN; E^*_x is an excitation energy of the nucleus formed at the xth step of the deexcitation cascade. It is clear that $\sigma^l_{(0)}(E^*_0) = \sigma^l_{fus}(E^*)$ at $E^*_{CN} = E^*_0 = E_{\rm c.m.} + Q_{\rm gg}$, where $Q_{\rm gg}$ is energy balance of reaction. The numbers of the being emitted neutrons, protons, α-particles, γ-quanta, $\nu(x)$n, $y(x)$p, $k(x)\alpha$, and $s(x)\gamma$, respectively, are functions of the step x. The emission branching ratio of these particles depends on the excitation energy and angular momentum of the being cooled intermediate nucleus $A = A_{\rm CN} - (\nu(x) + y(x) + 4k(x))$ and $Z = Z_{\rm CN} - (y(x) + 2k(x))$ [3]. We note that in Fig. 3 the maximum value of the excitation function of the total evaporation residues is about 0.12 mb and the complete spectrum is contributed mainly by the evaporation of charged particles which are accompanied with a small number of neutrons along the deexcitation cascade of CN.

4.1. Two regions of strong hindrance to complete fusion

Thus the experimental capture excitation function is decomposed into contributions of fusion-fission, quasifission, fast fission and total ER. One can see in Fig. 3 that at lowest energies $E_{\rm c.m.} < 125$ MeV, the fusion cross section is sufficiently lower than the quasifission contribution. The intense of hindrance to complete fusion is estimated by the fusion factor $P_{\rm CN}$ and it is determined by Eqs. (4) and (6). We present in Fig. 4 the calculated values of $P_{\rm CN}$ as a function of the beam energy. It is seen the hindrance to fusion is strong at very small and large values of the collision energy $E_{\rm c.m.}$.

The yield of quasifission is dominant at the sub-barrier beam energies leading to capture of deformed nuclei only with the small orientation angle of its symmetry axis relative to the beam direction. This phenomenon was found by Hinde and his colleagues [14] in the ^{16}O+^{238}U reaction where they observed the increase of the anisotropy of angular distribution of the fission fragments when beam energy decreases to the sub-barrier region. Its analysis has been discussed

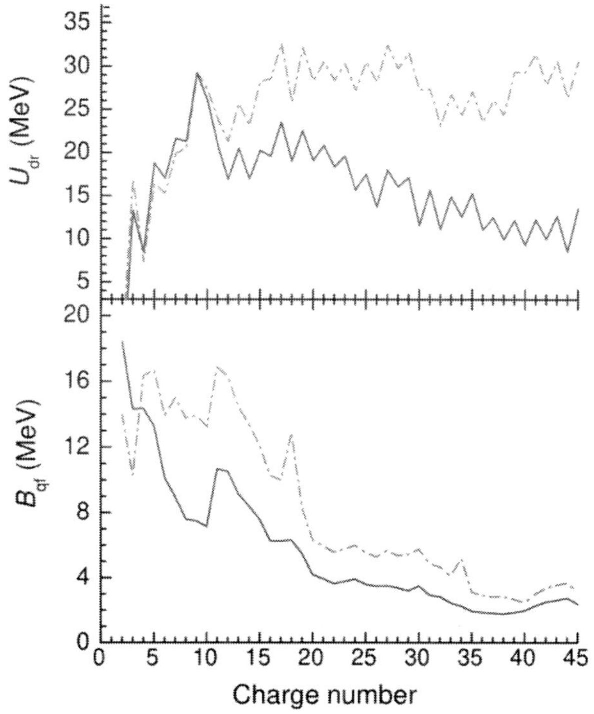

Figure 5. Driving potential calculated for the two different values of the orientation angles of target nuclei: $\alpha_T = 15°$ (solid line) and $\alpha_T = 45°$ (dashed line) for the $^{32}S+^{184}W$ reaction.

in Ref. [13] in connection with the explanation of the observed large angular anisotropy of the fission fragments. As an alternative suggestion for the origin of the observed anomaly, Liu *et al.* in Ref. [31] have put forward a new version of the preequilibrium fission model. The anomalous bump exists in the variation of the fragment anisotropy with the incident energy for the systems such as ^{19}F, ^{16}O $+^{232}Th$. Based on these studies, a new model of K pre-equilibrium fission was proposed, which can well explain the observed anomalous anisotropy. The authors of Ref. [14] analyzed in detail the angular anisotropy of fragments at low energies to show the dominant role of the quasifission in collisions of the projectile with the target nucleus when the axial symmetry axis of the latter is oriented along or near the beam direction. Large values of the anisotropy were obtained at low energies and these data were assumed to be connected with quasifission because a mononucleus or dinuclear system formed in the near tip collisions has an elongated shape.

This shape can be far from the one corresponding to the saddle point. The driving potential used in the DNS model depends on the shape of dinuclear system which is formed in collisions with the different orientation angles of the axial symmetry axis of deformed target. The small values of the fusion probability P_{CN} at lowest energies in the $^{32}S+^{184}W$ reaction are explained by the large values of B^*_{fus} for the dinuclear system formed at collisions of projectile with target oriented close to beam direction, *i.e.* when α_T is small. To demonstrate this conclusion we present in Fig. 5 the driving potential (upper panel) and quasifission barrier (lower panel) for the decay of dinuclear system as a function of the orientation angles of the target nucleus. It is seen from the upper panel of Fig. 5 that the value of driving potential corresponding to the projectile charge number Z=16 for the small orientation angle 15° (solid line) of the target is lower than that corresponding to the larger orientation angle 45° (dashed line). The fusion probability is larger if a value of the driving potential $U_{dr}(Z_{proj})$ corresponding to the initial charge number of the light fragment Z_L of dinuclear system is as possible at the same level or higher than its maximum value at Z=9 along the way to complete fusion which means the transfer of all nucleons of light nucleus into heavy nucleus: $Z_L \longrightarrow 0$ and $Z_H \longrightarrow Z_{CN}$.

The decrease of its values at collision energies $E_{c.m.} > 135$ MeV is explained by decreasing the quasifission barrier B_{qf} as a function of the orbital angular momentum ℓ. Because the depth

Journal of Physics: Conference Series **282** (2011) 012013 doi:10.1088/1742-6596/282/1/012013

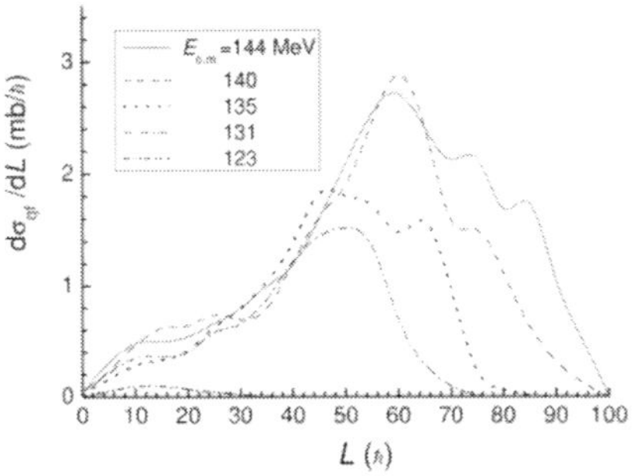

Figure 6. The partial quasifission excitation functions.

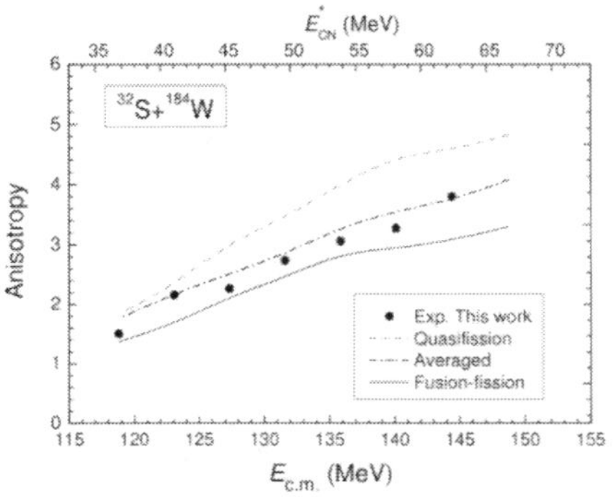

Figure 7. Comparison of the anisotropies measured (circles) with the theoretical results.

of the potential well being $B_{\rm qf}$ for a given charge asymmetry decreases due to the increase of the rotational energy $E_{\rm rot}$ of the dinuclear system (for details see [7]). At the same time the intrinsic fusion barrier $B^*_{\rm fus}$ increases by increasing ℓ. At small collision energies only small values of ℓ are populated because capture does not occur if the initial energy of projectile is not enough to overcome the Coulomb barrier of the entrance channel.

It is seen from Fig. 6 that the values $\ell > 45$ are populated at collision energies $E_{\rm c.m.} > 135$ MeV. Therefore, we conclude that the contribution of quasifission becomes dominant at $E_{\rm c.m.} > 135$ MeV and it has an effect on the anisotropy of angular distribution which increases by increasing the collision energy. This is connected with the fact that at these energies all orientation angles α_T can contribute to the formation of compound nucleus because the Coulomb barrier for large orientation angles may be overcomed. As we know, in collisions with large α_T the fusion probability is large. The decrease of $P_{\rm CN}$ at larger energies is explained by the decrease of the quasifission barrier B_{qf} by increasing ℓ in collisions with all orientation angles.

4.2. Anisotropy of fission fragment angular distribution and variance K_0^2 of K distribution

To clarify the role of quasifission fragments in the observed anisotropy $A_{\rm exp}$, we calculated the contributions of the quasifission ($A_{\rm qf}$) and fusion-fission ($A_{\rm CN}$) fragments. We used the expression of the approximated anisotropy of the fission fragment angular distribution suggested by Halpern and Strutinski in Ref. [32] and Griffin in Ref. [33]:

$$A \approx 1 + \frac{\langle l^2\rangle_i \hbar^2}{4\Im_{eff}{}^{(i)} T_i}, \text{ where } \frac{1}{\Im_{\rm eff}} = \frac{1}{\Im_{\parallel}} - \frac{1}{\Im_{\perp}}. \quad (13)$$

Here $\Im^{(i)}_{\rm eff}$ is the effective moment of inertia for the CN on the saddle point i = CN or it is the effective moment of inertia for the dinuclear system on the quasifission barrier i = DNS. In the last case, we calculate $\Im_{\rm eff}$ for the dinuclear system taking into account the possibility of different orientation angles of its constituent nuclei (see Appendix A of Ref.[13]), assuming that after capture the mutual orientations of the DNS nuclei do not change sufficiently. $\Im_{\parallel}$ and $\Im_{\perp}$ are the moments of inertia around the symmetry axis and a perpendicular axis, respectively.

Their values for $\Im^{(CN)}_{\rm eff}$ are determined in the framework of the rotating finite range model by Sierk [29]. The temperature of CN on the saddle point is found by the expression:

$$T_{\rm CN} = \left[\frac{E_{\rm c.m.} + Q_{\rm gg} - B_{\rm f}(\ell) - E_n}{A_{\rm CN}/12}\right]^{1/2}, \quad (14)$$

where $B_{\rm f}(\ell)$ is the fission barrier height. $B_{\rm f}(\ell)$ is calculated in terms of the rotating liquid drop model by Sierk [29]. $E_{\rm n}$ is the energy carried away by the pre-saddle fission neutrons. The

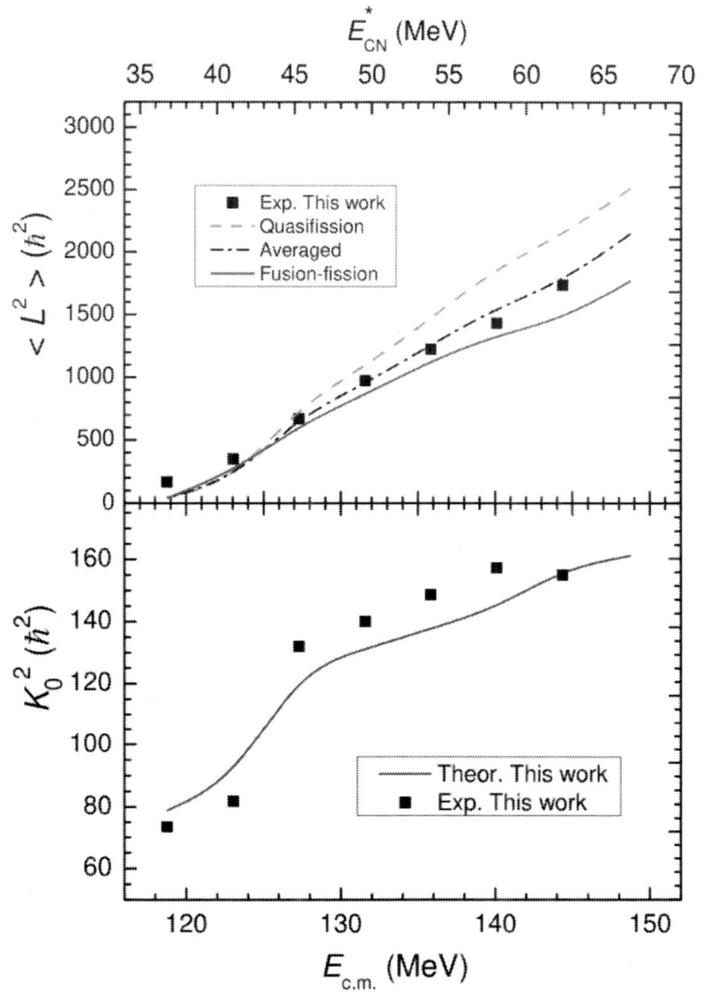

Figure 8. Mean square value of the angular momentum (upper panel) versus the collision energy $E_{\rm c.m.}$ for the ^{32}S+^{184}W reaction. The experimental data are shown in comparison with the results of DNS model which are obtained by averaging ℓ^2 by partial cross sections of the quasifission (dashed line) and complete fusion (solid line) events. In the lower panel the experimental and theoretical values of K_0^2 are compared.

temperature of DNS on the quasifission barrier is determined by the expression:

$$T_{\rm DNS} = \left[\frac{(E^*_{\rm DNS} - B_{\rm qf})}{A_{\rm CN}/12}\right]^{1/2}. \quad (15)$$

An important physical quantity in the formula (13) is the variance K_0^2 of the Gaussian distribution of the K projection :

$$K_0^2 = \frac{\Im_{\rm eff} T_{\rm saddle}}{\hbar^2}. \quad (16)$$

The experimental values of K_0^2 are used to fit the angular distribution of fission fragments by formula (1) (see Ref.[17, 34]).

In Fig. 7 we compare the anisotropy measured (circles) in this work with the theoretical results for the anisotropy of the quasifission (dashed line) and fusion-fission (solid line) fragments as a function of the center-of-mass energy (bottom axis) and excitation energy of the CN (top axis). The averaged theoretical anisotropy over the contributions of both mechanisms are presented by the dot-dashed line. It is seen that the averaged values of anisotropy are closer to the experimental data A_{exp}. Consequently we confirm that the measured cross section of the fission fragment formation and their angular distribution are results of mixing of the quasifission and fusion-fission products. In the upper panel of Fig.8, we compare experimental and theoretical values of mean square values of angular momentum. The experimental $\langle L^2\rangle$ values are obtained from the measured anisotropy A_{exp} and K_0^2 values used to fit measured angular distributions presented in Fig. 1. The theoretical values for fusion-fission and quasifission fragments are calculated by averaging ℓ^2 using the partial cross sections of the quasifission (dashed line) and complete fusion (solid line) events. The experimental data are well described with the averaged values of L^2 between the complete fusion and quasifission cross sections:

$$\langle L^2\rangle = \frac{(\sigma_{\text{fus}}\langle L^2\rangle_{\text{fus}} + \sigma_{\text{qf}}\langle L^2\rangle_{\text{qf}})}{\sigma_{\text{fus}} + \sigma_{\text{qf}}}. \quad (17)$$

In the lower panel of Fig.8, the experimental results of K_0^2 are compared with the theoretical values obtained from the description of A_{exp} (dashed line in Fig.7) and $\langle L^2\rangle$ (dot-dashed line in Fig.8) extracted from the experimental data of the angular distribution of fission fragments. This comparison shows again the dominancy of the quasifission fragments into measured data at low energies.

5. CONCLUSION

The fission angular distributions for the ^{32}S+^{184}W reaction have been measured. We obtained the experimental cross sections of fission-like, anisotropy A_{exp}, and $\langle l^2\rangle$ values compared with the results of the DNS model calculations. Assuming the measured fission cross sections is equal to the capture cross section they were decomposed into fusion-fission, quasifission and fast fission contributions by the dinuclear system model. The hindrance to complete fusion at small collision energies increases due to the increase of the quasifission events and it is explained by the elongated shape of the dinuclear system which is formed in collisions with small orientation angle of the symmetry axis of ^{184}W to the beam direction. Because of collisions with small orientation angles the intrinsic fusion barrier B^*_{fus} is larger and quasifission barrier B_{qf} is smaller in comparison with the corresponding barriers in the case of collision with large orientation angles. An increase of the quasifission contribution at large beam energies is connected by the angular momentum dependence of the quasifission B_{qf} and intrinsic fusion B^*_{fus} barriers: at large angular momentum of dinuclear system B_{qf} decreases and B^*_{fus} increases. The large quasifission barrier increases the life-time of dinuclear system allowing it to transform into a CN [18, 7]. It is concluded that the effects of competition between fusion and quasifission in the reaction play an important role in the dynamics process. The total evaporation residue and fusion-fission excitation functions are calculated in the framework of the advanced statistical model. The maximal value of the total cross section of the evaporation residue is 0.12 mb and fusion-fission cross section is comparable with the quasifission cross section. The contribution of the fast fission appears at $E_{\text{c.m.}} > 131$ MeV and it is sufficiently small in comparison with contributions of fusion-fission and quasifission reactions.

ACKNOWLEDGMENTS

This work was supported by the Major State Basic Research Development Program under Grant No. G2000077400 and the National Natural Science Foundation of China under Grant Nos. 10375095, 10735100, and 10811120019. A.A. Nasirov is grateful to the Istituto Nazionale di Fisica Nucleare and Department of Physics of the University of Messina for the support received in the collaboration between the Dubna and Messina groups.

References

[1] B.B. Back, *et al.*, Phys. Rev. C**32**, 195 (1985)

[2] V. V. Volkov, N. A. Antonenko, E. A. Cherepanov, A. K. Nasirov, V. P. Permjakov, Phys. Lett B **319** 425 (1993); Phys. Rev. C51, 2635 (1995)
[3] G. Fazio, G. Giardina,. G. Mandaglio, F. Hanappe, A. I. Muminov, A. K. Nasirov, U. T. Yakhshiev, F. Hanappe, T. Materna, and L. Stuttge, Eur. Phys. Jour. A **19** 89 (2004).
[4] G.G. Adamian, N.V. Antonenko, W. Scheid, Phys. Rev. C **68**, 034601 (2003).
[5] H. Q. Zhang, C. L. Zhang, C. J. Lin, Z. H. Liu, F. Yang, A. K. Nasirov, G. Mandaglio, M. Manganaro and G. Giardina, Phys. Rev. C **81**, 034611 (2010).
[6] G. Giardina, S. Hofmann, A. I. Muminov, A. K. Nasirov, Eur. Phys. J. A **8** 205 (2000).
[7] G. Fazio, G. Giardina,. G. Mandaglio, R. Ruggeri, A. I. Muminov, A. K. Nasirov, Yu. Ts. Oganessian, A. G. Popeko, R. N. Sagaidak, A. V. Yeremin, S. Hofmann, F. Hanappe, C. Stodel, Phys.Rev. C **72** 064614 (2005).
[8] A. D' Arrigo, G. Giardina, M. Herman, and A. Taccone, Phys. Rev. C **46**, 1437 (1992).
[9] A. D' Arrigo, G. Giardina, M. Herman, A. V. Ignatyuk, A. Taccone, J. Phys. G **20**, 365 (1994).
[10] R. N. Sagaidak, V. I. Chepigin, A. P. Kabachenko, J. Rohac, Yu. Ts. Oganessian, A. G. Popeko, A. A. Yeremin, A. D'Arrigo , G. Fazio, G. Giardina, M. Herman, R. Ruggeri, and R. Stiricle, J. Phys. G **24**, 611 (1998).
[11] A. Sobiczewski, M. Kowal, Int. Jour. of Mod. Phys. E **18**, No: 4, 869 (2009)
[12] I. Muntian, Z. Patyk, A. Sobiczewski, Phys. of At. Nuclei **66** Issue: 6, 1015 (2003).
[13] A.K. Nasirov, A.I. Muminov, R.K. Utamuratov, G. Fazio, G. Giardina, F. Hanappe, G. Mandaglio, M. Manganaro, and W. Scheid, Eur. Phys. J. A **34**, 325 (2007).
[14] D.J. Hinde, M. Dasgupta, J.R. Leigh, J.C.Mein, C.R.Morton, J.O. Newton, H. Timmers, Phys. Rev. C **53**, 1290 (1996).
[15] M. G. Itkis et al., Nucl. Phys. A**734**, 134 (2004).
[16] G. N. Knyazheva, E. M. Kozulin, R. N. Sagaidak, A. Yu. Chizhov, M. G. Itkis, N. A. Kondratiev, V. M. Voskressensky, M. Stefanini, B. R. Behera, L. Corradi, Phys. Rev. C **75**, 064602 (2007).
[17] J. R. Huizenga, A. N. Behkami, and L. G. Moretto, Phys. Rev. **177** 1826 (1969)
[18] Avazbek Nasirov, Akira Fukushima, Yuka Toyoshima, Yoshihiro Aritomo, Akhtam Muminov, Shuhrat Kalandarov, Ravshanbek Utamuratov, Nucl. Phys. A **759**, 342 (2005).
[19] C. Gregoire, C. Ngo, E. Tomasi, B. Remaud, F. Scheuter, Nucl.Phys. **A387**, 37c (1982).
[20] V. E. Viola et, al Phys. Rev. C **31** 1550 (1985)
[21] A. K. Nasirov, G. Giardina, G. Mandaglio, and M. Manganaro, F. Hanappe, S. Heinz and S. Hofmann, A. I. Muminov, W. Scheid Phys.Rev. C **79**, 024606 (2009).
[22] G. Fazio, G. Giardina, F. Hanappe, G. Mandaglio, M. Manganaro, A. I. Muminov, A. K. Nasirov, and C. Sacca, J. Phys. Soc. Jpn. **77**, 124201 (2008).
[23] V.V. Volkov, in Contributed Papers of Nucleus-Nucleus Collision II, Visby, 1985, edited by B. Jakobson and K. Aleclett (North-Holland, Amsterdam, 1985), Vol.1, p.54; Izv. Akad. Nauk SSSR Ser. Fiz. 50, 1879 (1986); in Proceedings of International School-Seminar on Heavy Ion Physics, Dubna, 1986, D7-87-68 (Dubna, 1987), p. 528; in Proceedings of the 6th International Conference on Nuclear Reaction Mechanisms, Varenna, 1991, edited by E. Gadioli (Ricerca Scientifica ed Educasione Permanente Supplemento n.84, 1991), p.39. on Heavy Ion Physics
[24] P. Moller and J. R. Nix, Atomic Data and Nuclear Data **39**, 213 (1988).
[25] S. Raman, C. H. Malarkey, W. T. Milner, C. W. Nestor, Jr., and P. H. Stelson, Atomic Data and Nuclear Data **36**, 1 (1987).
[26] R. H. Spear, Atomic Data and Nuclear Data **42**, No. 1, 55 (1989).
[27] G. Fazio, G. Giardina, G. Mandaglio, F. Hanappe, A. I. Muminov, A. K. Nasirov, W. Scheid, L. Stittge, Mod. Phys. Lett. A Vol. **20** No. 6 (2005) 391-405.
[28] K. Pomorski, B. Nerlo-Pomorska, A. Surowiec, M.Kowal, J. Bartel, K. Dietrich, J. Richert, C. Schmitt, B. Benoit, E. de Goes Brennand, L. Donadille, C. Badimong, Nucl. Phys. A **679** (2000) 2553
[29] A. J. Sierk, Phys. Rev. C **33**, 2039 (1988).
[30] G. Fazio, G. Giardina, A. Lamberto, A. I. Muminov, A. K. Nasirov, F. Hanappe and L. Stuttge, Eur. Phys. Jour. A **22** 75 (2004).
[31] Z. H. Liu, H.Q. Zhang, J.C. Xu, Y. Qiao, X. Qian, and C.J. Lin, Phys. Lett. B **353**, 173 (1995).
[32] I. Halpern and V. M. Strutinski, Proceeding of the Second United National International Conference on the Peaceful Uses of Atomic Energy, Geneva, 1958 (United Nations, Geneva, 1958), Vol. 15,p.408.
[33] J. J. Griffin, Phys. Rev **116**, (1959).
[34] D.J. Hinde, A. C. Berriman, M. Dasgupta, J. R. Leigh, J. C. Mein, C. R. Morton, and J. O. Newton, Phys. Rev. C **60**, 054602 (1999).

Halo effects on fusion cross section in $^{4,6}He+^{64}Zn$ collision around and below the coulomb barrier

M Fisichella[1,2], V Scuderi[2,3], A Di Pietro[2], P Figuera[2], M Lattuada[2,3], C Marchetta[2], M Milin[4], A Musumarra[2,3], M G Pellegriti[2,3], N Skukan[5], E Strano[2,3], D Torresi[2,3], M Zadro[5]

[1] Dipartimento di Fisica, Università di Messina, Messina, Italy
[2] INFN- Laboratori Nazionali del Sud and sezione di Catania, Catania, Italy
[3] Dipartimento di Fisica ed Astronomia, Università di Catania, Catania, Italy
[4] Department of Physics Faculty of Science University of Zagreb, Zagreb, Croatia
[5] Ruđer Bošković Institute, Zagreb, Croatia

Abstract. The structure of the halo nuclei is expected to influence the fusion mechanism at energies around and below the Coulomb barrier. Here new data of $^{4}He+^{64}Zn$ at sub-barrier energies are presented which cover the same energy region of previous measurements of $^{6}He+^{64}Zn$. The fusion cross section was measured by using an activation technique where the radioactive evaporation residues produced in the reaction were identified by the X-ray emission which follows their electron capture decay. By comparing the two system, we observe an enhancement on the fusion cross section in the reaction induced by ^{6}He, at energy below the Coulomb barrier. It is shown that this enhancement seems to be due to static properties of halo 2n ^{6}He nucleus.

1. Introduction

In recent years, great effort have been made to investigate the effect of the projectile structure on the fusion reaction mechanism in collisions induced by halo nuclei, at energies around and below the Coulomb barrier. In typical halo nuclei, such as ^{6}He, ^{11}Li, ^{11}Be, ^{19}C or ^{8}B, the outer nucleons are characterized by low binding energy (0.1-1 MeV) and they occupy single-particle states having low angular momentum (*s* or *p*). The corresponding single-particle wave function of these valence nucleons has a long tail, which extends mostly outside the potential well. The structure of halo nuclei is expected to strongly affect the reaction mechanisms, especially at energies near the Coulomb barrier. Direct processes are expected to be favored due to the low binding energies of valence nucleons. Concerning the effects that halo structure could generate on the fusion mechanism one should take into account two different categories: static and dynamic effects. Static effects are due to the long range of the radial wave-functions associated with the valence orbitals. The presence of the diffuse halo affects the shape of projectile-target potential, thus reducing the Coulomb barrier and therefore enhancing the fusion cross section around and below the barrier. Dynamic effects are due to the coupling of the relative motion of projectile and target to their intrinsic excitations or to the other reactions channels. In particular, due to the very low break-up threshold, coupling with the break-up may be important. A lot of theoretical studies have been done in order to understand the role of these dynamic effects on the fusion cross section with halo nuclei. However, until now, contradictory effects on fusion cross section have been predicted by different theoretical models. The role of the break-up process of halo and weakly bound projectile in fusion have been investigated via Coupled Channel

(CC) calculations. In order to properly describe the reaction dynamics, in collisions involving these nuclei, the coupling not only to bound but also to continuum states (break-up states) must be included in CC calculations. Continuum-Discretised Coupled-Channel (CDCC) calculations have been, hence, developed [1,2]. These calculations predict that dynamical effects enhance the sub-barrier total fusion (complete + incomplete) cross-section with respect to the nocoupling case. In addition, due to the break-up, a suppression of the total fusion cross-section is predicted by CDCC calculations at energies above the barrier [2]. However, the calculated cross-sections and hence the amount of sub-barrier enhancement, depend upon the adopted approximations and the phase space in the continuum considered (range of energies, energy bin and partial waves) [3]. Opposite results have been obtained with a different approach [4] based on a time dependent wave-packet formalism which uses a three body model (core, halo and target). According to this model, the cross-section in the neutron halo case is slightly suppressed compared to the non-halo case.

2. Reaction induced by halo nuclei

The experimental investigation on the reaction processes, using radioactive beams, is quite difficult owing to the very low intensity (10^5-10^7pps) of the available halo beams, coupled to the difficulty of measuring fusion cross-sections at low energies for reactions induced by light projectiles. Fusion reactions induced by the 2n halo ^{6}He nucleus in the energy region around the Coulomb barrier are the most studied of any involving light radioactive beams. This is because ^{6}He beams are available in several radioactive beam facilities with good intensities (up to 10^7pps). The only low energy reaction studies involving the 1n halo ^{11}Be, existing in the literature, concerns the system ^{11}Be+^{209}Bi. Also from the experimental point of view different authors did not reach similar conclusion about the enhancement/suppression effect on the fusion cross section due to the projectile halo structure since most of the existing data do not really explore the region below the barrier with reasonable errors. Moreover is not always clearly discussed which is the role played in the observed final result by different static and dynamic effects. However, a remarkable feature of reactions involving halo nuclei is also the very large near and sub-barrier direct reaction cross-sections, dominating fusion by orders of magnitude. Different papers have evidenced that direct reactions completely dominate the total reaction cross-section at these energies [5,6,7,8].

The first measurement of fusion with ^{6}He was performed at the Flerov laboratory in Dubna for the system ^{6}He+^{209}Bi at energies above the barrier [9]. In this experiment, a very large fission cross-section was measured and it was attributed to fusion-fission processes. The same reaction was successively measured at the TwinSol facility at the University of Notre Dame by Kolata and collaborators [10]. The fusion cross-section was extracted using an activation technique by detecting the α particles emitted in the on-line and off-line decay of the evaporation residues. An enhancement of the sub-barrier fusion with respect to a statistical model calculation and to a single barrier penetration calculations [10], but also to fusion induced by ^{4}He onto the same ^{209}Bi target was found. Usually the fusion cross section of the reactions induced by ^{6}He are compared with those obtained using ^{4}He beam on the same target. The ^{4}He represents the ^{6}He core and by comparing the two systems one can observe the effect due to the presence of valence neutrons.

The reaction ^{4,6}He+^{238}U were measured twice at the CRC in Louvain la Neuve by the same group [11,12] . They determined the fusion cross section by the detection of the fission fragments. A large enhancement of fission around and below the barrier in the ^{6}He case respect the ^{4}He ones was observed. The analysis of the light particle, emitted in coincidence with the fission fragments, allowed to conclude that the observed fission enhancement is due to transfer-fission events and no effect on fusion due to the structure of ^{6}He are observed.

The fusion reaction ^{6}He + ^{206}Pb was performed at the ISOL facility DRIBs at the Flerov Laboratory in Dubna [13]. The 2n evaporation channel is compared to the measured 1n evaporation channel of ^{4}He+^{208}Pb. In this paper an enhancement of the sub-barrier fusion cross-section for the ^{6}He induced reaction is observed with respect to the ^{4}He ones. In a model of "sequential fusion" proposed by Zagrabaev [14], the transfer of the two "valence" neutrons in the ^{6}He (with positive Q-value) causes a gain in the energy of the relative motion of the two nuclei and thus facilitates fusion. Therefore, the enhancement is attributed to the coupling to the transfer channel having positive Q-value. However, the two reactions ^{6}He+^{206}Pb and ^{4}He+^{208}Pb form the same compound nucleus, but owing to the very different reaction *Q*-value this, for the same $E_{c.m.}$, is formed at a different excitation energy. In particular at the lowest measured beam energy the excitation energy of the compound nucleus ^{212}Po, formed in the reaction with ^{4}He beam, is below the particle threshold, therefore it would be necessary to compare the measured total fusion cross-section for the two reactions ^{4,6}He + ^{208}Pb rather than the excitation function for one particular evaporation channel.

Navin et al. [19] measured neutron-transfer and fusion cross-sections for the ^{6}He +^{63,65}Cu, at energies around the Coulomb barrier at the SPIRAL ISOL facility at GANIL. They compared this measurement with the ones obtained using same targets and ^{4}He beam. The last measurement was made at the 14UD BARC-TIFR Pelletron Accelerator Mumbai. In the case of ^{6}He beam they performed coincident measurement between γ-rays and light charged particle, too, which allowed to disentangle fusion and neutron transfer mechanisms. In particular they observed ^{66}Cu residue cross-sections much larger than the ones calculated using the statistical model. This result suggested that the dominant production mechanism was other than fusion-evaporation, most probably neutron transfer. Charged particle- γ-ray coincidence events were consistent with this hypothesis.

The only fusion reaction, present in literature, induced by ^{11}Be was performed at RIKEN, Japan, in two successive experiment [15,16]. Also in this case the evaporation residues were identified from their delayed α activity. The cross-section data, for the weakly bound one-neutron halo ^{11}Be, were compared with those for the well bound ^{10}Be nucleus. In addition, the fusion of the stable but weakly bound ^{9}Be nucleus with ^{209}Bi was measured by the same group at the Munich Tandem accelerator [Sig99]. The complete fusion cross-section reported in [16] for the three reaction ^{9,10,11}Be+^{209}Bi show no difference within the errors.

2.1. ^{6}He+^{64}Zn experiment

In [6,18] we had performed the ^{6}He+^{64}Zn reaction at the Centre of Research of Cyclotron in LLN to measure elastic scattering, direct reaction and fusion cross section at energies around the Coulomb barrier. The same quantities were also measured using a ^{4}He beam. To measure the fusion cross-section we used an activation technique based on the off-line measurement of the atomic X-ray emission following the electron capture decay of the evaporation residues produced in the reaction. This technique will be discussed, in details, in the next paragraph. Stacks of four ^{64}Zn targets, each followed by a ^{93}Nb catcher, were irradiated with ^{6}He beam in order to measure different points in the fusion excitation function with a single ^{6}He energy. The fusion cross-section could be obtained by summing up the cross-sections of all evaporation channels. The same nuclei can be produced by fusion-evaporation and by other reaction processes. For these reason we had performed a comparison of the measured cross-section for the different channels with a statistical model calculation, performed with the CASCADE code. This comparison showed a good agreement except for the ^{65}Zn residue, were a large enhancement with respect to the calculation was observed. The excess in the yield measured for this channel was attributed to one and two neutron transfer reactions. Therefore, in order to obtain the fusion excitation function we have subtracted the contribution due to the transfer by replacing the measured value for ^{65}Zn with the one calculated using the statistical model calculations. The results are shown in figure 1. No evident effects were observed for ^{6}He+^{64}Zn fusion with respect to ^{4}He+^{64}Zn within the measured energy range. However the ^{6}He data extended lower energies than the ^{4}He ones. Therefore, new data with ^{4}He beam are needed in order to investigate the role of valence neutrons at energies below to the Coulomb barrier.

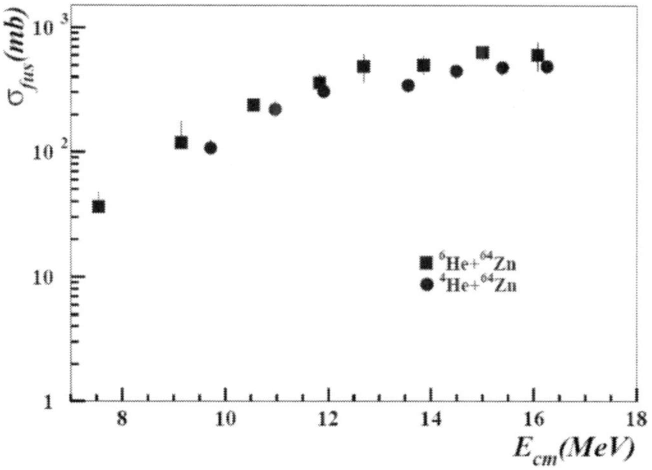

Figura 1. $^4He+^{64}Zn$ (circles) and $^6He+^{64}Zn$ (squares) fusion cross-sections [6]

3. New $^4He+^{64}Zn$ experiment

3.1 Experimental technique

The new $^4He+^{64}Zn$ experiment was performed at Ruder Boskovic Institute (Zagreb) with an average ^4He beam current of about 10^{11} pps. The reaction was measured at E_{lab}=8.1 MeV, 9.2 MeV and 10.1 MeV. As in our previous measurement [6,18] the fusion excitation function was performed using the activation technique. The direct detection of evaporated residues (E.R.), produced in the collision of a low energy light projectile onto a medium mass target is not always possible since the largest fraction of E.R. will not come out from the target owing to the their low kinetic energy. However by choosing a suitable target, as in our case, almost all the produced radioactive residues decay by electron capture (E.C). It is possible to obtain E.R. unstable against E.C. decay and "detect" the E.R. by looking at X-ray emitted in their decay. This technique consists of two steps. The first is the target activation. During this time the Zn target (540μg/cm^2) is first irradiated with the ^4He beam. The fusion-evaporation reactions take place within the target and the largest fraction of E.R. produced will not come out from it. The small fraction of residues emerging from the target is stopped in the ^{93}Nb catcher (1000μg/cm^2), placed behind the target. In figure 2 a schematic view of experimental set-up is shown.

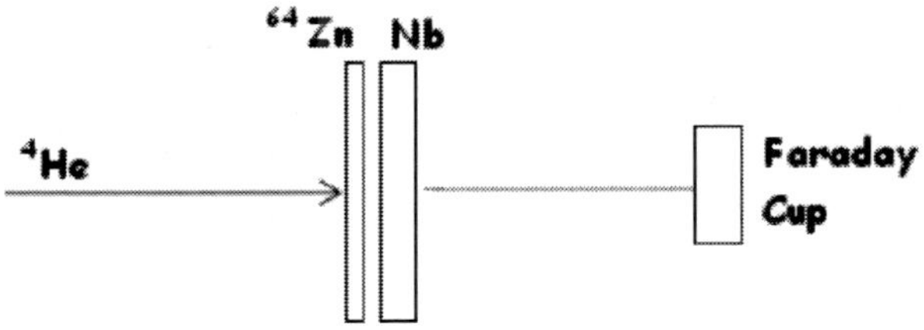

Figura 2. Experimental set-up for the fusion measurement. The faraday cup was used to measure the beam current.

Possible reactions induced by the beam on the ^{93}Nb catchers do not represent a problem since the X-ray energies are different to the ones corresponding to reactions on ^{64}Zn. At the end of the irradiation time the activated target and catcher are placed very close to the Si(Li) detector to measure the residues activity. The irradiated target were measured at Laboratori Nazionali del Sud for about twenty days. One of the advantages of this technique is the 100% intrinsic detection efficiency of Si(Li) detector in the energy region of interest (7-11 MeV). Moreover the atomic X-ray in this energy region can be detected with extremely low background .

3.2 Experimental result

Typical X-ray spectra measured off-line for the reaction ^{4}He+^{64}Zn are shown in figure 3, where the two peaks correspond to the K_α and K_β X-ray emission of Zn. The K_β emission represents about 15% of the total k X-rays emission. In the present experiment, the analysis was performed only on the K_α lines.

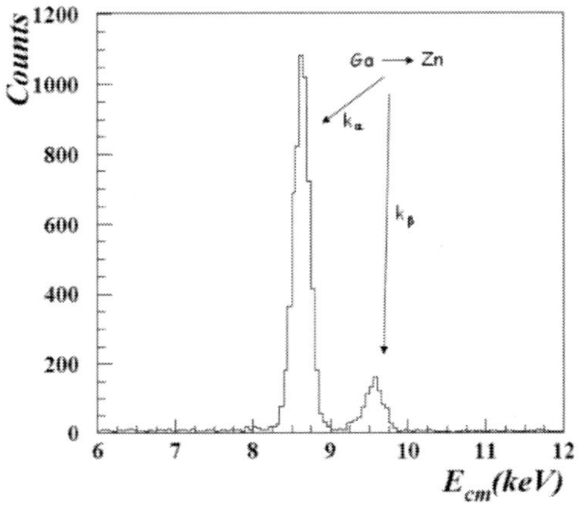

Figure 3. Typical X-ray spectra measured off-line for the reaction ^{4}He+^{64}Zn. It is possible to distinguish two peaks which correspond to the Kα and K_β X-ray emission of Zn

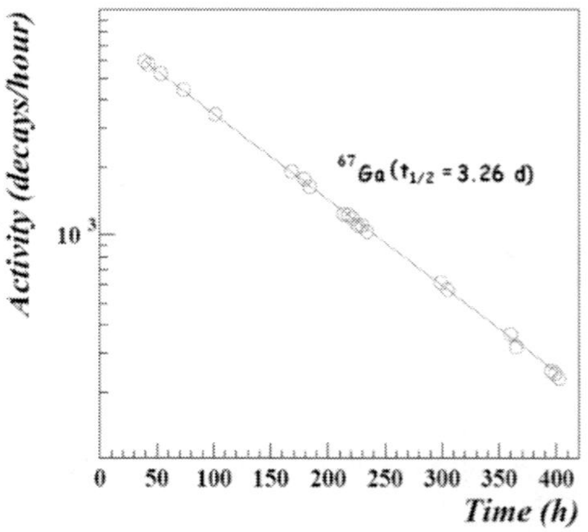

Figure 4. Activity curve for the ^{67}Ga isotope extracted in the activation run.

From the X-ray energies we can only identify different elements but not different isotopes. However it is possible to discriminate the different isotope contributions by following the X-ray activity as a function of time. In the ^{4}He+^{64}Zn reaction we had estimated with statistical model calculations that the only residues produced could be ^{67}Ge and ^{67}Ga. However, the ^{67}Ge is a shortlived nucleus ($t_{1/2}$= 18 minutes) and it decays by E.C. 100% into ^{67}Ga. For this reason after one day we expect to observe only the ^{67}Ga contribution. In figure 4 the activity curve for the Ga isotope is shown. As one can see from the slope of the activity curve only the contribution of ^{67}Ga isotope is present, as expected. Therefore, the total fusion cross section corresponds to the longer lived ^{67}Ga ($T_{1/2}$=3.26 d) cross section extracted by using the following formula:

$$\sigma = \frac{A_{0exp}}{N_i N_t \lambda_0 K_\alpha \varepsilon_T} \quad (1)$$

A_{0exp} is the experimentally measured activity at the end of the irradiation as obtained from the fit of the activity curve (figure 4). This term is then corrected for the known K_α fluorescence probability and for the detector efficiency (ε_T); λ_0 is the deacy constant; N_t is the number of target atoms per cm^2

and N_i is the integrated beam current (i.e. the number of incident particles). This was determined by using a faraday cup. A correction for the X-ray absorption by ^{64}Zn and ^{93}Nb foils has also been made. In figure 5 both the preliminary data of ^{4}He+^{64}Zn (open circles) and the previous data of ^{4,6}He+^{64}Zn [6] are shown. By observing in particular the points around 9.5 MeV, we notice that the new data are in good agreement with the previous ones.

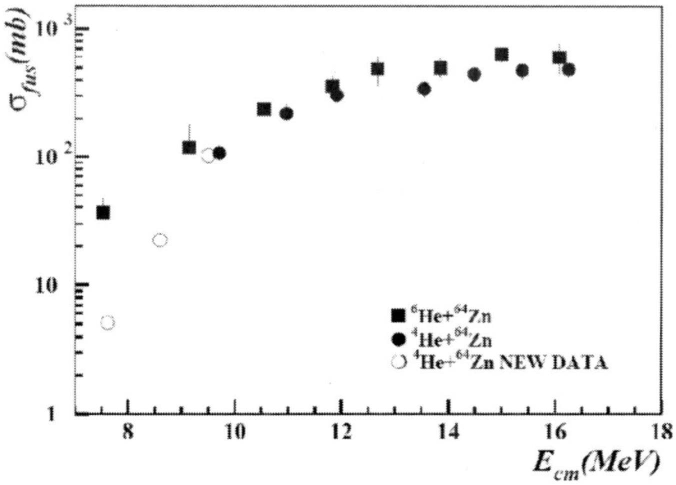

Figura 5. ^{6}He+^{64}Zn (squares) and ^{4}He+^{64}Zn (circles) fusion excitation function. The ^{6}He and ^{4}He at upper energies (closed circles) are the data of the previous experiment [6]. With the open circles are plotted the new preliminary ^{4}He data.

Concerning the comparison of the two fusion excitation function an enhancement of the fusion excitation functions at the energies below the Coulomb barrier in the case of ^{6}He with respect the ^{4}He one is observed. The enhancement is attributed to the halo structure of ^{6}He, however this comparison includes both the static and the dynamic effect, which are respectively due to the size of the system and the coupling to other channels, such as break-up.

4. Conclusion

In the last years different experiments have been performed to study the fusion mechanisms in collisions induced by halo nuclei. Performing a comparison of $\sigma_{fus}(E_{cm})$ between halo and corresponding well bound isotopes a clear systematic behavior for all systems is not yet observed.
Concerning our data the comparison of the ^{4,6}He+^{64}Zn we previously measured has been extended to lower energies by using an activation technique. It has been observed an enhancement of the ^{6}He+^{64}Zn fusion cross section with respect to ^{4}He+^{64}Zn at energies around and below the Coulomb barrier.
In general we underline that most of the existing data do not really explore the region below the barrier with reasonable errors and it is not always clearly discussed which is the role played by static and dynamic effects in the fusion mechanism. Therefore we believe that new fusion cross-section measurements at sub-barrier region are needed.

5. References

[1] Hagino K et al. 2002 *Phys. Rev.* **C61** 037602
[2] Diaz-Torres A, Thompson J J 2002 *Phys. Rev. C* **65** 024606.
[3] Canto L F et al. 2006 *Phys. Rep.* **424**, 1.

[4] Ito M, et al. 2006 *Phys. Lett. B* **37**, 53
[5] Aguilera E F et al. 2001 *Phys. Rev. C* **63**, 061603(R).
[6] Di Pietro A et al. 2004 *Phys. Rev. C* **69** 044613.
[7] Bychowski J P et al. *2004 Phys. Rev. Lett.* **596** 26.
[8] De Young P Aet al., 2005 *Phys. Rev. C* **71** 051601(R).
[9] Fomichev A S et al. 1995 *Z. Phys. A* **351** 129.
[10] Kolata J Jet al. 1998 *Phys.Rev.Lett.* **81** 4580.
[11] Trotta M et al. 2000 *Phys. Rev. Lett.* **84** 2342
[12] Raabe R et al. 2004 Nature **431** 823
[13] Penionzhkevich Y Eet al. 2006 *Phys. Rev. Lett.* **96** 162701.
[14] Zagrebaev V I 2003 *Phys. Rev. C* **67** 061601(R)
[15] Signorini C et al.1998 *Eur. Phys. J. A* **2** 227
[16] Signorini C et al.2004 *Nucl. Phys. A* **735** 329.
[17] Signorini C et al.1999 *Eur. Phys. J. A* **5** 7.
[18] Di Pietro A et al. 2007 *Eur.Phys.J. Special Topics* **150** 15.
[19] Navin A et al., 2004 *Phys. Rev. C* **70** 044601

Journal of Physics: Conference Series **282** (2011) 012015 doi:10.1088/1742-6596/282/1/012015

Decay studies of rare-earth nuclei to superheavy elements and the associated shell effects

Sushil Kumar

Department of Physics, Chitkara University, solan-174103, H.P., India

E-mail: sushilk17@gmail.com

Abstract. The effects of shell closure in nuclei via cluster decay have been investigated. In this context,the Preformed Cluster Model (*PCM*) based on Quantum Mechanical Fragmentation Theory is used. The key point in the cluster radioactivity is that it involves the interplay of close shell effects of parent and daughter nucleus. Small half life for a parent indicates shell stabilized daughter and long half life indicates the stability of the parent against the decay. Cluster decays of rare-earth nuclei are studied with a view to look for neutron magic shells for the $_{50}Sn$ nucleus as the daughter product always, to find the most probable cluster decays and the possibility, if any, of new neutron shells. The recently observed α-decay chains $^{293-294}117$ which were produced by the fusion reactions with target ^{249}Bk and projectile ^{48}Ca at Dubna, in Russia. The α-decay calculations are performed for all the parents of the two alpha decay chains and compared with the experimental data and with other theoretical model.

1. Introduction

Since the discovery of ^{14}C-decay from ^{223}Ra by Rose and Jones [1] in 1984, many other ^{14}C-decays from other radioactive nuclei (^{221}Fr, ^{221,222,224,226}Ra, ^{223,225}Ac and ^{226}Th) and some 12 to 13 neutron-rich clusters, such as ^{20}O, ^{23}F, $^{22,24-26}$Ne, ^{28,30}Mg and ^{32,34}Si, have been observed experimentally for the ground-state decays of translead ^{226}Th to ^{242}Cm parents [2, 3, 4, 5], which all decay with the doubly closed shell daughter ^{208}Pb (Z=82, N=126) or its neighboring nuclei. Theoretically, such an exotic natural radioactivity of emitting particles (nuclei) heavier than α-particle was already predicted in 1980 by Sandulescu, Poenaru and Greiner [6] on the basis of the Quantum Mechanical Fragmentation Theory (QMFT). Todate, ^{34}Si is the heaviest cluster observed with the longest decay half-life ever measured ($log_{10}T_{1/2}(s) = 29.04$) from ^{238}U parent, and the smallest branching ratio of cluster w.r.t. α-decay ($B = \lambda_{cluster}/\lambda_{\alpha} \sim 10^{-17}$ for ^{28,30}Mg decay of ^{238}Pu [4].

Keeping in mind the doubly magic nature of the ^{208}Pb daughter, a second island of heavy-cluster radioactivity was predicted by Poenaru et. al. [7] on the basis of analytical superasymmetric fission model (ASAFM). in the decays of some neutron-deficient rare-earth nuclei in to ^{100}Sn (Z=N=50) daughter or a neighboring nucleus. Recently, Sarkar et al., [8] predicted a new shell closure at $N = 90$ for the Sn isotopes on the basis of shell model calculations. Experimentally, several unsuccessful attempts [9, 10, 11, 12] have been made to measure the ^{100}Sn-daughter radioactivity from the ^{114}Ba parent nucleus produced in ^{58}Ni+^{58}Ni reaction. Instead, a new phenomenon of intermediate mass fragments (IMFs, with $3 \leq Z \leq 9$), also referred to as 'clusters' or 'complex fragments', emitted from the excited compound nucleus, was observed at various incident energies [13, 14, 15, 16], with ^{12}C as one the fragment or cluster emitted. However, the above study of ground-state cluster-decays of rare-earth nuclei still remains of interest from the point of view of associated shell effects.

In this paper, the heavy cluster emissions to many more rare-earth parents (329 cases) with $_{50}Sn$ always as the daughter product is considered. Specifically, we have considered the emission of various isotopes of C, O, Ne, Mg, Si, S, Ar, Ca, Ti, Cr, Fe and Ni respectively, from neutron-deficient to neutron-rich Ba, Ce, Nd, Sm, Gd, Dy, Er, Yb, Hf, W, Os and Pt parents, with a view to look for ^{100}Sn and ^{132}Sn radioactivities, as well as any other new Sn radioactivity with new shell closures in neutrons. Since the cluster decays are more probable with daughters as magic nuclei, the decay half-lives are expected to drop (be minimum) for the magic daughters. The same idea was utilized earlier for the (spherical) sub-shell closed $_{40}Zr$ daughter [17, 18], including also a brief report of the results on $_{50}Sn$ daughter [17].
Next, the superheavy mass region extend effectively due to the availability and advancement in the radioactive nuclear beam technology. The ^{48}Ca beam is most prominent at present for the synthesis of superheavy elements (SHE). The neutron rich superheavy elements Z= 112-116, and 118 have been produced by (the doubly magic nucleus) ^{48}Ca induced reactions with actinide targets ^{238}U, ^{237}Np, ^{244}Pu, ^{243}Am, ^{248}Cm and ^{248}Cf respectively at low excitation energies at FLNR in Dubna, Russia [19, 20, 21]. In addition to the above mentioned SHE, recently two isotope $^{293}117$ and $^{294}117$ are observed of a new superheavy element Z=117, as a result of the fusion reaction between ^{48}Ca projectiles and radioactive ^{249}Bk target nuclei [22]. In this reaction, the excitation energy of the compound nucleus $^{297}117$ is reported to be E^*= 39 MeV and E^*= 35 MeV, which as a result de-excite by the emission of four and three neutrons, respectively. Recently observed two isotope of the element Z=117 with mass number 293 and 294 are studied for the alpha decay characteristics. Calculations for the $^{293,294}117$ alpha decay chains are compared with the experimental results [22] and the macroscopic- micrscopic(MM) model dependent calculations of A. Sobiczewski [23].

The calculations are based on the preformed cluster model (PCM)[24, 25, 26], described briefly in Sect. 2. The results of calculation are presented in Sect. 3 and a summary of results is given in Sect. 4.

2. The Preformed Cluster Model (PCM)

The preformed cluster model (PCM) [24, 25, 26] uses the dynamical collective coordinates of mass and charge asymmetries

$$\eta = (A_1 - A_2)/(A_1 + A_2)$$

and

$$\eta_Z = (Z_1 - Z_2)/(Z_1 + Z_2),$$

first introduced in the Quantum Mechanical Fragmentation Theory [27, 28, 29]. These are in addition to the usual coordinates of relative separation R and deformations β_i $(i = 1, 2)$. Then, in the standard approximation of decoupled R- and η-motions [24, 25], The decay constant λ(the decay half-life $T_{1/2}$)in PCM is defined as

$$\lambda = \frac{ln2}{T_{1/2}} = P_0 \nu_0 P. \tag{1}$$

Here P_0 is the cluster (and daughter) preformation probability and P the barrier penetrability which refer, respectively, to the η and R motions. The ν_0 is the barrier assault frequency. The

P_0 are the solutions of the stationary Schrödinger equation in η,

$$\{-\frac{\hbar^2}{2\sqrt{B_{\eta\eta}}}\frac{\partial}{\partial\eta}\frac{1}{\sqrt{B_{\eta\eta}}}\frac{\partial}{\partial\eta}+V_R(\eta)\}\psi^{(\nu)}(\eta)=E^{(\nu)}\psi^{(\nu)}(\eta), \tag{2}$$

which on proper normalization are given as

$$P_0=\sqrt{B_{\eta\eta}}\mid\psi^{(\nu)}(\eta(A_i))\mid^2(2/A), \tag{3}$$

with i=1 or 2 and ν=0,1,2,3.... (ν=0 coresspond to the ground state and ν=1,2,3...for the higher excited states.) Eq. (2) is solved for the ground state (ν=0) and at a fixed $R=R_a=C_t(=C_1+C_2)$. The C_i are Süssmann central radii $C_i=R_i-(1/R_i)$, with the radii $R_i=1.28A_i^{1/3}-0.76+0.8A_i^{-1/3}fm$. The fragmentation potential $V_R(\eta)$ in (2) is calculated simply as the sum of the Coulomb interaction, the nuclear proximity potential [30] and the ground state binding energies of two nuclei,

$$V(R_a,\eta)=-\sum_{i=1}^{2}B(A_i,Z_i)+\frac{Z_1Z_2e^2}{R_a}+V_P, \tag{4}$$

with B.E.'s taken from the 2003 experimental compilation of Audi and Wapstra [31] and from the 1995 calculations of Möller et al. [32] whenever not available in [31]. Thus, full shell effects are contained in our calculations that come from the experimental and/or calculated [32] binding energies.
The charges Z_1 and Z_2 in (4) are fixed by minimizing the potential in η_Z coordinate. The Coulomb and proximity potentials in (4) are for spherical nuclei. The mass parameters $B_{\eta\eta}(\eta)$, representing the kinetic energy part in (2), are the classical hydrodynamical masses [33].
The WKB tunnelling probability, calculated is $P=P_iP_b$ with

$$P_i=exp[-\frac{2}{\hbar}\int_{R_a}^{R_i}\{2\mu[V(R)-V(R_i)]\}^{1/2}dR] \tag{5}$$

$$P_b=exp[-\frac{2}{\hbar}\int_{R_i}^{R_b}\{2\mu[V(R)-Q]\}^{1/2}dR]. \tag{6}$$

These integrals are solved analytically [25] for R_b, the second turning point, defined by $V(R_b)=Q$-value for the ground-state decay. The assault frequency ν_0 in (1) is given simply as

$$\nu_0=(2E_2/\mu)^{1/2}/R_0, \tag{7}$$

with $E_2=(A_1/A)Q$, the kinetic energy of the lighter fragment, for the Q-value shared between the two products as inverse of their masses.

3. Results and Discussion

In the present study, however, as other most probable clusters (isotopes of O, Ne, Mg, Si, S, Ar, Ca, Ti, Cr, Fe and Ni) from heavier neutron-deficient and neutron rich rare-earth parents ($^{118-170}$Ce, $^{118-176}$Nd, $^{122-184}$Sm, $^{132-190}$Gd, $^{132-194}$Dy, $^{138-200}$Er, $^{148-200}$Yb, $^{154-208}$Hf, $^{156-208}$W, $^{160-210}$Os and $^{168-210}$Pt) are also considered, interestingly, ^{12}C remains to be the most favorable cluster-decay from ^{112}Ba parent with ^{100}Sn-daughter, but for ^{132}Sn-daughter the most favorable cluster is ^{78}Ni from ^{210}Pt. This is illustrated in Fig. 2 for both the ^{100}Sn and ^{132}Sn daughters, where the most probable clusters emitted from Ba to Pt parents are plotted. The

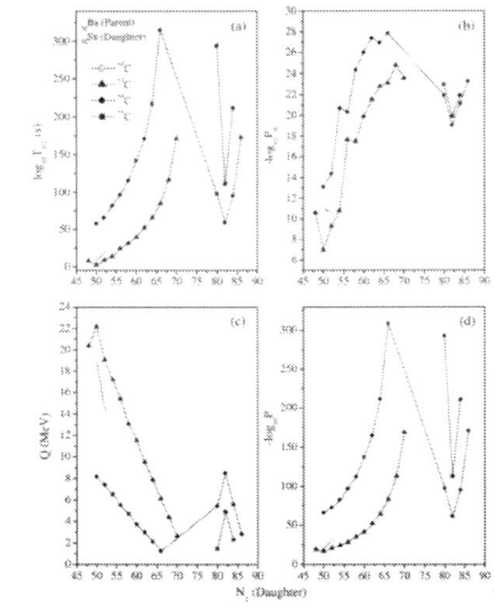

Figure 1. The decay half-lives $T_{1/2}$(s) and other characteristic quantities like preformation factors P_0, Q-values (in MeV), and penetrabilities P of different Carbon clusters emitted with $_{50}$Sn daughters from various isotope of Ba nuclei, plotted as a function of daughter neutron number N_1.

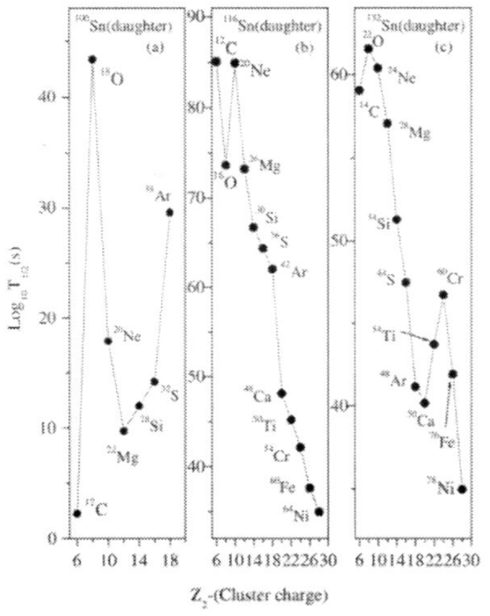

Figure 2. $log_{10}T_{1/2}(s)$ for the most probable clusters emitted from various Ba to Pt parents with (a) ^{100}Sn (b) ^{116}Sn and (c) ^{132}Sn daughter, calculated on the basis of the PCM, plotted as a function of cluster proton number Z_2. Note the different ordinate-scales used in these figures.

fact that the most probable cluster ^{78}Ni, arising from Pt parents, occur, at $N_1 = 82$ of the $_{50}$Sn daughter, is illustrated in Fig. 3 for $T_{1/2}$ alone. Interesting enough, in addition to the strong minima at daughter neutrons $N_1 = 82$, a minimum is also shown to be persent at $N_1 = 66$ for the $_{50}$Sn daughter, with ^{64}Ni as the emitted cluster from ^{180}Pt parent. Thus, a new possibility of ^{116}Sn-daughter radioactivity is indicated.
Finally, Fig. 4 gives a complete histogram of the decay half-lives $log_{10}T_{1/2}$(s) as a function of the neutron number N_1 of the emitted $_{50}$Sn-daughters with the most probable clusters (minimum $T_{1/2}$ values) from some 329 parents taken from Ba to Pt with mass numbers $A = 110 - 210$. Note that here $_{50}$Sn-daughter is kept fixed and all possible clusters are considered from different parents (total 1617 combinations with $_{50}$Sn-daughter and probable cluster), and then the one with minimum half-life time is plotted. Apparently, the shortest half-life time $log_{10}T_{1/2}(s) = 2.27$ (with Q-value=22.16 MeV) is obtained for ^{12}C decay of ^{112}Ba. The role of magic $N_1 = 82$ is also evident with a minimum in the histogram at ^{132}Sn-daughter due to the emission of ^{78}Ni cluster from ^{210}Pt parent. The predicted half-life $log_{10}T_{1/2}(s) = 34.974$ (with a Q-value=119.292 MeV), which is beyond the limit of present day experiments. Thus, as expected, the strongest shell effects occur at $N_1 = 50$ and 82. In addition, another minimum due to ^{64}Ni cluster emitted from ^{180}Pt parents could also be of interest for a closed shell (either spherical and/or deformed) at $N_1 = 66$. This minima are comparable to $N_1 = 82$, with a predicted decay half-life of $log_{10}T_{1/2}(s) = 34.975$(with a Q-value=124.192 MeV), which by all means is very large for experiments.

In the second section of superheavy elements decay study, Fig.5 contains two α-decay chains

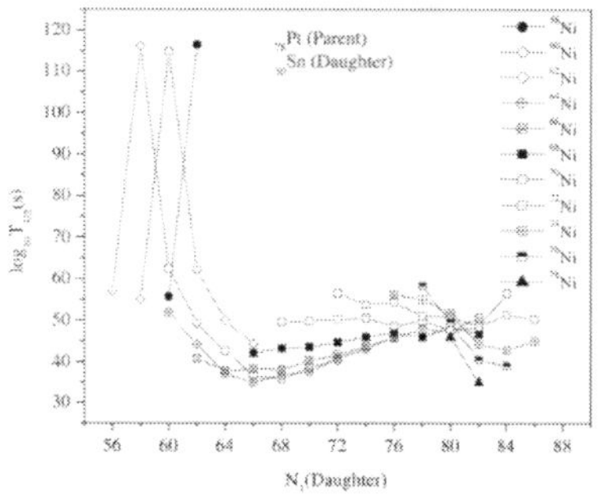

Figure 3. The decay half-lives $T_{1/2}$(s) plotted as a function of daughter neutron number N_1 and for different Ni clusters emitted from various Pt parents with $_{50}$Sn daughters.

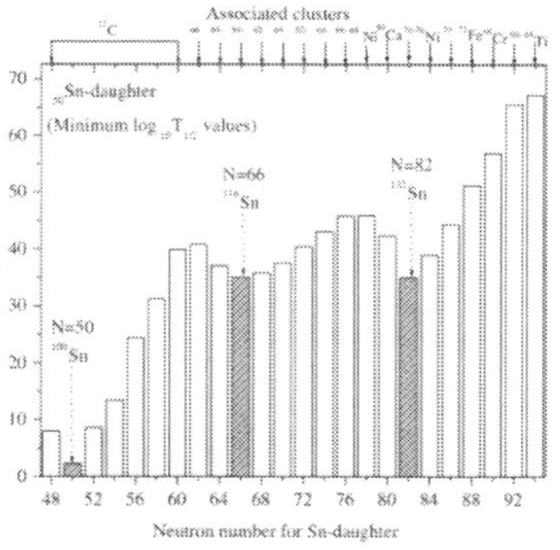

Figure 4. Histogram of $log_{10}T_{1/2}(s)$ versus $_{50}$Sn-daughter neutron number N_1 for the most probable clusters emitted from various Ba to Pt parents with $_{50}$Sn as a daughter nucleus always, calculated on the basis of the PCM. The associated clusters are also shown on the top panel.

of the Z=117 element, $^{294}117$ and $^{293}117$ are shown for the alpha decay half-lives and Q-value calculations. These two decay chains are calculated for the two different Q-values i.e., Q^{PCM} and Q^{Exp} to see the effect of the Q-value on the alpha decay half-lives. The Q-value of $^{293-294}117$ parents in PCM is more than two order less than the experimental and the MM values, and hence the correspond change in half-life. This value is calculated by using with B.E.'s taken from the 2003 experimental compilation of Audi and Wapstra [31] and from the 1995 calculations of Möller et al. [32] whenever not available in [31]. The experimental alpha decay half-lives are maximum at $^{274}107$ (Z=107,N=167) and $^{286}113$ (Z=113,N=173) parent nuclei, in $^{294}117$ α-decay chain. While in $^{293}117$ α-decay chain it is maximum at $^{285}113$ (Z=113,N=172) parent nucleus. Maximum half-lives at these nuclei can be attributed either due to the magicity of protons at Z= 114 or neutrons at N=172 or due to both. Interesting, the smaller Q-value means larger half-life, which is followed by one α-decay chain $^{293}117$ but not completely followed by the second $^{294}117$ α-decay chain. This discrepancy is noticed at $^{290}115$ and $^{282}111$ parent nuclei. Since both the theoretical model (PCM and MM) follow the same trend.

In Fig.5 the PCM model calculations for the alpha decays are compared with the experimental results and the macroscopic-microscopic (MM) model calculations. The PCM model calculations for the alpha decay half-lives shows the same trend as the MM calculations and experimental results. The differences in half lives may be either due to the consideration of alpha decays in these decay chains from ground state of the parent to the ground state of the daughter. Whereas the possibility of alpha-decay from the excited state also exists [34, 35, 36]. Also, the penetration and preformation probabilities are calculated for both the alpha decay chains $^{293,294}117$. The preformation probability factor P_0 calculated from eq.(3) is larger for the $^{290}115$ and $^{293}117$ parents in the ground state i.e., ν=0. Here, decay constant λ defined in eq. (1) is a combined effect of the P_0 the cluster (and daughter) preformation probability, P the barrier penetrability and the ν_0 which is the barrier assault frequency.

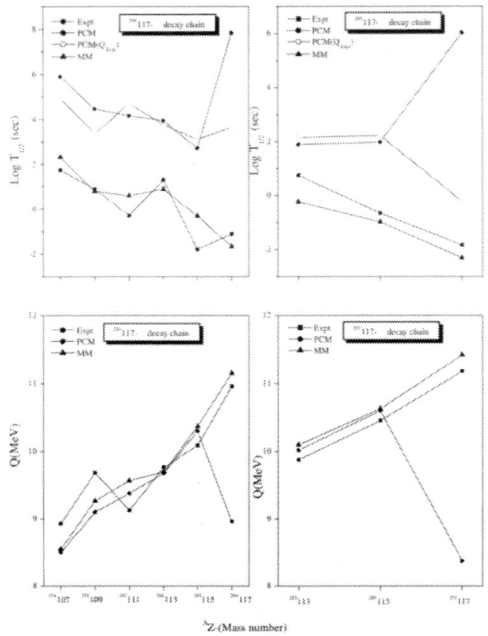

Figure 5. The α-decay half-lives calculated on the basis of PCM and comparision with experimental data[22] and those calculated on the basis of the MM [23] plotted as a function of the parent nucleus mass for the α-decay chains of $^{294-293}117$.

4. Summary

As already stated in the introduction, the cluster decays of various isotopes of $_{56}$Ba to $_{78}$Pt parents are calculated for the daughter nucleus to be always an isotope of $_{50}$Sn nucleus. For example, for the neutron-deficient $^{110-132}$Ba and neutron-rich $^{144-150}$Ba parents considered here, different isotopes of Carbon cluster would give rise to various isotopes of $_{50}$Sn daughter. This is illustrated in Fig. 1 for the Q-values of various C-decays, together with the logarithms of penetrability P, preformation factor P_0, and the decay half-life $T_{1/2}$ as a function of N_1, the neutron number of $_{50}$Sn daughter. The impinging frequency ν_0 is nearly constant $\sim 10^{21}(s^{-1})$. All the four quantities Q, P, P_0, and $T_{1/2}$ show the shell effects at magic N_1=50 and 82; the Q, P and P_0 being large and $T_{1/2}$ small at these numbers. Thus, the most favorable decay is ^{12}C from ^{112}Ba nucleus in the $48 \leq N_1 \leq 70$ region, leaving behind ^{100}Sn as the daughter product, and the ^{14}C cluster from ^{146}Ba in the $72 \leq N_1 \leq 86$ region with ^{132}Sn as the daughter product. The calculated alpha decay half-lives are compared with the experimental results and the other theoretical model (MM) calculations. The PCM calculation results shows the same trend as in experimental data and MM calculations. In the alpha decay study, role of the Q-value is also studied. A small change in the Q-value produce a large variation in the half-life.

5. Acknowledgments

The author is greatful to the Organizing Committee of International Symposium Quasifission Process in Heavy Ion Reactions at University of Messina, (Italy) for the invitation to present this talk and kind hospitality during the conference. The author also very much thankful to the Chitkara University and Department of Science and Technology, Govt of India, for financial support, File No. SR/ITS/02851/2010-2011.

6. References

[1] Rose H J and Jones H J 1984 *Nature (London)* **307**, 245
[2] Gupta R K and Greiner W, 1994 *Int. J. Mod. Phys.* E **3**, 335.
[3] Bonetti R and Guglielmetti A, in *Heavy Elements and Related New Phenomena*, edited by W. Greiner and R. K. Gupta, (World Scientific, Singapore, 1999), Vol. II, p. 643.
[4] Bonetti R and Guglielmetti A 2007 *Romanian Reports in Physics* **59**, 301.
[5] Guglielmetti A et al. 2008 *Journal of Phys.: Conference Series* **111**, 012050.
[6] Săndulescu A, Poenaru D N and Greiner W 1980 *Sov. J. Part. Nucl.* **11**, 528.
[7] Poenaru D N, Greiner W and Gherghescu R 1993 *Phys. Rev.* C **47**, 2030.
[8] Sarkar S and Saha Sarkar M 2010 *Phys. Rev.* C **81**, 064328.
[9] Oganessian Yu Ts *et al.* 1994 *Z. Phys.* A **349**, 341.
[10] Guglielmetti A, *et al.* 1995 *Phys. Rev.* C **52**, 740.
[11] Guglielmetti A, *et al.* 1997 *Phys. Rev.* C **56**, R2912.
[12] Mazzocchi C, *et al.* 2002 *Phys. Lett.* B **532**, 29.
[13] Gomez del Campo J, *et al.* 1988 *Phys. Rev. Lett.* **61**, 290.
[14] Gomez del Campo J, *et al.* 1991 *Phys. Rev.* C **43**, 2689.
[15] Gomez del Campo J,*et al.* 1998 *Phys. Rev.* C **57**, R457.
[16] La Commara M,*et al.* 2000 *Nucl. Phys.* A **669**, 43.
[17] Kumar S, Balasubramaniam M and Gupta R K, Symp. on Nucl. Phys., Mumbai (India), Dec. 8-12, Vol. **46B**, (Published by the Department of Atomic Energy (DAE)-Board of Research in Nuclear Sciences (BRNS), Govt. of India, 2003), p. 30.
[18] Kumar S, *et al.* 2009 *J. Phys. G: Nucl. Part. Phys.* **36**, 015110.
[19] Oganessian Yu Ts *et al.* 2004 *Phys. Rev.* C **69**, 021601 (R).
[20] Oganessian Yu Ts *et. al.* 2006 *Phys. Rev.* C **74**, 044602.
[21] Oganessian Yu Ts *et. al.* 2007 *J. Phys. G: Nucl. Part. Phys.* **34**, R165, and references therein.
[22] Oganessian Yu Ts *et. al.* 2010 *Phys. Rev. Lett.* **104**, 142502.
[23] Sobiczewski A 2010 *Acta Phy. Pol.* B **41**, 157.
[24] Gupta R K , in *Proceedings of the 5th International Conference on Nuclear Reaction Mechanisms*, Varenna, Italy, Editor: E. Gadioli, (Ricerca Scientifica ed Educazione Permanente, Milano, 1988), p. 416.
[25] Malik S S and Gupta R K, Phys. Rev. **C39**, 1992 (1989).
[26] Kumar S and Gupta R K 1997 *Phys. Rev.* C **55**, 218.
[27] Maruhn J and Greiner W 1974 *Phys. Rev. Lett.* **32**, 548.
[28] Gupta R K , Scheid W, and Greiner W 1975 *Phys. Rev. Lett.* **35**, 353.
[29] Gupta R K 1977 *Sovt. J. Part. Nucl.* **8**, 289; 1978 *Nucl. Phys. and Solid St. Phys. Symp.* (India) **21A**, 171.
[30] Blocki J, Randrup J, Swiatecki W J, and Tsang C F 1977 *Ann. Phys. (NY)* **105**, 427.
[31] Audi G, Wapstra A H and Thibault C 2003 *Nucl. Phys.* A **729**, 337.
[32] Möller P, Nix J R, Myers W D and Swiatecki W J 1995 *At. Data Nucl. Data Tables* **59**, 185.
[33] Kröger H and Scheid W 1980 *J. Phys. G* **6**, L85.
[34] Delion D S *et al* 2007 *Phys. Rev.* C **76**, 044301.
[35] Jianmin Dong, Hongfei Zhang, Yanzhao Wang, Wei Zuo, Junqing Li 2010 *Nuclear Physics* A **832**,198.
[36] Dongdong Ni1 and Zhongzhou Ren 2010 *Phys. Rev.* C **81**, 024315.

Journal of Physics: Conference Series **282** (2011) 012016 doi:10.1088/1742-6596/282/1/012016

Bremsstrahlung emission of photons accompanying ternary fission of ^{252}Cf

S P Maydanyuk[1,◇] **and V S Olkhovsky**[1,†],
G Mandaglio[2,3], **M Manganaro**[2,3], **G Fazio**[2,3], **G Giardina**[2,3,⋆] **and C Saccá**[4]

[1] Institute for Nuclear Research, National Academy of Science of Ukraine, Kiev, 03680, Ukraine
[2] Dipartimento di Fisica dell'Università di Messina, Messina, 98166, Italy
[3] Istituto Nazionale di Fisica Nucleare, Sezione di Catania, Catania, 95123, Italy
[4] Dipartimento di Scienze della Terra dell'Università di Messina, Messina, 98166, Italy

E-mail: ◇ maidan@kinr.kiev.ua

E-mail: † olkhovsk@kinr.kiev.ua

E-mail: ⋆ giardina@nucleo.unime.it

Abstract. We present the first results on the bremsstrahlung emission of photons accompanying ternary spontaneous fission of the ^{252}Cf nucleus. We also compare our calculations on the basis of quantum model with preliminary experimental data and find a good agreement between theory and experiment for photon energies up to 500 keV, when the α-particle emission is in presence of the field of two fission fragments of the daughter nucleus.

1. Introduction

Nuclear fission accompanied by light charged particle emission (often called ternary fission) has been widely studied [1–4]. The main interest about this process is related to the possibility of obtaining new additional information about the fission dynamics. The study of the bremsstrahlung emission of photons accompanying ternary fission allows us other opportunities in understanding of this process. This process has never been studied elsewhere. In this paper we present our model and calculations of the bremsstrahlung probabilities in the ternary fission of the ^{252}Cf nucleus.

After review in Sect. 2 of our results in determination of bremsstrahlung probabilities during the spontaneous fission of ^{252}Cf, in Sects. 3 and 4 we present our first results in the theoretical study of bremsstrahlung photons emitted during the ternary fission of the ^{252}Cf nucleus. In Sect. 5 we summarize the paper.

2. Model of the bremsstrahlung emission accompanying the spontaneous fission

2.1. Shape of the nuclear system undergoing fission

In the fission process it is necessary to describe in a continuous way the sequence of shapes of a fissioning nucleus from its ground state, through its saddle and scission configurations, to the separated fragments at infinity. In a study of ground state shapes one has to be able to describe a sphere, oblate and prolate spheroids, octupole deformations, and positive and

Journal of Physics: Conference Series **282** (2011) 012016 doi:10.1088/1742-6596/282/1/012016

negative hexadecapole deformations. Since many questions studied in the fission process do not sensitively depend upon deviations from axial symmetry, we shall initially apply the model to shapes of nuclear systems that are axially symmetric.

We chose to specify the nuclear shape in terms of smoothly joined portions of three quadratic surfaces of revolution: two spheroids connected by a hyperboloidal neck (see Fig. 3 of Ref. [6]). In terms of the cylindrical coordinate system, we shall use the equation for the drop's surface written explicitly as [6]

$$\rho^2 = \begin{cases} a_1^2 - (a_1^2/c_1^2)\,(z - l_1)^2, & \text{for } l_1 - c_1 \le z \le z_1 \\ a_2^2 - (a_2^2/c_2^2)\,(z - l_2)^2, & \text{for } z_2 \le z \le l_2 + c_2 \\ a_3^2 + (a_3^2/c_3^2)\,(z - l_3)^2, & \text{for } z_1 \le z \le z_2. \end{cases} \tag{1}$$

Here, the quantity l_i specifies the position of the center of the ith quadratic surface, c_i is its symmetry axis, and a_i is its transverse semiaxis ($i = 1, 2, 3$). There are nine coordinates in the specification of a nuclear shape. However, following the description reported in the paper of Bolsterli *et al.* [6], one coordinate is eliminated by the assumption that the volume of the nuclear system remains constant during its evolution, and other two coordinates are eliminated by the requirement that the middle surface joins smoothly with both surfaces of the forming fragments (at points denoted by z_1 and z_2). This introduces three relations between the original nine degrees of freedom, and reduces the number to six. Elimination of the center of mass coordinate finally reduces the number of shape coordinates to five. In Fig. 1 (a) one can see what shapes we have obtained by such a way at different distances between centers of the ^{12}C fragment and the daughter nucleus formed in fission of the ^{252}Cf nucleus.

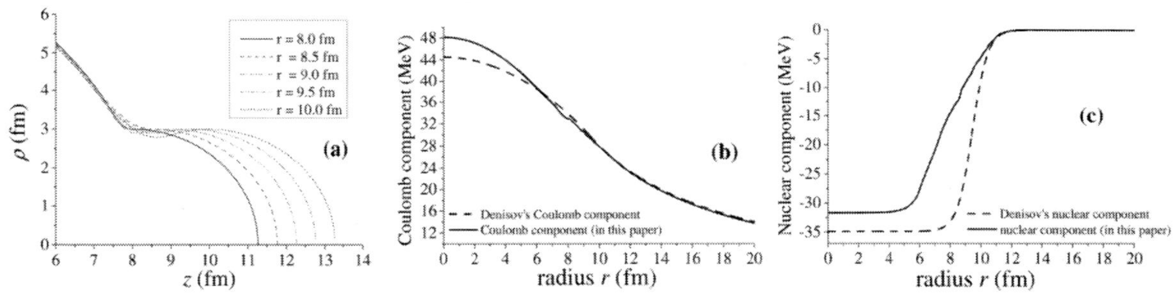

Figure 1. (a) Shapes of the nuclear surface for emission of the ^{12}C fragment calculated for different distances r between centers of the daughter nucleus and the fragment during spontaneous fission process of the ^{252}Cf nucleus (axes z and ρ are defined from eq. (1)). Potential obtained in this paper and spherically symmetric potential proposed by Denisov and Ikezoe in [8] for the α decay of the ^{252}Cf nucleus for (b) the Coulomb components and (c) the nuclear components .

2.2. Interaction potential between daughter nucleus and emitted fragments

Once the nuclear shape has been specified, the next step is to generate an interaction potential between the daughter nucleus and fragment with respect to the shape and distance between such nucleus and fragment. If such a procedure is useful in the general theory of fission, it has to be capable of handling very deformed shapes and the evolution from one parent nucleus to two final fragments (the daughter nucleus and the related fission fragment). The interaction potential between the daughter nucleus and the fragment could be written as

$$V_{\text{total}}(\mathbf{r}) \;=\; V_{\text{N}}(\mathbf{r}) + V_{\text{C}}(\mathbf{r}). \tag{2}$$

We define the nuclear component $V_{\rm N}(\mathbf{r})$ and the Coulomb component $V_{\rm C}(\mathbf{r})$ as

$$V_{\rm N}(\mathbf{r}) = E_{\rm N,\,nucleus}(\mathbf{r}) - E_{\rm N,\,fragment}, \qquad V_{\rm C}(\mathbf{r}) = E_{\rm C,\,nucleus}(\mathbf{r}) - E_{\rm C,\,fragment}, \tag{3}$$

where $E_{\rm N,\,nucleus}(\mathbf{r})$ and $E_{\rm C,\,nucleus}(\mathbf{r})$ are nuclear and Coulomb energies for the total nuclear system, $E_{\rm N,\,fragment}(\mathbf{r})$ and $E_{\rm C,\,fragment}(\mathbf{r})$ are nuclear and Coulomb energies for that fragment only, defined in the folding form (see also Ref. [6]):

$$\begin{array}{ll} E_{\rm N,\,nucleus}(\mathbf{r}) = -\lambda_N \displaystyle\int\limits_V \frac{d\mathbf{r}^3}{1+\exp(|\mathbf{r}-\mathbf{r}'|/a)}, & E_{\rm N,\,fragment} = -\lambda_N \displaystyle\int\limits_V \frac{d\mathbf{r}^3}{1+\exp(|\mathbf{r}|/\mathbf{a})}, \\ E_{\rm C,\,nucleus}(\mathbf{r}) = \lambda_C \displaystyle\int\limits_{V,\,\mathbf{r}\neq\mathbf{r}'} \frac{d\mathbf{r}'^3}{|\mathbf{r}'-\mathbf{r}|}, & E_{\rm C,\,fragment} = \lambda_C \displaystyle\int\limits_{V,\,r\neq 0} \frac{d\mathbf{r}^3}{|\mathbf{r}|}. \end{array} \tag{4}$$

The integration is over the volume V of the shape defined relatively given distance r between the daughter nucleus and the fragment emitted. Introduced parameters $\lambda_C = Z_1 Z_2 e^2/V_p$ and $\lambda_N = M_p/V_p$ (where V_p and M_p are volume and mass of the parent nucleus) give us practically very precise coincidence between the potential defined by (2)–(4) and the potential obtained by procedure of parametrization proposed by Denisov and Ikezoe in Ref. [8] for α-decay (see Fig. 1 (b, c) for ^{252}Cf).

The total potential depends on the difference between centers of the daughter nucleus and the emitted fragment, and we use the spherically symmetric approximation for further calculation of the wave functions.

The Q value concerning each emitted fragment in fission is calculated by standard procedure as

$$Q = M_p - M_d - M_f, \tag{5}$$

where M_p, M_d and M_f are the masses of the parent, daughter and specific fragment nuclei, respectively (see Ref. [7]).

2.3. Photon emission probability accompanying the spontaneous fission

We define the photon emission probability formed by a separate fragment during fission in terms of the transition matrix elements for the composite quantum system (the daughter nucleus and the fragment) from its state before the photon emission (initial state i) into its state after the photon emission (finale state f), as [9]

$$\frac{dP(w,\vartheta)}{dE_\gamma} = N_0\, w\, |p\,(w,\vartheta)|^2, \qquad k_{i,f} = \sqrt{2m\,E_{i,f}}, \qquad w = E_i - E_f, \tag{6}$$

where

$$p(w,\vartheta) = -\sqrt{\frac{1}{3}} \sum_{l=0}^{+\infty} i^l (-1)^l\,(2l+1)\,P_l(\cos\vartheta) \sum_{\mu=-1,1} h_\mu J_{m_f}(l,w), \qquad h_\pm = \mp\frac{1}{\sqrt{2}}(1\pm i). \tag{7}$$

Here, the vector $\mathbf{k}$ represent the photon impulse in the direction of its propagation, the vector $\mathbf{r}$ is the radius vector marking the position of center of the emitted fragment relative to the center of the daughter nucleus, and θ is the angle between the direction $\mathbf{n}_1 = \mathbf{r}/r$ of the fragment motion and the propagation direction $\mathbf{n}_2 = \mathbf{k}/k$ of the photon emitted, where $k = |\mathbf{k}|$ and $r = |\mathbf{r}|$. $E_{i,f}$ and $k_{i,f}$ are total energy and wave vector of the system in the initial state i or in the final state f, $\psi_i(\mathbf{r})$ and $\psi_f(\mathbf{r})$ are the wave functions of the system in the initial and final state i and f, $w = k = |\mathbf{k}|$ is the photon frequency (energy) (see Refs. [10–13] for a detailed description of all parameters, calculation of wave functions, boundary conditions etc.). The radial integral $J_{m_f}(l,w)$ is

$$J_{m_f}(l,w) = \int_0^{+\infty} r^2\, R_f^*(r,E_f)\, \frac{\partial R_i(r,E_i)}{\partial r}\, j_l(kr)\, dr, \tag{8}$$

where $R_i(r)$ and $R_f(r)$ are the radial components of the total wave functions $\psi_i(\mathbf{r})$ and $\psi_f(\mathbf{r})$, $j_l(kr)$ is the spherical Bessel function of order l. N_0 is a coefficient calculated by (see Ref. [14])

$$N_0 = \frac{Z_{\rm eff}^2 e^2}{(2\pi)^4 m}, \tag{9}$$

where $Z_{\rm eff}$ and m are the effective charge and reduced mass of the daughter-fragment system, respectively.

2.4. Calculation of bremsstrahlung spectra for the spontaneous fission of ^{252}Cf

We applied the above-described method to calculate the spectrum of photons emitted during the spontaneous fission of the ^{252}Cf nucleus [9]. At first, we had to check our model and calculations on a simple problem: the estimation of the bremsstrahlung radiation formed by only one fragment with arbitrary mass and charge numbers A and Z, respectively which participates in the complex process of fission.

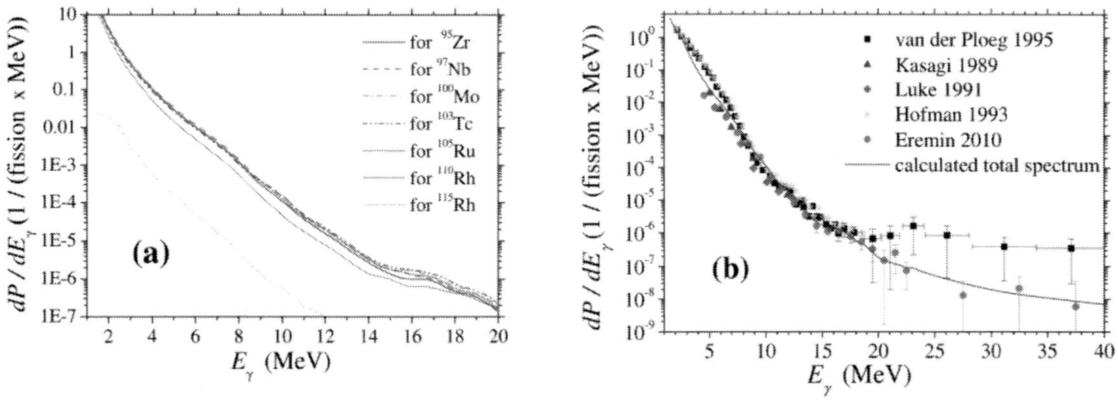

Figure 2. (a) Calculation of the bremsstrahlung photon emission probabilities caused by heavy fragments included in mass region $A = 95 - 115$ during the spontaneous fission of ^{252}Cf [9]; (b) Total bremsstrahlung photon probability in the spontaneous fission of ^{252}Cf [9]: calculation of the spectrum (solid line) obtained by averaging the spectra for contributions of all fragments; in comparison with the sets of experimental data given by squares [16], triangles [17], diamonds [18], stars [19], and circles [20].

Calculations for heavy fragments are the most difficult. Fragments with large masses participate in fission with larger Q-values. In order to achieve stable spectra, it is necessary to take a larger number of oscillations in the integrant function into account for the integration of the matrix element. Moreover, the high energies of emitted photons reinforce such difficulty. After explicit integration of the matrix element in far asymptotic region, we have improved our code that allows us to achieve such stability for different heavy fragments. Results of such calculations for some fragments within the 95-115 mass region are presented in Fig. 2 (a).

The total spectrum of photons which are emitted during the spontaneous fission of ^{252}Cf was obtained by averaging the spectra for contributions of all fragments, and normalizing the resulting spectrum to the experimental data at lower energies. Such obtained spectrum is presented in Fig. 2 (b) by solid line. Our calculations were obtained in the framework of a fully quantum approach and agree well with the experimental data in literature concerning bremsstrahlung photon emission in the spontaneous fission of ^{252}Cf in the E_γ energy range up to about 20 MeV. In the 20-38 MeV range, however, the calculations agree only with Eremin *et al.*'s experimental data [20]. So, this model could be a convenient basis of developing the study of photon emission in the ternary fission of the same ^{252}Cf nucleus.

3. Model for the bremsstrahlung emission accompanying ternary fission

In order to generalize the above-described model for the emission probability in the ternary fission it is necessary to solve the following steps:

(i) to define the geometry of the surface of nuclear system undergoing ternary fission;
(ii) to calculate the interaction potential between the α-particle and the other two fragments of fissioning nucleus;
(iii) to calculate the matrix element of the α-emission by taking the motion of the two fragments into account;
(iv) to reach the condition that the final spectra have to be stable for light and heavy fragments of the fissioning nucleus, for different geometries of fragments in the fissioning process.

3.1. Geometry of the fissioning system and interaction potential

In the first step we define the geometry of the nuclear system which decays into three fragments. In Fig. 3 (a) are reported some parameters which we use.

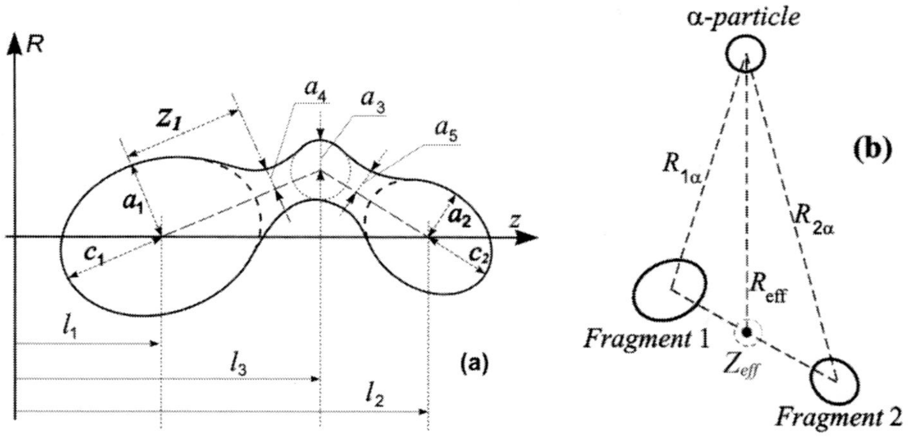

Figure 3. Geometry of the fissioning system: (a) separation into three fragments; (b) transition to the effective charge Z_{eff} at far distances.

In order to study the ternary fission process, as emitted fragment we shall choose α-particle. By using the surface of the nuclear system reported in Fig.3 (a), we calculate the interaction potential between the α-particle and daughter nucleus which is composed of two fragments. Here, we use standard folding procedure. The new potential V_{total} depends on the geometry of the relative positions of the two fragments and the α-particle. Here, as the main parameter, we consider the relative distance R_{12} between the centers of masses of two fragments. So, we write the total potential as

$$V_{\text{total}}(\mathbf{r}, \mathbf{R}_{12}) = V_{\text{N}}(\mathbf{r}, \mathbf{R}_{12}) + V_{\text{C}}(\mathbf{r}, \mathbf{R}_{12}), \tag{10}$$

where the nuclear component $V_{\rm N}(\mathbf{r})$ and the Coulomb component $V_{\rm C}(\mathbf{r})$ are

$$\begin{aligned} V_{\rm N}(\mathbf{r},\mathbf{R}_{12}) &= V_{\rm N,\,nucleus}(\mathbf{r},\mathbf{R}_{12}) - E_{\rm N,\,fragment}, \\ V_{\rm C}(\mathbf{r},\mathbf{R}_{12}) &= V_{\rm C,\,nucleus}(\mathbf{r},\mathbf{R}_{12}) - E_{\rm C,\,fragment}. \end{aligned} \quad (11)$$

Here, $V_{\rm N,\,nucleus}(\mathbf{r},\mathbf{R}_{12})$ and $V_{\rm C,\,nucleus}(\mathbf{r},\mathbf{R}_{12})$ are the nuclear and Coulomb potential components of the total daughter nuclear system, while $E_{\rm N,\,fragment}(\mathbf{r})$ and $E_{\rm C,\,fragment}(\mathbf{r})$ are the nuclear and Coulomb potential components for the separated fragments, defined by the folding form:

$$\begin{aligned} V_{\rm N,\,nucleus}(\mathbf{r},\mathbf{R}_{12}) &= -\lambda_N \int\limits_{V(\mathbf{R}_{12})} \frac{d\mathbf{r}^3}{1+\exp(|\mathbf{r}-\mathbf{r}'|/a)}, & E_{\rm N,\,fragment} &= -\lambda_N \int\limits_{V} \frac{d\mathbf{r}^3}{1+\exp(|\mathbf{r}|/a)}, \\ V_{\rm C,\,nucleus}(\mathbf{r},\mathbf{R}_{12}) &= \lambda_C \int\limits_{V(\mathbf{R}_{12}),\,\mathbf{r}\neq\mathbf{r}'} \frac{d\mathbf{r}'^3}{|\mathbf{r}'-\mathbf{r}|}, & E_{\rm C,\,fragment} &= \lambda_C \int\limits_{V,\,r\neq 0} \frac{d\mathbf{r}^3}{|\mathbf{r}|}. \end{aligned} \quad (12)$$

3.2. Calculations of the Coulomb component at far distances

In principle, by using the above-described procedure , one can calculate the wave functions and bremsstrahlung probabilities. However, in order to realize such a procedure we have to solve some particular problems: the time of calculations of spectra on the basis of the folding procedure is very long, and it is very difficult to obtain convergent calculations and stable spectra because we have to take far distances into account where it is needed to calculate the wave function starting from the corresponding folding potential.

For such reasons, in order to calculate the Coulomb component at far distances we change two fragments (located at distances $\mathbf{R}_{1\alpha}$ and $\mathbf{R}_{2\alpha}$ from the α-particle) with a new effective fragment of $Z_{\rm eff}$ charge located at the distance $\mathbf{R}_{\rm eff}$ from the α-particle with the aim to have the same influence on the α-particle as the two original fragments (see Fig. 3 (b)). Therefore, in eqs. (10)–(12) we use $\mathbf{R}_{\rm eff}$ instead of the variable $\mathbf{r}$ for a fixed angular geometry. We determine this $Z_{\rm eff}$ charge from the following condition:

$$Z_{\rm eff}(R_{\rm eff}) = R_{\rm eff}\left\{\frac{Z_1}{R_{1\alpha}} + \frac{Z_2}{R_{2\alpha}}\right\}, \quad (13)$$

where $R_{\rm eff} = |\mathbf{R}_{\rm eff}|$, $R_{1\alpha} = |\mathbf{R}_{1\alpha}|$, $R_{2\alpha} = |\mathbf{R}_{2\alpha}|$. It turns out that such an approach allows us to achieve stability in calculations of the bremsstrahlung spectra which are convergent. Another important result of this procedure is that the spectra are sensitive to the chosen geometry of the fissioning nucleus.

3.3. Results

We shall start from considerations of calculations obtained for the α-decay of ^{252}Cf. In Fig. 4 (a) we present our results for the bremsstrahlung emission of photons (full line) caused by the α-particle of E_α=6.2 MeV when the daughter nucleus is not separated into two fragments. As one can see this spectrum has the similar order as in our previous results for the spectra of the α-decay of the ^{214}Po (red dashed line) and ^{226}Ra (blue dashed line) nuclei. In Fig. 4 (b) we present new calculation for the emission of photons caused by the α-particle emitted from the same ^{252}Cf nucleus but when the daughter nucleus is separated into two fragments (dash-dotted line) but in this case is E_α=16 MeV [21] .In this last figure we also report for a comparison the result (full line) of the previous Fig. 4 a of the photon spectrum for the α-decay at E_α=6.2 MeV. As one can see the relevant difference with the last spectrum is partially explained by the higher energy E_α=16 MeV of the emitted α-particle in presence of the two fragments.

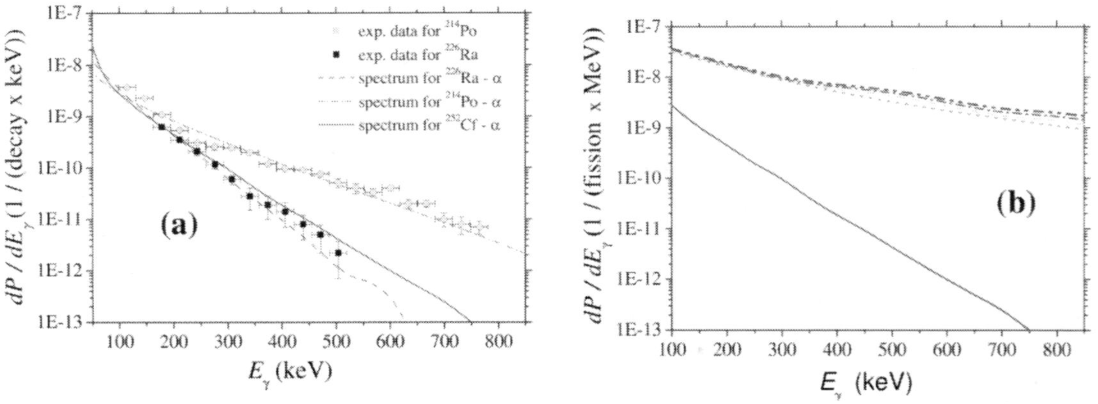

Figure 4. The bremsstrahlung photon emission probabilities caused by α-particle emitted from the ^{252}Cf nucleus: (a) the spectrum caused by α-particle emitted from nucleus (solid blue line) in comparison with the spectra due to α decay of the ^{214}Po and ^{226}Ra nuclei (red dash-dotted and green dashed lines, respectively); (b) spectra caused by α-particle when we take into account fission of the daughter nucleus into two fragments for different kind of fragmentation and a fixed neck length value of 7 fm (A_1=120, Z_1=45(dash-double dotted line); A_1=80, Z_1=30(dash dotted line);A_1=40, Z_1=15(dashed line);A_1=8, Z_1=4(dotted line);) we also add for a comparison the bremsstrahlung spectrum accompanying the α-decay (solid line) already presented in left figure when the daughter nucleus is not separated.

In Fig. 5 we present the calculated spectra obtained by the above-described approach in comparison with the preliminary experimental data [22] of photon emission in presence of the ternary fission. These data were obtained analyzing the event of the photon emission in coincidence with the α-particle, emitted from the ^{252}Cf nucleus, having a kinetic energy $E_\alpha > 6.2$ MeV corresponding to motion of α-particles in presence of the field of the two fission fragments of daughter nucleus. By comparing the experimental data with the calculated spectra, we can conclude that: (1) the general tendency of the calculated spectra is similar to the one of the experimental data; (2) there is some deviation between calculation and data of emitted photons in the 200–400 keV energy region of E_γ. On such a basis we can rise the following question: *what mechanism could improve the description of the experimental data?*

4. Improvement of the model

In order to resolve the above-mentioned question, let us collect such aspects of the model which influence in an appreciable mode the photon emission spectra.

(i) In calculation of the photon emission probability caused by the α-particle in presence of the binary nucleus, we consider:
  (a) different separations of the same binary nucleus into two fragments;
  (b) the geometry of the fissioning nucleus;
  (c) the neck of the binary nucleus at formation of the α-particle;
  (d) the dynamics of the relative motion between α-particle and fragments of the binary nucleus;
  (e) the angles between emitted photons and fragments.

(ii) One can assume that the two fragments of the binary nucleus are in motion during the fission in order to give an additional contribution to the emission probability of the bremsstrahlung

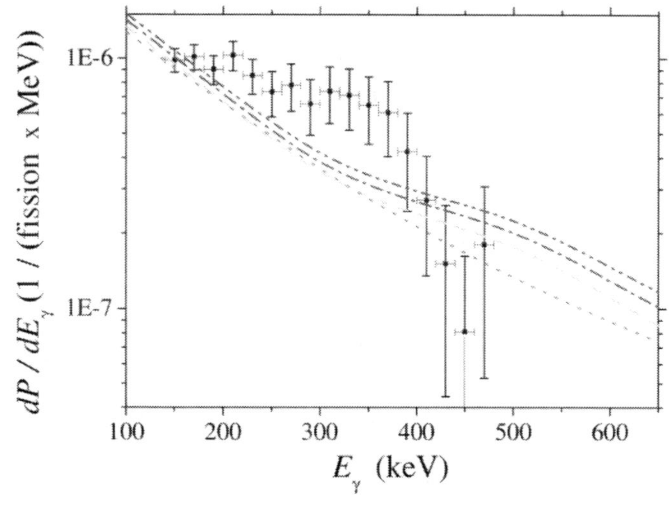

Figure 5. Bremsstrahlung photon emission probabilities in the ternary fission of the ^{252}Cf nucleus: calculated spectra accompanying of α-emission in presence of different kind of fragmentation of the daughter nucleus as presented in Fig 4 (b), in comparison with the preliminary experimental data [22] (full squares).

photons.

The last point seems to have the largest influence on the spectra. For this reason, we shall continue our analysis on this item.

4.1. Contribution of the binary nucleus

Let us consider the operator of cluster emission from the total nuclear system (α-particle and daughter nucleus) where both α-particle and residual nucleus are composed by nucleons (we indicate nucleons of α-particle by the index i, and nucleons of the residual nucleus by the index j):

$$\hat{H}_\gamma = -e\sqrt{\frac{2\pi\hbar}{w}} \sum_{n=1,2} \mathbf{e}^{(n),*} \Big\{ \sum_{i=1}^{4} \frac{Z_i}{m_i} e^{-i\mathbf{k}\mathbf{r}_i} \hat{\mathbf{p}}_i + \sum_{j=1}^{A} \frac{Z_j}{m_j} e^{-i\mathbf{k}\mathbf{r}_j} \hat{\mathbf{p}}_j \Big\}. \qquad (14)$$

We pass to new variables, where we take into account centers of masses of the α-particle and of the two fragments in the daughter nucleus. We define coordinates of centers of masses of the α-particle by $\mathbf{r}_\alpha$, of the daughter nucleus by $\mathbf{R}_A$, and of the total nuclear system by $\mathbf{R}$, as

$$\mathbf{r}_\alpha = \frac{1}{m_\alpha} \sum_{i=1}^{4} m_i \, \mathbf{r}_{\alpha i}, \quad \mathbf{R}_A = \frac{1}{M} \sum_{j=1}^{A} m_j \, \mathbf{r}_{Aj}, \quad \mathbf{R} = \frac{M\mathbf{R}_A + m_\alpha \mathbf{r}_\alpha}{M + m_\alpha}. \qquad (15)$$

Introducing the new relative coordinates ρ_{α_i}, ρ_{A_j} and $\mathbf{r}$:

$$\mathbf{r}_i = \mathbf{r}_\alpha + \rho_{\alpha_i}, \quad \mathbf{r}_j = R_A + \rho_{A_j}, \quad \mathbf{r} = \mathbf{r}_\alpha - \mathbf{R}_A \qquad (16)$$

we obtain:

$$\mathbf{r}_i = \mathbf{R} + \frac{M}{M + m_\alpha} \mathbf{r} + \rho_{\alpha_i}, \quad \mathbf{r}_j = \mathbf{R} - \frac{m_\alpha}{M + m_\alpha} \mathbf{r} + \rho_{A_j}. \qquad (17)$$

As a result, we obtain the matrix element in the following form (see Appendix A):

$$\begin{aligned}\langle f|\,\hat{H}_\gamma|\,i\rangle \;&=\; -e\sqrt{\frac{2\pi\hbar}{w}}\sum_{n=1,2}\mathbf{e}^{(n),*}\,\delta(\mathbf{K}_f-\mathbf{k})\cdot\bigg\{\Big\langle f_A,\,f_\alpha\Big|\,Z_{\rm eff}(\mathbf{r})\,e^{-i\,\mathbf{kr}}\,\mathbf{p}\,\Big|\,i_A,\,i_\alpha\Big\rangle+\\ &+\Big\langle f_\alpha\Big|\,e^{i\,\mathbf{kr}\frac{m_\alpha}{M+m_\alpha}}\Big\langle f_A\Big|\,Z_{\rm A}(\mathbf{k})\,\mathbf{p}_{Aj}\,\Big|\,i_A\Big\rangle\Big|\,i_\alpha\Big\rangle\bigg\}\;,\end{aligned} \tag{18}$$

where the indexes i and f denote the initial state (the state before photon emission) and the final state (after photon emission), $|\,s_A\rangle$ is wave function, describing the internal states of the daughter nucleus, s_α is wave function describing relative motion (with possible tunnelling) of the α-particle concerning the daughter nucleus. Here, $Z_{\rm eff}(\mathbf{r})$, $Z_\alpha(\mathbf{k})$ and $Z_{\rm A}(\mathbf{k})$ are, respectively, the effective charge of the system composed by the α-particle and daughter nucleus, the charge form-factor of the α-particle and the charge form-factor of the daughter (binary) nucleus, as defined in Appendix A.

Calculation of various contributions are presented in Fig. 6 (a). In order to analyze how the various contributions influence the spectra, we made calculation for the same fragmentation of the binary nucleus (we chose one fragment with $A_1 = 120$ and $Z_1 = 45$) and normalized the spectra to the experimental data. Such calculations are presented in Fig. 6 (b). This last figure

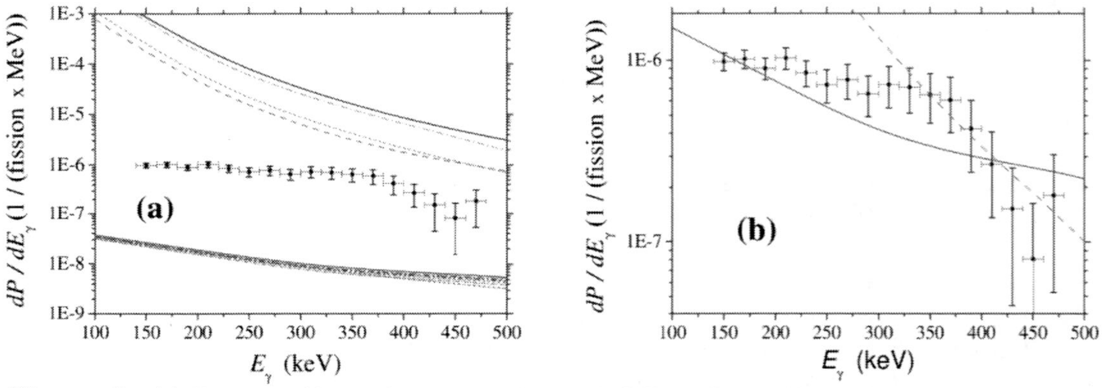

Figure 6. (a) Bremsstrahlung photon emission probabilities due to the nuclear system composed by α-particle and daughter binary nucleus (dawn lines) composed by two kind of fragments (full blue line for A_1=80 and Z_1=30; red dashed line for A_1=120, Z_1=45; blue dashed line for A_1=40, Z_1=15; red dotted line for A_1=8, Z_1=4), in comparison with the data of [22]. In the same figure are reported the bremsstrahlung probabilities of photon emission due to the fission fragments of ithe binary daughter nucleus for the above.mentioned kinds of fragments (upper lines). (b)As in the previous left figure, the contribution of bremsstrahlung due to the α-emission in presence of the binary daughter nucleus is reported by blue solid line; while, the spectrum due to only one kind of fragmentation of the binary nucleus for the neck length value of 7 fm is reported by red dashed line ($A_1 = 120$ and $Z_1 = 45$), after normalization of both lines to the experimental data.

shows that the component of emission caused by fragments of the binary nucleus well describes the experimental data at energies higher than 300 keV, while at lower energies the emission is mainly caused by the motion of the α-particle in the field of the daughter nucleus.

4.2. Delay in the emission of the binary nucleus

Following the previous assumption, the total matrix element of emission can be written in two parts:

$$\big\langle f\big|\hat{H}_\nu\big|i\big\rangle = \big\langle f\big|\hat{H}_\nu\big|i\big\rangle_\alpha + \big\langle f\big|\hat{H}_\nu\big|i\big\rangle_{\rm binary}. \tag{19}$$

Here, the second term represents the emission of photons caused by the relative motion of two charged fragments of the daughter nucleus; it can be written as

$$\begin{aligned}\left\langle f\left|\hat{H}_\gamma\right|i\right\rangle_{\text{binary}} &= -e\sqrt{\frac{2\pi\hbar}{w}}\sum_{n=1,2}\mathbf{e}^{(n),*}\left\langle f\left|e^{i\,\mathbf{kr}\,\frac{m_\alpha}{M+m_\alpha}}\cdot\left\langle f_A\right|Z_{\text{eff,A}}(\rho)\,\mathbf{p}_\rho\left|i_A\right\rangle\right|i\right\rangle = \\ &= -e\sqrt{\frac{2\pi\hbar}{w}}\sum_{n=1,2}\mathbf{e}^{(n),*}\left\langle f\left|e^{i\,\mathbf{kr}\,\frac{m_\alpha}{M+m_\alpha}}\right|i\right\rangle\cdot\left\langle f_A\right|Z_{\text{eff,A}}(\rho)\mathbf{p}_\rho\left|i_A\right\rangle.\end{aligned} \tag{20}$$

In this expression calculation of the first matrix element $\left\langle f\right|\exp[i\,\mathbf{kr}\,\frac{m_\alpha}{M+m_\alpha}]\left|i\right\rangle$ represents a relevant difficulty. In order to estimate it, in first approximation we use the following assumption:

$$\left\langle f\left|e^{i\,\mathbf{kr}\,\frac{m_\alpha}{M+m_\alpha}}\right|i\right\rangle \simeq N_i^2 = \frac{\mu}{k_i}, \tag{21}$$

where N_i is the normalization factor of the wave function for the state before emission of photon (see Ref.s [10–13, 15] for details). In particular, applying such approximation, we obtained the spectra for the contributions of the binary nucleus presented in Fig. 6 (a) by lines which beyond the experimental data.

In order to find the next correction to the approximation (21), we should take into account dependence of this matrix element both on the wave vector $\mathbf{k}$ of the emitted photon, and on the angle between vectors $\mathbf{k}$ (i.e. direction of the emission of photon) and $\mathbf{r}$ (i.e. direction of motion of α-particle). On such a basis, we write the following (second) approximation as

$$\left\langle f\left|e^{i\,\mathbf{kr}\,\frac{m_\alpha}{M+m_\alpha}}\right|i\right\rangle \simeq \frac{\mu}{k_i}\cdot F\left[k,\cos(\mathbf{k},\mathbf{r})\right]. \tag{22}$$

where $F\left[k,\cos(\mathbf{k},\mathbf{r})\right]$ is a new unknown function. We find this function by such a way that it should include better description of the experimental data at low energies by the emission of the α-binary nucleus and at high energies by the contribution of the fragment emission from the binary nucleus. In particular, such situation can be present in the following dynamical scheme of the ternary fission process:

(i) In the starting stage of ternary fission, α-particle leaves a region of the nuclear system. It moves and emits photons at low energies. The fragments of binary nucleus are moved very slowly, and emission of photons can be neglected.

(ii) After, in the next stage of ternary fission, two fragments of the binary nucleus begin to move outside with acceleration. They give emission of photons at higher energies, which becomes prevailing.

Taking into account such a scheme, we assume for the function $F\left[k,\cos(\mathbf{k},\mathbf{r})\right]$ the form $F = 1 - f_0/(1+\exp((w-w_0)/w_1))$. Then, we obtain the unknown parameters by the fitting procedure where its inclusion into calculations should describe effectively experimental data in the whole energy region of the emitted photons (we have $f_0 = 1+\exp(-w_0/w_1)$, $w_0 = w_1\cdot\ln(\exp(w_2/w_1)-2)$, $w_1 = 0.057$, $w_2 = 0.24$, and w is the energy of the emitted photons).

Results of calculation of spectra on basis of such assumptions are presented in Fig. 7 (a). As one can see, the new spectrum, (presented by solid green line) constructed on the basis of the second approximation (22) and the function $F\left[k,\cos(\mathbf{k},\mathbf{r})\right]$, is in agreement with the experimental data [20] for the spontaneous fission of the ^{252}Cf nucleus.

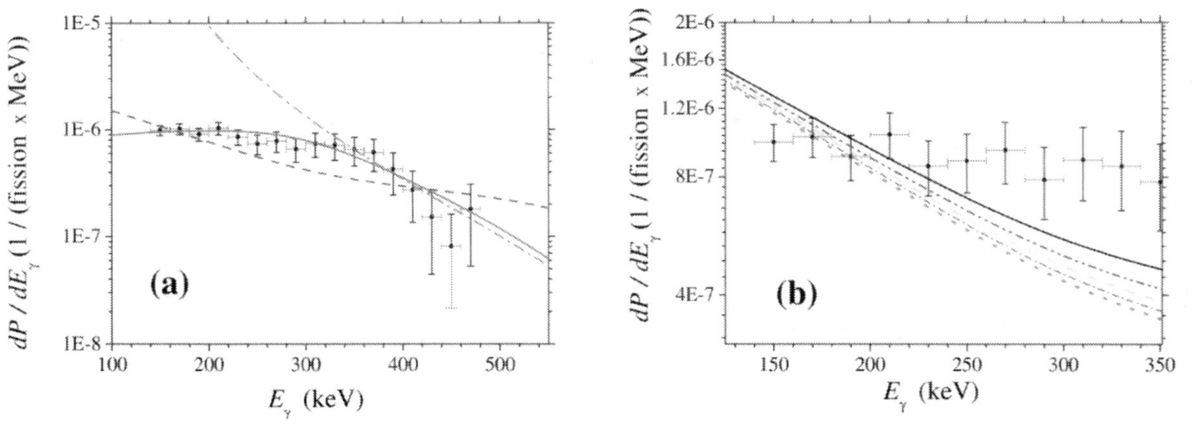

Figure 7. Bremsstrahlung photon emission probabilities in the ternary fission of ^{252}Cf and experimental data [22]. In calculation we use a chosen separation of the binary nucleus into two fragments with $A_1 = 120$, $Z_1 = 45$ and $A_2 = 128$, $Z_2 = 51$; (a) the total spectrum (full line) including the emission of α-particle in the field of the daughter binary nucleus (dashed line) and contribution of photon emission due to the motion of the fission fragments of daughter nucleus (red dashed line, when we assume the neck length of 7 fm); (b) spectra for different lengths of the neck (11 fm, full line; 9 fm, dash-double dotted; 7 fm, dashed line; 5 fm, dash-dotted; 3 fm, dotted line) between the two fragments of the fissioned daughter nucleus.

According to this description, at higher energies (above 300 keV) the emission of photons is mainly determined by the contribution of the binary nucleus. So, at lower energies we have mainly the emission caused by the α-particle in the field of the binary nucleus. By other words, in the energy range smaller than 300 keV, the analysis of the dependence of the spectrum on the length of the neck is useful. The results of calculations are presented in Fig. 7 (b). From this result we deduce that the optimal value of the neck is about of 11 fm, when the calculated curve is closest to the experimental data, in the assumed case of separation of the binary nucleus into two fragments with $A_1 = 120$, $Z_1 = 45$, and $A_2 = 128$, $Z_2 = 51$.

5. Conclusions

We have elaborated a new model describing the emission of the bremsstrahlung photons accompanying the ternary fission of heavy nuclei. For the chosen ^{252}Cf nucleus, the general tendency of our calculated spectra of bremsstrahlung photon emission is in good agreement with the preliminary experimental data [22] at lower and higher photon energies in the energy range up to 500 keV (see Fig. 5). However, there is some disagreement between theory and experiment in region 200–400 keV of the emitted photons.

In order to improve the description of experimental data, also in this intermediate energy range we suggest a new way for the dynamics of the ternary fission process:

(i) when the α-particle leaves the region of the ternary nuclear system, the binary nucleus is separated very slowly and gives an unimportant contribution to the total spectrum of the emitted photons;

(ii) at this stage, α-particle moves fast and emits photons of low energies;

(iii) then, the binary nucleus begins to separate quickly, and fragments move with acceleration giving appreciable photon emission at high energies.

This model and mechanism allows us to achieve better results.

By comparing our spectra with the preliminary experimental data we find that: (1) the α-particle is emitted when the length of the neck of the binary nucleus is 11 fm; (2) the probable separation of the binary nucleus is into two similar fragments.

Appendix A.

Let us write the operator of photon emission of the nuclear system decaying in the laboratory system by α-particle and daughter nucleus, where α-particle and daughter nucleus are composed of nucleons (we describe nucleons of α-particle by index i, and nucleons of the nucleus by index j):

$$\hat{H}_\gamma = -\sum_{i=1}^{4} \frac{e\,Z_i}{m_i\,c}\,\mathbf{A}(\mathbf{r}_i,t)\,\hat{\mathbf{p}}_i - \sum_{j=1}^{A} \frac{e\,Z_j}{m_j\,c}\,\mathbf{A}(\mathbf{r}_j,t)\,\hat{\mathbf{p}}_j. \tag{A.1}$$

By using the following form for the potential of the electromagnetic field:

$$\mathbf{A}(\mathbf{r},t) = \sqrt{\frac{2\pi\hbar c^2}{w}} \sum_{n=1,2} \mathbf{e}^{(n),*} e^{-i\mathbf{kr}}, \tag{A.2}$$

we rewrite the operator of emission (A.1) as the used expression (14).

by using the relative coordinates ρ_{α_i}, ρ_{A_j} and $\mathbf{r}$ defined in relation (16) we find the corresponding momenta:

$$\begin{aligned} &\mathbf{p}_i = -\,i\hbar\,\frac{\mathbf{d}}{\mathbf{d}\mathbf{r}_i} = \mathbf{p}_\alpha + \mathbf{p}_{\alpha_i}, \quad \mathbf{p}_\alpha = -\,i\hbar\,\frac{\mathbf{d}}{\mathbf{d}\mathbf{r}_\alpha}, \quad \mathbf{p}_{\alpha_i} = -\,i\hbar\,\frac{\mathbf{d}}{\mathbf{d}\rho_{\alpha_i}}, \\ &\mathbf{p}_j = -\,i\hbar\,\frac{\mathbf{d}}{\mathbf{d}\mathbf{r}_j} = \mathbf{P}_A + \mathbf{p}_{A_j}, \quad \mathbf{P}_A = -\,i\hbar\,\frac{\mathbf{d}}{\mathbf{d}\mathbf{R}_A}, \quad \mathbf{p}_{A_j} = -i\hbar\,\frac{\mathbf{d}}{\mathbf{d}\rho_{A_j}}, \quad \mathbf{p} = \mathbf{p}_\alpha - \mathbf{P}_A. \end{aligned} \tag{A.3}$$

Taking into account the relative coordinate (16), we obtain

$$\mathbf{R}_A = \mathbf{R} - \frac{m_\alpha}{M+m_\alpha}\,\mathbf{r}, \quad \mathbf{r}_\alpha = \mathbf{R} + \frac{M}{M+m_\alpha}\,\mathbf{r} \tag{A.4}$$

and we can rewrite the first and second expressions in (16) as

$$\mathbf{r}_i = \mathbf{R} + \frac{M}{M+m_\alpha}\,\mathbf{r} + \rho_{\alpha_i}, \quad \mathbf{r}_j = \mathbf{R} - \frac{m_\alpha}{M+m_\alpha}\,\mathbf{r} + \rho_{A_j}. \tag{A.5}$$

Substituting these expressions in relation (14), we find:

$$\begin{aligned} \hat{H}_\gamma \;=\; & -e\sqrt{\frac{2\pi\hbar}{w}} \sum_{n=1,2} \mathbf{e}^{(n),*}\, e^{-i\mathbf{k}\left[\mathbf{R} - \frac{m_\alpha}{M+m_\alpha}\mathbf{r}\right]} \times \\ & \times \Biggl\{ \Bigl[e^{-i\mathbf{kr}} \sum_{i=1}^{4} \frac{Z_i}{m_i}\, e^{-i\mathbf{k}\rho_{\alpha_i}} + \sum_{j=1}^{A} \frac{Z_j}{m_j}\, e^{-i\mathbf{k}\rho_{A_j}} \Bigr] . \mathbf{P} + \\ & + \Bigl[e^{-i\mathbf{kr}} \frac{M}{M+m_\alpha} \sum_{i=1}^{4} \frac{Z_i}{m_i}\, e^{-i\mathbf{k}\rho_{\alpha_i}} - \frac{m_\alpha}{M+m_\alpha} \sum_{j=1}^{A} \frac{Z_j}{m_j}\, e^{-i\mathbf{k}\rho_{A_j}} \Bigr]\, \mathbf{p} + \\ & + e^{-i\mathbf{kr}} \sum_{i=1}^{4} \frac{Z_i}{m_i}\, e^{-i\mathbf{k}\rho_{\alpha_i}}\, \mathbf{p}_{\alpha_i} + \sum_{j=1}^{A} \frac{Z_j}{m_j}\, e^{-i\mathbf{k}\rho_{A_j}}\, \mathbf{p}_{A_j} \Biggr\}. \end{aligned} \tag{A.6}$$

Now we present the wave function of the total system in the following expression (see Ref. [23]):

$$|s\rangle = e^{-i\,\mathbf{K}_s\cdot\mathbf{R}}\, s_\alpha\,|\,s_A\rangle, \tag{A.7}$$

where $s = i$ or f are the indexes denoting the initial state (before emission of photon) and the final state (after emission of photon)), $\hbar\,\mathbf{K}_s$ is the full momentum of the total system, $|\,s_A\rangle$ is the wave function describing internal states of the daughter nucleus, s_α is the wave function describing the relative motion (with possible tunneling) of the α-particle concerning the daughter nucleus. Assuming $\mathbf{K}_i = 0$, we calculate the matrix element:

$$\begin{aligned}
\langle f|\,\hat{H}_\gamma|\,i\rangle \;=\; & -e\sqrt{\frac{2\pi\hbar}{w}}\sum_{n=1,2}\mathbf{e}^{(n),*}\Biggl\{\Biggl\langle f_\alpha,\,f_A\Biggl|\,e^{i\,(\mathbf{K}_f-\mathbf{k})\cdot\mathbf{R}}\,e^{i\,\mathbf{kr}\,\frac{m_\alpha}{M+m_\alpha}}\;\times \\
& \times\Bigl[e^{-i\mathbf{kr}}\sum_{i=1}^{4}\frac{Z_i}{m_i}\,e^{-i\mathbf{k}\rho_{\alpha_i}}+\sum_{j=1}^{A}\frac{Z_j}{m_j}\,e^{-i\mathbf{k}\rho_{A_j}}\Bigr]\,\mathbf{P}\,\Biggr|\,i_\alpha,\,i_A\Biggr\rangle+ \\
& +\Biggl\langle f_\alpha,\,f_A\Biggl|\,e^{i\,(\mathbf{K}_f-\mathbf{k})\cdot\mathbf{R}}\,e^{i\,\mathbf{kr}\,\frac{m_\alpha}{M+m_\alpha}}\Bigl[e^{-i\mathbf{kr}}\,\frac{M}{M+m_\alpha}\sum_{i=1}^{4}\frac{Z_i}{m_i}\,e^{-i\mathbf{k}\rho_{\alpha_i}}\,- \\
& \qquad -\frac{m_\alpha}{M+m_\alpha}\sum_{j=1}^{A}\frac{Z_j}{m_j}\,e^{-i\mathbf{k}\rho_{A_j}}\Bigr]\,\mathbf{p}\,\Biggr|\,i_\alpha,\,i_A\Biggr\rangle+ \\
& +\Biggl\langle f_\alpha,\,f_A\Biggl|\,e^{i\,(\mathbf{K}_f-\mathbf{k})\cdot\mathbf{R}}\,e^{i\,\mathbf{kr}\,\frac{m_\alpha}{M+m_\alpha}}\,e^{-i\mathbf{kr}}\sum_{i=1}^{4}\frac{Z_i}{m_i}\,e^{-i\mathbf{k}\rho_{\alpha_i}}\,\mathbf{p}_{\alpha_i}\Biggr|\,i_\alpha,\,i_A\Biggr\rangle+ \\
& +\Biggl\langle f_\alpha,\,f_A\Biggl|\,e^{i\,(\mathbf{K}_f-\mathbf{k})\cdot\mathbf{R}}\,e^{i\,\mathbf{kr}\,\frac{m_\alpha}{M+m_\alpha}}\sum_{j=1}^{A}\frac{Z_j}{m_j}\,e^{-i\mathbf{k}\rho_{A_j}}\,\mathbf{p}_{A_j}\Biggr|\,i_\alpha,\,i_A\Biggr\rangle\Biggr\}.
\end{aligned} \tag{A.8}$$

In our calculation we shall not use the first term of expression (A.8), since we study process of fission in the center-of-mass system, and the third term because we assume that α-particle is not deformed in the studied process of fission. Now, introducing the effective charge of the system (α-particle and daughter nucleus) as

$$Z_{\text{eff}}(\mathbf{r}) \;=\; e^{i\,\mathbf{kr}}\left\{e^{-i\,\mathbf{kr}\,\frac{M}{M+m_\alpha}}\,\frac{M\,Z_\alpha}{M+m_\alpha}-e^{i\,\mathbf{kr}\,\frac{m_\alpha}{M+m_\alpha}}\,\frac{m_\alpha\,Z_{\text{A}}}{M+m_\alpha}\right\}, \tag{A.9}$$

the charged form-factor of α-particle as

$$Z_\alpha(\mathbf{k}) = \Bigl\langle \beta_f\Bigl|\sum_{i=1}^{4}\frac{Z_i}{m_i}\,e^{-i\mathbf{k}\rho_{\alpha_i}}\Bigr|\,\beta_i\Bigr\rangle\,, \tag{A.10}$$

and the charged form-factor of the daughter (binary) nucleus as

$$Z_{\text{A}}(\mathbf{k}) = \Bigl\langle \beta_f\Bigl|\sum_{j=1}^{A}\frac{Z_j}{m_j}\,e^{-i\mathbf{k}\rho_{A_j}}\Bigr|\,\beta_i\Bigr\rangle\,, \tag{A.11}$$

and by using (A.9) we obtain

$$\langle f|\, \hat{H}_\gamma |\, i\rangle = -e\sqrt{\frac{2\pi\hbar}{w}} \sum_{n=1,2} \mathbf{e}^{(n),*}\, \delta(\mathbf{K}_f - \mathbf{k}) \cdot \biggl\{ \Bigl\langle f_A,\, f_\alpha \Bigl|\, Z_{\rm eff}(\mathbf{r})\, e^{-i\,\mathbf{kr}}\, \mathbf{p} \,\Bigr|\, i_A,\, i_\alpha \Bigr\rangle + \\ + \Bigl\langle f_\alpha \Bigl|\, e^{i\,\mathbf{kr}} \frac{m_\alpha}{M+m_\alpha} \bigl\langle f_A \bigl|\, Z_{\rm A}(\mathbf{k})\, \mathbf{p}_{A_j} \,\bigr|\, i_A \bigr\rangle \Bigr|\, i_\alpha \Bigr\rangle \biggr\}. \tag{A.12}$$

In particular, in the first approximation (so called dipole) we obtain:

$$Z_{\rm eff}^{(\rm dip)} = \frac{M\, Z_\alpha - m_\alpha\, Z_{\rm A}}{M + m_\alpha}. \tag{A.13}$$

References

[1] Daniel A V *et al.* 2004 *Phys. Rev. C* **69** 041305(R)
[2] Ramayya A V, Hamilton J H, Hwang J K 2007 *Romanian Reports in Physics* **59** 595
[3] Ramayya A V *et al.* 1998 *Phys. Rev. C* **57** 2370
[4] Kopach Yu N, Singer P, Mutterer M, Klemens M, Hotzel A, Schwalm D, and Thirolf P 1999 *Phys. Rev. Lett.* **82** 303.
[5] Nix J R 1969 *Nucl. Phys.* A **130** 241
[6] Bolsterli M, Fiset E O, Nix J R, and Norton J L 1972 *Phys. Rev. C* **5** 1050
[7] Audi G, Wapstra A H, and Thibault C 2003 *Nucl. Phys. A* **729** 337
[8] Denisov V Yu and Ikezoe H 2005 *Phys. Rev. C* **72** 064613
[9] Maydanyuk S P *et al.* 2010 *Phys. Rev. C* **82** 014602
[10] Maydanyuk S P and Olkhovsky V S 2003 *Prog. Theor. Phys.* **109** 203
[11] Maydanyuk S P and Olkhovsky V S 2006 *Eur. Phys. J. A* **28** 283
[12] Giardina G *et al.* 2008 *Eur. Phys. J. A* **36** 31
[13] Giardina G *et al.* 2008 *Mod. Phys. Lett. A* **23** 2651
[14] S. P. Maydanyuk, The Open Nucl. Part. Phys. Journ. **2**, 1 (2009).
[15] Maydanyuk S P *et al.* 2009 *Nucl. Phys. A* **823** 3
[16] Van der Ploeg H *et al.* 1995 *Phys. Rev.* C **52** 1915
[17] Kasagi J *et al.* 1989 *Journ. Phys. Soc. Jpn.* **58** 620
[18] Luke S J *et al.* 1991 *Phys. Rev. C* **44** 1548
[19] Hofman D J *et al.* 1993 *Phys. Rev.C* **47** 1103
[20] Eremin N V *et al.* 2010 *Int. J. Mod. Phys. E* **19** 1183
[21] Hwang J K *et al.* 2000 *Phys. Rev. C* **61** 047601
[22] Eremin N V *et al.* 2010 private communication
[23] Kopitin I V, Dolgopolov M A, Churakova T A and Kornev A S 1997 *Yad. Fiz.* **60** 869

Journal of Physics: Conference Series **282** (2011) 012017 doi:10.1088/1742-6596/282/1/012017

Investigation of high energy γ-rays accompanying spontaneous fission of ^{252}Cf in double and triple neutron-γ coincidences

N V Eremin[1,†], S S Markochev[2,◇], A A Paskhalov[1], G Mandaglio[3], M Manganaro[3], G Fazio[3], G Giardina[3,⋆] and M V Romaniuk[3]

[1] Skobeltsin Institute of Nuclear Physics, Lomonosov Moscow State University, 119899, Moscow, Russia

[2] Moscow Institute of Physics and Technology, 141700, Dolgoprudny, Russia

[3] Dipartimento di Fisica dell'Universitá di Messina, Messina, 98166, Italy and Istituto Nazionale di Fisica Nucleare, Sezione di Catania, Catania, 95123, Italy

E-mail: † eremin@spels.ru

E-mail: ◇ sergey.markochev@gmail.com

E-mail: ⋆ giardina@nucleo.unime.it

Abstract. The high energy bremsstrahlung γ-rays accompanying the spontaneous fission of ^{252}Cf were measured in the 10-70 MeV energy range. The photons were detected by two BGO scintillator detectors ($\oslash 7.6cm \times 7.6cm$) in coincidence with neutrons detected by plastic scintillator detector, in the 90° and 180° geometry of the two BGO detectors with respect to the axis of the plastic scintillator. The distance from the ^{252}Cf source to the BGO detectors was 10 cm, and the one to the plastic detector was 50 cm. The fast digital shape analysis technique and the time-of-flight method were used to reject pile-up effects and cosmic ray background. The γ-ray emission probability of $3 \times 10^{-8} photon/(MeV \times fission)$ at $E_\gamma = 70$ MeV was obtained.

1. Introduction

One of the intriguing questions in physics of spontaneous fission of heavy nuclei remains the possibility of emission of high energy γ-rays ($E_\gamma > 20$ MeV) accompanying spontaneous fission. The coherent nucleus-nucleus bremsstrahlung mechanism may be responsible for this phenomenon. A similar process of bremsstrahlung emission was observed for the α-decay of heavy nuclei. The photon bremsstrahlung spectra for the α-decay of the ^{226}Ra and ^{214}Po nuclei were measured up to E_γ energy of about 1 MeV[1, 2]. In the case of spontaneous fission of heavy nuclei one can expect the emission of high energy γ-rays because of the bigger Q-value (Q $\sim$ 200 MeV). The preliminary results of the high energy bremsstrahlung photon emission probability accompanying the binary spontaneous fission of ^{252}Cf were presented in Ref. [3]. The calculation by the quantum mechanical model presented in Ref. [4] describe in good way the experimental bremsstrahlung spectrum obtained in [3] for binary spontaneous fission of ^{252}Cf. Moreover, in Fig. 3 (b) of Ref.[4] were also presented the theoretical results of the bremsstrahlung spectrum for the α-decay of the ^{226}Ra, ^{214}Po, and ^{252}Cf nuclei, in comparison with the available experimental data of ^{226}Ra and ^{214}Po.

The first experimental investigation of high energy γ-rays was made by the group of Kasagi[5]. They selected the γ-γ-coincidences among the emitted γ-rays accompanying the fission fragments. The observed spectrum had a high energy "tail" which began from E_γ about 20 MeV up to about 160 MeV. Authors connected these γ-rays with the coherent nucleus-nucleus bremsstrahlung emission, affirming instead that at E_γ lower than 20 MeV the cascade of γ-rays and the contribution of γ-rays due to the giant dipole resonances of excited fission fragments were dominant.

The Kasagi report was so unexpected that many attempts have been made in order to repeat this result. There were two approaches in experimental investigations: *i*) to measure the exclusive γ-ray spectrum by means of big volume scintillators with plastic anticoincidence shielding[6, 7]; *ii*) to measure the high energy γ-rays by using the γ-γ[5, 7] or γ-fission fragments[7, 8] coincidence techniques. In the energy range $E_\gamma < 20$ MeV there is a reliable agreement between all authors, but above 20 MeV the results are contradictory.

Yu.N. Pokotilovskii[6] measured high energy γ-rays by using a big volume NaJ(Tl) detector with an anticoincidence plastic scintillator covering. The upper limits of the γ-ray emission probability in the $E_\gamma = 20-140$ MeV energy region of about 6×10^{-9} $photon/(fission \times MeV)$ at $E_\gamma = 40$ MeV and 1×10^{-9} $photon/(fission \times MeV)$ at $E_\gamma = 100$ MeV were obtained.

The γ-γ-coincidence method was used by the group of Kasagi[5], and later by the group of Pandit[9]. Their experiments got the same results,in the E_γ = 8-80 MeV energy range. The relative probability of γ-ray emission at 20 MeV was about 7×10^{-6} $photon/(fission \times MeV)$ and at 80 MeV was about 3×10^{-7} $photon/(fission \times MeV)$.

S.J. Luke *et al.*[7] used the γ-γ and γ-fission fragment coincidence methods to measure high energy γ-rays emitted from ^{252}Cf. The high energy γ-rays were detected by the NaJ(Tl) scintillator with an active anticoincidence shielding. By the first method the integrated yield was $(2.4 \pm 1.3) \times 10^{-6} photon/fission$ and by the second one the upper limit of about 1.8×10^{-6} $photon/fission$ was obtained.

H. van der Ploeg *et al.*[8] used the γ-fission fragment coincidence technique. The γ-ray spectrum was measured by big volume BaF_2 and NaJ(Tl) detectors. The NaJ(Tl) detector was covered by a plastic anticoincidence shielding. By the NaJ detector, the determinations of upper limits were 5.7×10^{-7} and 2.0×10^{-8} $photon/(fission \times MeV)$ in the 20-30 and 30-40 MeV regions, respectively.

In such experiments, the difficulties related to the identification and selection of events included the influence of cosmic ray background events, pile-up effects of gamma-neutron emission multiplicity events and other kind of events on the rate of the true interested events. With the aim to reduce the cosmic ray background we propose a new experimental method based on the neutron-γ (n-γ) coincidence. We strongly rejected the pile-up effects by using the digital storage oscilloscope technique and digital processing of signals.

2. Experiment

2.1. Experimental setup

The experimental setup consisted of two BGO scintillator detectors and one plastic scintillator detector connected with a digital storage oscilloscope Tektronix-TDS 7704B. The BGO cylindrical scintillator crystals ($\oslash 76mm \times 76mm$) and plastic scintillator ($\oslash 60mm \times 20mm$) were coupled with PMTs XP 4312(Photonis Co. Prod.). BGO detectors were used to measure γ-ray energy. The distance between the ^{252}Cf source and each of the two BGO detectors was chosen 10 cm in order to avoid multiple taking of γ emitted from source. The plastic scintillator detector was able to detect γ-rays as well as neutrons. The distance between the source and the plastic detector was chosen 50 cm with the aim to distinguish neutrons from photons by the time-of-flight (TOF) method. The axes of the two BGO detectors in respect to the axis of the plastic detector formed the angles 90° and 180°, respectively.

The software code developed for the Tektronix-TDS 7704B allowed us to detect double and triple coincidences within the 200 ns time window between the signal from plastic detector and one (in double coincidence mode) or two signals (in triple coincidence mode) from the BGO detectors, and to save the shapes of pulses from all detectors in the oscilloscope memory. We used signals from plastic detector as "start" and the signals from BGO detectors as "stop". Experimental data were collected in a continuous mode during the total interval of 78 days of data taking. The off-line processing of signals was made by a personal computer. The activity of the ^{252}Cf source was $6.1 \times 10^6 neutron/s$.

The absolute efficiency of the γ-ray detecting system was calculated by GEANT4. In simulation, the source of γ-rays was modelled in the geometry of our experimental setup. We assume the absolute efficiency of detector as the number of photons which lost almost all its energy (with accuracy of the energy averaging interval) in the BGO detector per the total number of γ-rays emitted from the source. For γ-rays of 70 MeV leaving 65-70 MeV of its energy the absolute detecting efficiency of BGO was about 0.2%.

2.2. Data processing
The goal of data processing was to determine the energy of pulses in both BGO detectors and to estimate the delay time between the pulses of the plastic detector and the BGO detectors in the coincidence modes. We used the delay time to distinguish the n-γ coincidences from the γ-γ ones and the cosmic shower coincidences.

The data processing included the rejection of pile-up events to be sure that we measured the γ-ray energy accurately. The registered pulses (voltage versus time) were smoothed and local extremums were found for each pulse. If in neighbourhood of such local extremum the pulse area at the right side of the considered extremum was greater than the one at the left side, then this pulse was considered as a superposition of pulses and rejected.

To calculate the delay time we found the time marker position for each pulse. The marker position was found by the constant fraction discriminator method at 1/3 of the pulse amplitude. The amplitudes of pulses were found for the smoothed shapes of signals. To obtain the energy of BGO pulses, we calculated the pulse area by using the coefficients of the calibration of the BGO detectors.

2.3. Calibration
To calibrate each of the BGO detectors we used standard radioactive sources as ^{137}Cs, ^{60}Co and Pu(Be), and cosmic muons. The experimental setup for the calibration by muons is presented in Fig.1 (left side), and the BGO detector was in coincidence with the plastic detector. We calculated by GEANT4 the energy lost by cosmic muons in the BGO scintillator, and the results of calculation (triangles) in comparison with experimental data (circles) are presented in Fig.1 (right side). The experimental and theoretical spectra have the same maximum near 70 MeV.

3. Results and discussion
3.1. Time-of-flight spectrum
The main idea of our experiment was the use of the TOF method to reduce the influence of cosmic background by means of the delay time between the pulses in the plastic detector and the BGO detectors, so that we could choose the n-γ coincidences from the ones which also comprised the cosmic shower coincidences. The typical shape of the TOF spectrum is presented in Fig.2. The time resolution of the γ-γ peak was about 2 ns. Cosmic rays gave a contribution in two ways. Near the peak of γ-γ coincidences there is a peak of cosmic shower coincidences which shows a Gaussian-like distribution. The shift between the top of γ-γ peak and the top of cosmic peak is about 2 ns. This is needed time to a γ-photon to pass 50 cm from the ^{252}Cf source to

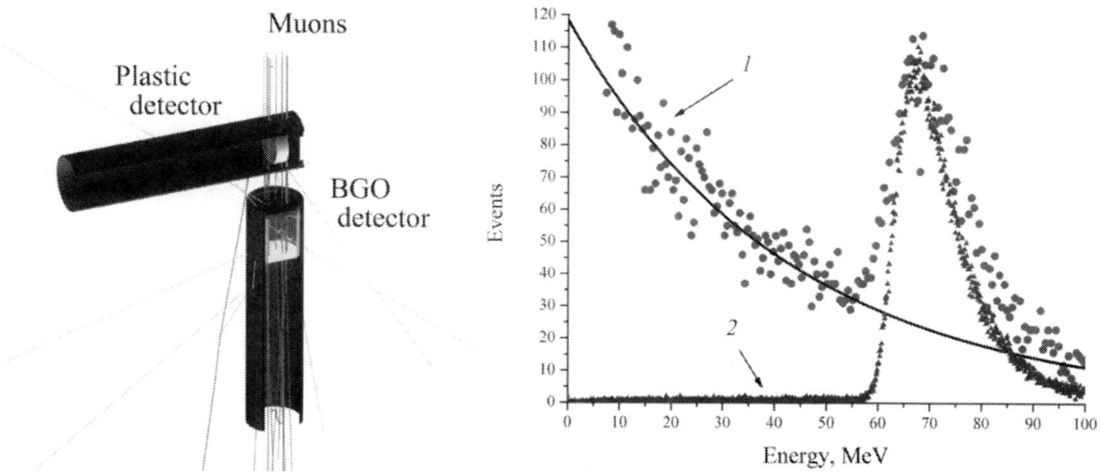

Figure 1. Experimental setup for the calibration of the BGO detectors (left side), and energy loss by cosmic muons in the BGO crystal (right side). Circles are the experimental data (cosmic background was approximated by an exponential function), triangles are the GEANT4 calculations.

the plastic detector. Also there were random coincidences between cosmic rays and neutrons emitted from the source, and between cosmic rays and γ-rays emitted from the source. These events show an uniform distribution.

To estimate the influence of cosmic ray events we made measurements with the same geometry but without source. The total rate of high energy γ-rays ($E_\gamma > 30$ MeV) was $1.7 \times 10^{-5} photon/(MeV \times s)$ in the region of γ-γ coincidence peak. In the experiment with the ^{252}Cf source the rate of γ-rays ($E_\gamma > 30$ MeV) from n-γ coincidences was $1.5 \times 10^{-6} photon/(MeV \times s)$. But the peak of n-$\gamma$ coincidences cross only the tail of the peak of cosmic coincidences, so cosmic events may give a contribution lower than about 1%. The final estimation of the rate of cosmic coincidences was lower than $1.7 \times 10^{-7} photon/(MeV \times s)$ in the region of n-γ peak.

3.2. Emission probability and result

In the TOF spectrum we selected one time window in the region of n-γ coincidences and another one of the same time duration is symmetrically chosen at the left side with respect to the peak of cosmic shower coincidences. In the second region there are only random coincidences. In order to subtract the contribution of cosmic ray background from the n-γ coincidences, we subtracted the spectrum of the random coincidences (left window) from the true n-γ coincidences plus the random spectrum (right window). The probability of γ-ray emission accompanying spontaneous fission of ^{252}Cf was calculated by formula:

$$P_\gamma(E_\gamma) = \frac{N_{n-\gamma}(E_\gamma)}{\epsilon_\gamma(E_\gamma) \cdot T \cdot N_n \cdot \Delta E_\gamma} \quad (1)$$

where $P_\gamma(E_\gamma)$ is the probability of photon emission with energy E_γ; $N_{n-\gamma}(E_\gamma)$ is the number of the registered n-γ coincidence events with photons of energy E_γ; $\epsilon_\gamma(E_\gamma)$ is the absolute efficiency of the BGO detector which registers photons with energy E_γ, calculated by GEANT4; T is the total time of the collected coincidence events; N_n is the number of registered pulses by the plastic detector in single mode; ΔE_γ is the interval of energy for the averaging procedure.

The obtained probability is presented in Fig.3. For 90° and 180° geometry of BGO, the

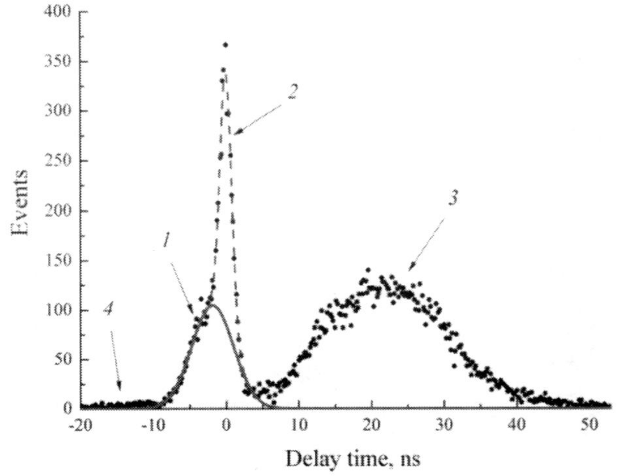

Figure 2. TOF spectrum of ^{252}Cf. The labels denote: *1* - the peak of cosmic showers; *2* - the peak of γ-γ coincidences; *3* - the peak of the n-γ coincidences; *4* - the random coincidences. Dashed line represents the approximation of peaks *1* and *2* by sum of two Gaussian curves. The Gaussian representing the peak of the cosmic showers is denoted by full line.

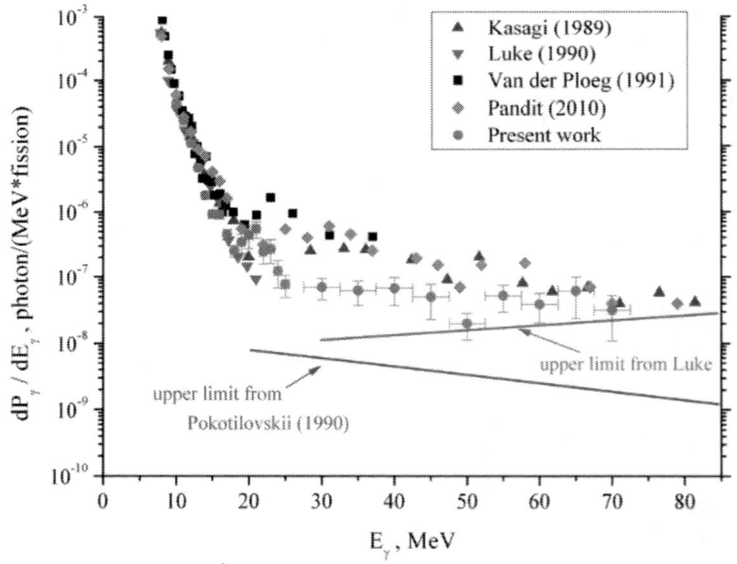

Figure 3. Photon emission probability of the high energy γ-rays accompanying spontaneous fission of the ^{252}Cf nucleus.

emission probability values at high energy γ-rays were of the same order. In comparison with the results of Kasagi[5] and Pandit[9] groups, our obtained probability values are about one order lower at $E_\gamma \sim 30$ MeV and become equal at $E_\gamma \sim 70$ MeV. The relevant difference between the obtained probabilities at $E_\gamma \sim 30$ MeV of other groups and our presented results may be due to their incomplete cut off of the cosmic ray background leading to an overestimation of the γ-emission probability by the γ-γ coincidence method.

Pokotilovskii[6] and Luke[7] used complicated active and passive shieldings in order to reduce cosmic ray background. For our opinion, they may reduce also true coincidence events in the case of high energy γ-rays, because such γ-rays do not lost all its energy in the γ-detector since they lost some part in the anticoincidence shielding. In Fig.3 our results reproduce the estimated upper limits for the high energy γ-rays presented in Refs. [6, 7].Therefore, only at $E_\gamma \sim 70$ MeV our results are in good agreement with the determinations of Kasagi and Pandit, and with the upper limit given by Luke.

3.3. Triple coincidences

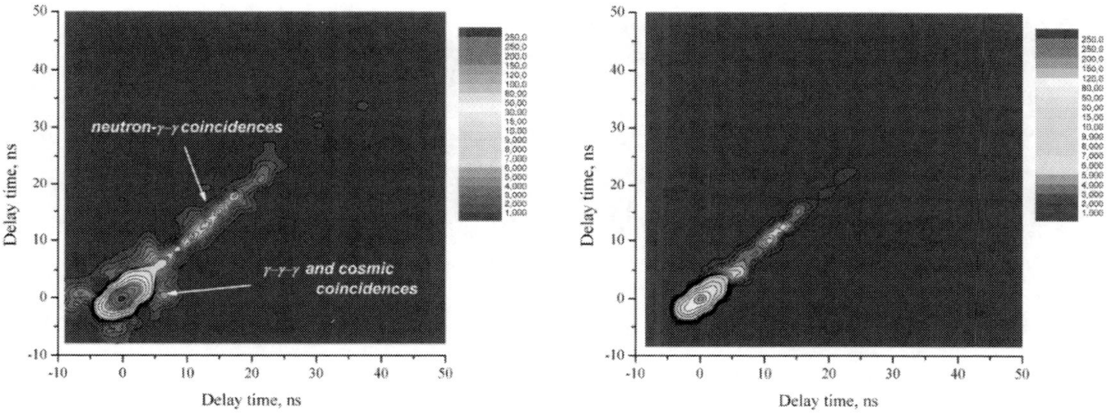

Figure 4. Number of triple coincidences (denoted by color) as a function of delay time between pulses of two BGO detectors and plastic detector. In the left side of figure, the down arrow indicates the region of triple $\gamma - \gamma - \gamma$ and cosmic coincidences; the triple n–$\gamma - \gamma$ coincidence distribution is clearly resolved in comparison to the triple $\gamma - \gamma - \gamma$ coincidences. In the right side of figure are presented the events of triple coincidences with $E_\gamma > 10$ MeV only.

Fig.4 shows our preliminary results of triple coincidences between pulses of the two BGO detectors and plastic detector. The peak of the γ-γ-γ coincidences and the distribution of the n-γ-γ coincidences are clearly resolved. The n-γ-γ coincidence distribution presents a peak (at the typical delay time of about 20 ns) corresponding to neutrons with energy of about 5 MeV. Also a local maximum (indicated in figure by the upper arrow) along the n-γ-γ coincidence distribution, is observed at the delay time of about 10 ns, corresponds to neutrons with energy of about 10 MeV. In the right side of figure are reported only the events of triple coincidences with $E_\gamma > 10$ MeV.

References

[1] Giardina G, Fazio G, Mandaglio G, Manganaro M, Maydanyuk S P, Olkhovsky V S, Eremin N V, Paskhalov A A, Smirnov D A, Sacca C 2008 *Mod. Phys. Lett.* A **23** 2651

[2] Giardina G, Fazio G, Mandaglio G, Manganaro M, Sacca C, Eremin N V, Paskhalov A A, Smirnov D A, Maydanyuk S P, Olkhovsky V S 2008 *Europ. Phys. Journ.* A **36** 31

[3] Eremin N V, Paskhalov A A, Markochev S S , Tsivetkov E A, Mandaglio G, Manganaro M, Fazio G, Giardina G, and Romaniuk M V 2010 *Int. J. Mod. Phys. E* **19** 1183

[4] Maydanyuk S P , Olkhovsky V S, Mandaglio G, Manganaro M, Fazio G and Giardina G 2010 *Phys. Rev. C* **82** 014602
[5] Kasagi J, Hama H, Yoshida K, Sakurai M, Ishii K 1989 *J. Phys. Soc. Jpn. Suppl.* **58** 620
[6] Pokotilovskii Yu N 1990 *Sov. J. Nucl. Phys.* **52** 942
[7] Luke S J, Gossett C A, Vandenbosh R 1991 *Phys. Rev.* C **44** 1548
[8] van der Ploeg H, Postma R, Bacelar J C, Berg T, Iacob V E, Jongman J R, Woude A 1992 *Phys. Rev. Lett.* **68** 3145
[9] Pandit D, Mukhopadhyay S, Bhattacharya S, Pal S, De A, Banerjee S R 2010 *Phys. Lett.* B **690** 473

Journal of Physics: Conference Series **282** (2011) 012018 doi:10.1088/1742-6596/282/1/012018

Symmetry Energy Effects on Low Energy Dissipative Heavy Ion Collisions

C Rizzo[1 2], V Baran[3], M Colonna[1], M Di Toro[1 2], M Odsuren[4]

[1] Laboratori Nazionali del Sud INFN, I-95123 Catania, Italy
[2] Physics and Astronomy Dept., University of Catania
[3] Physics Faculty, Univ. of Bucharest and NIPNE-HH, Romania
[4] Nuclear Research Center, National University of Mongolia, Ulaan-Baatar

E-mail: ditoro@lns.infn.it

Abstract.
We investigate the reaction path followed by Heavy Ion Collisions with exotic nuclear beams at low energies. We focus on the interplay between reaction mechanisms, fusion vs. break-up (fast-fission, deep-inelastic), that in exotic systems is expected to be influenced by the symmetry energy term at densities around the normal value. The method described here, based on the event by event evolution of phase space quadrupole collective modes, will nicely allow to extract the fusion probability at relatively early times, when the transport results are reliable. Fusion probabilities for reactions induced by ^{132}Sn on $^{64,58}Ni$ targets at 10 AMeV are evaluated. We obtain larger fusion cross sections for the more n-rich composite system, and, for a given reaction, with a soft symmetry term above saturation. A collective charge equilibration mechanism (the Dynamical Dipole Resonance, DDR) is revealed in both fusion and break-up events, depending on the stiffness of the symmetry term just below saturation. Finally we investigate the effect of the mass asymmetry in the entrance channel for systems with the same overall isospin content and similar initial charge asymmetry. As expected we find reduced fusion probabilities for the more mass symmetric case, while the DDR strength appears not much affected. This is a nice confirmation of the prompt nature of such collective isovector mode.

1. Introduction

Production of exotic nuclei has opened the way to explore, in laboratory conditions, new aspects of nuclear structure and dynamics up to extreme ratios of neutron (N) to proton numbers (Z). An important issue addressed is the density dependence of the symmetry energy term in the nuclear Equation of State (EOS), of interest also for the properties of astrophysical objects [1, 2, 3, 4]. By employing Heavy Ion Collisions (HIC), at appropriate beam energy and centrality, the isospin dynamics at different densities of nuclear matter can be investigated [3, 4].

We focus the attention on the interplay of fusion vs. deep-inelastic mechanisms for dissipative HIC with exotic nuclear beams at low energies, just above the Coulomb Barrier (between 5 and 20 AMeV), where unstable ion beams with large asymmetry will be soon available.

We show that the reaction dynamics undergoes a fusion/break-up path bifurcation at very early times due to coulomb and angular momentum effects. In such critical transition stage the final outcome is essentially ruled by dynamical fluctuations and it will be rather sensitive to differences in the neutron-repulsive symmetry term.

Moreover, it is now well established that in the same energy range, for fusion reactions between nuclei with different N/Z ratios, the charge equilibration process has a collective character, the prompt Dynamical Dipole Resonance (DDR), a kind of large amplitude Giant Dipole Resonance (GDR), which is thermally excited in the final equilibrated residues, see the recent [5] and refs. therein. The gamma yield resulting from the decay of such pre-equilibrium isovector mode can encode information about the early stage of the reaction [6, 7, 8, 9]. This collective response is appearing in the intermediate neck region, while the system is still in a highly deformed dinuclear configuration with large surface contributions, and so it will be sensitive to the density dependence of symmetry energy below saturation [5]. Here we will show that this mode is present also in break-up events, provided that a large dissipation is involved. In fact we see that the strength of such fast dipole emission is not much reduced passing from fusion to very deep-inelastic mechanisms. This can be expected from the fact that such excitation is related to an entrance channel collective oscillation. Thus we suggest the interest of a study of the prompt gamma radiation, with its characteristic angular anisotropy [5], even in deep-inelastic collisions with radioactive beams.

2. Reaction Dynamics

The reaction dynamics is described by a Stochastic Mean-Field (SMF) approach, extension of the microscopic Boltzmann-Nordheim-Vlasov transport equation [3], where the time evolution of the semi-classical one-body distribution function $f(\mathbf{r},\mathbf{p},t)$ is following a Boltzmann-Langevin evolution dynamics (see [10] and refs. therein).

In the SMF model the fluctuating term $\delta I[f]$ is implemented in an approximate way, through stochastic spatial density fluctuations [11]. Stochasticity is essential to get distributions, as well as to allow the growth of dynamical instabilities. In order to map the particle occupation at each time step, gaussian phase space wave packets (test particles) are considered. In the simulations 100 test particles per nucleon have been employed for an accurate description of the mean field dynamics. In the collision integral, I_{coll}, an in-medium depending nucleon-nucleon cross section, via the local density, is employed [12, 13].

The mean field is built from Skyrme forces:

$$\begin{aligned} U_{n,p} &= A\frac{\rho}{\rho_0} + B(\frac{\rho}{\rho_0})^{\alpha+1} + C(\rho)\frac{\rho_n-\rho_p}{\rho_0}\tau_q + \\ &+ \frac{1}{2}\frac{\partial C}{\partial \rho}\frac{(\rho_n-\rho_p)^2}{\rho_0} \end{aligned} \tag{1}$$

where $q = n,p$ and $\tau_n = 1, \tau_p = -1$. The coefficients A, B and the exponent α, characterizing the isoscalar part of the mean-field, are fixed requiring that the saturation properties of symmetric nuclear matter ($\rho_0 = .145 fm^{-3}$, $E/A = -16MeV$), with a compressibility modulus around 200 MeV, are reproduced. The function $C(\rho)$ will give the potential part of the symmetry energy:

$$\frac{E_{sym}}{A}(\rho, T=0) = \frac{E_{sym}}{A}(kin) + \frac{E_{sym}}{A}(pot) \equiv \frac{\epsilon_F}{3} + \frac{C(\rho)}{2\rho_0}\rho \tag{2}$$

For the density dependence of the symmetry energy, we have considered two different parametrizations [14, 15], that are presented in Fig.1. In the Asysoft EOS choice, $\frac{C(\rho)}{\rho_0} = 482 - 1638\rho$, the symmetry energy has a weak density dependence close to the saturation, being almost flat around ρ_0. For the Asystiff case, $\frac{C(\rho)}{\rho_0} = \frac{32}{\rho_0}\frac{2\rho}{\rho+\rho_0}$, the symmetry energy has a clear increasing slope at normal density. Aim of this work is to show that fusion probabilities, fragment properties in break-up events, as well as properties of prompt collective modes, in collisions induced by neutron-rich exotic beams, are sensitive to the different slopes of the symmetry term around saturation.

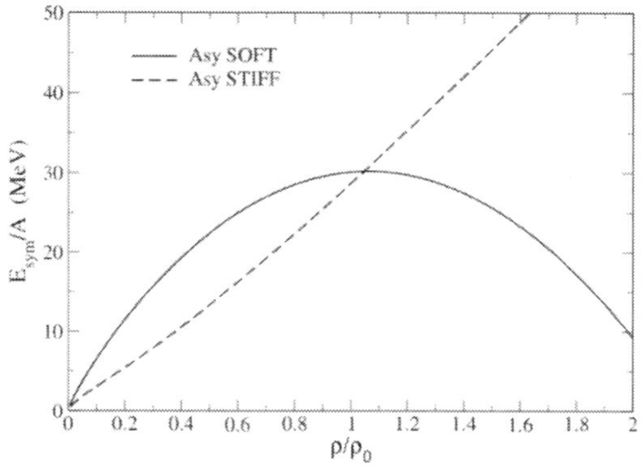

Figure 1. Density dependence of the symmetry energy for the two parametrizations. Solid line: Asysoft. Dashed line: Asystiff

3. Fusion dynamics for ^{132}Sn induced reactions

In order to study isospin and symmetry energy effects on the competition between fusion and break-up (deep-inelastic) we consider the reactions ^{132}Sn + $^{64,58}Ni$ at 10 AMeV, having in mind that ^{132}Sn beams with good intensities in this energy range will be soon available in future Radioactive Ion Beam facilities. In particular, we have performed collision simulations for semi-peripheral impact parameters (from b =4.5 fm to b = 8.0 fm, with Δb= 0.5 fm), to explore the region of the transition from fusion to break-up dominance. We have found a reliable criterion that can indicate when the reaction mechanism is changing. This will also allow to evaluate the corresponding absolute cross sections and the symmetry energy effects..

The method is based on a phase space analysis of quadrupole collective modes. The information on the final reaction path is deduced investigating the fluctuations of the system at early times (200-300 fm/c), when the formation of composite elongated configurations is observed and phenomena associated with surface metastabity and/or instability may take place, [16].

We start considering the time evolution, in each event, of the quadrupole moment in coordinate space which is given by:

$$Q(t) =< 2z^2(t) - x^2(t) - y^2(t) >,$$

averaged over the space distribution in the composite system. At the same time-steps we construct also the quadrupole moment in momentum space:

$$QK(t) =< 2p_z^2(t) - p_x^2(t) - p_y^2(t) >,$$

in a spatial region around the center of mass. The z-axis is along the rotating projectile-like/target-like direction, the x-axis is on the reaction plane.

We run 200 events for each set of macroscopic initial conditions and we take the average over this ensemble.

In Fig.2 we present the time evolution of the mean space quadrupole moment at various centralities in the fusion-break-up transition region for the two reactions and for the two choices of the symmetry term. We notice the difference in Q(t) between the behavior corresponding to more peripheral impact parameters and that obtained for b=5-6 fm, where we have still a little oscillation in the time interval between 100 and 300 fm/c, good indication of a fusion contribution.

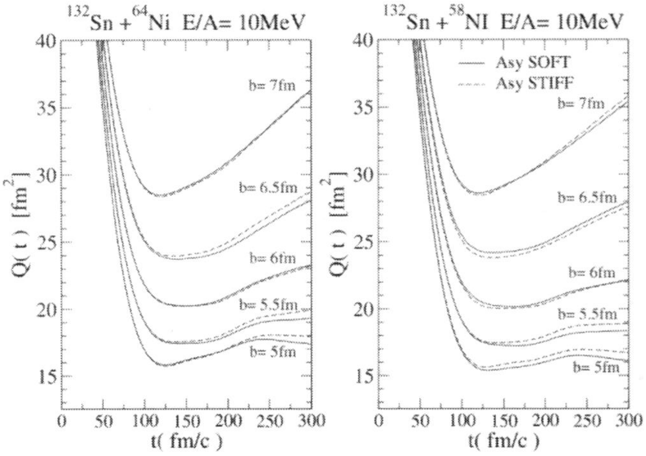

Figure 2. Time evolution of the space quadrupole moments in the angular momentum transition region, between b=5.0 and 7.0 fm. Solid line: Asysoft. Dashed line: Asystiff.

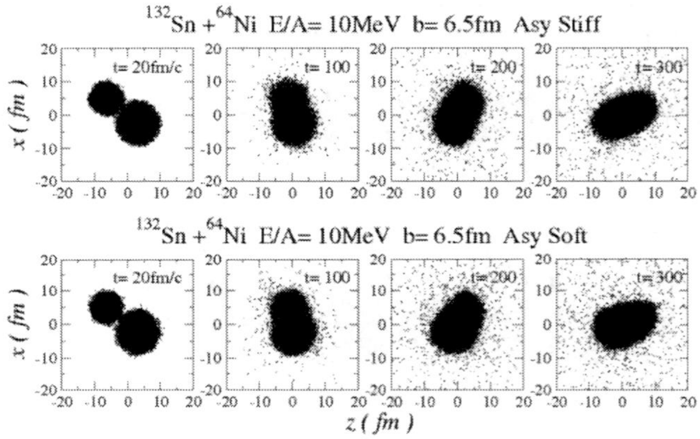

Figure 3. Time evolution of the space density distributions for the reaction $^{132}Sn +^{64} Ni$ (n-rich systems), 10 AMeV beam energy, for semicentral collisions, b=6.5 fm impact parameter (average over 20 events). Upper Panel: Asystiff. Lower Panel: Asysoft.

We can interpret these observations assuming that starting from about b = 5 fm, we have a transition from fusion to a break-up mechanism, like deep-inelastic. Positive values of the Q(t)-slope should be associated with a quadrupole deformation velocity of the dinuclear system that is going to a break-up exit channel. We notice a slight systematic difference, especially in the most neutron-rich system, with a larger deformation velocity in the Asystiff case. Hence, just from this simple analysis of the average space quadrupole "trajectories" we can already appreciate that the Asysoft choice seems to lead to larger fusion cross sections.

The latter point can also be qualitatively seen from the time evolution of the space density distributions projected on the reaction plane, as shown in Fig.3. The formation of a more compact configuration in the Asysoft case can be related to a larger fusion probability.

It is very instructive to look also at the time evolution of the quadrupole deformations in momentum space. In Fig.4 we present the time evolution of the average p-quadrupole moments at various centralities for the two systems and the two choices of the symmetry term. We notice

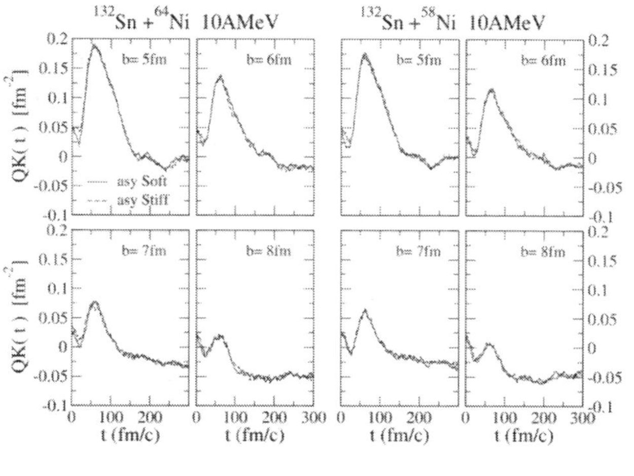

Figure 4. Time evolution of the momentum quadrupole moments, in a sphere of radius 3 fm around the c.o.m., for different centralities and for the two systems. Solid line: Asysoft. Dashed line: Asystiff.

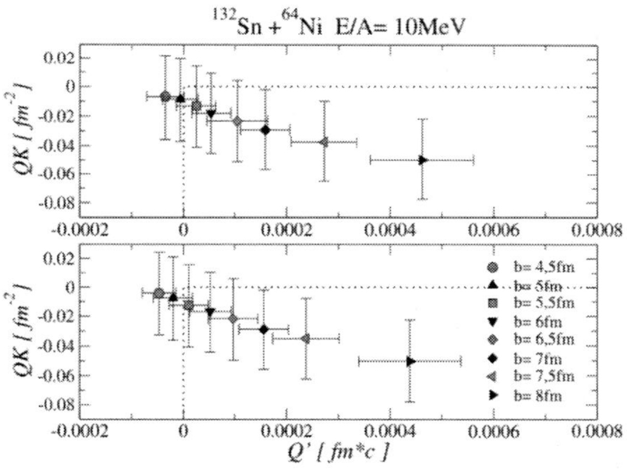

Figure 5. ^{132}Sn + ^{64}Ni system. Mean value and variance of QK vs Q', averaged over the 100-300 fm/c time interval, at various centralities in the transition region. The box limited by dotted lines represents the break-up region. Upper panel: Asystiff. Bottom Panel: Asysoft.

a difference between the plots corresponding to peripheral or central collisions. With increasing impact parameter the quadrupole QK(t) becomes more negative in the time interval between 100 and 300 fm/c: the components perpendicular to the symmetry axis, that is rotating in reaction plane, are clearly increasing eventually leading to a separation of the deformed dinuclear system. Thus the break-up probability will be larger if the quadrupole moment in p-space is more negative.

From Fig.s 2, 4 one can see that there is a region of impact parameter (b = 5-6.5 fm) where the derivative of the quadupole moment in coordinate space, Q', and the quadrupole moment in momentum space, QK, are both rather close to zero. This is the region where we expect that fluctuations of these quantities should play an important role in determining the fate of the reaction and an event-by-event analysis is essential to estimate fusion vs. break-up probabilities.

The Q'-QK correlation plots for the two systems and the two asy-EOS are presented in

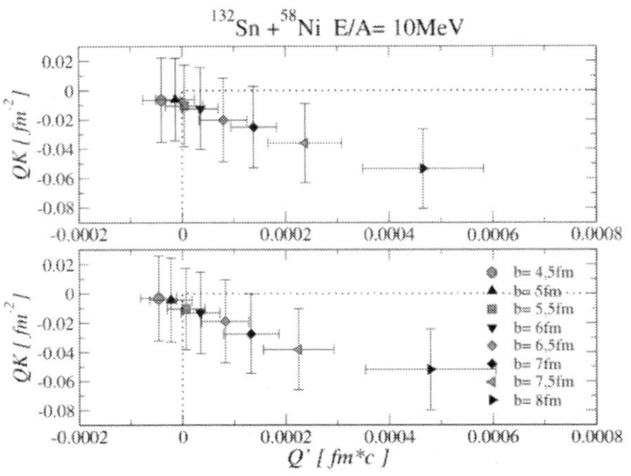

Figure 6. Like in Fig.5 but for the ^{132}Sn + ^{58}Ni system.

Figs.5 and 6, respectively. The displayed quantities are mean value and variance of the two extracted properties of the phase space moment evolution in the "bifurcation" time interval $t = 200 - 300\ fm/c$. We can evaluate the normal curves and the relative areas for each impact parameter in order to select the events: break-up events will be located in the regions with both positive slope of Q(t) and negative QK. In this way, for each impact parameter we can evaluate the fusion events by the difference between the total number of events and the number of break-up cases. Finally the fusion cross section is obtained (in absolute value) by

$$\frac{d\sigma}{dl} = \frac{2\pi}{k^2} l \frac{N_f}{N_{tot}}, \tag{3}$$

where l is the angular momentum calculated in the semiclassical approximation, k is the relative momentum of the collision, N_f the number of fusion events and N_{tot} the total events in the angular momentum bin.

In Fig.7 we present the fusion spin distribution plots. We note that just in the centrality transition region there is a difference between the σ-fusion corresponding to the two different asy-EOS, with larger values for Asysoft.

In fact, the total cross sections are very similar: the difference in the area is about 4-5 % in the neutron rich system, 1128 mb (Asysoft) vs. 1078 mb (Asystiff), and even smaller, 1020 mb vs. 1009 mb, for the ^{58}Ni target. However, through a selection in angular momentum, $130 \leq l \leq 180$ ($\hbar$), we find that the Asysoft curve is significantly above the Asystiff one, and so in this centrality bin the fusion cross section difference can reach a 10% in the case of the more neutron-rich system. Then it can be compared to experimental data as an evidence of sensitivity to the density dependence of the symmetry energy.

From the comparison of the total areas for the two systems we can also estimate isospin effects on the total fusion cross section, with a larger value in the more neutron-rich case, as also recently observed in fusion reactions with Ar + Ni [17] and Ca + Ca isotopes [18].

Finally we like to note that for the neutron-rich case, $^{132}Sn + ^{64}Ni$, our absolute value of the total fusion cross section presents a good agreement with recent data, at lower energy (around 5 AMeV), taken at the ORNL [19].

In Fig.7 for the same system (left panel) we show also the results obtained with the macroscopic fusion probability evaluation code $PACE4$, [20, 21] obtained with different l-diffuseness parameters, fixing, as input parameters, our total fusion cross section and maximum angular momentum. We see that in order to have a shape similar to our $\sigma(l)$ distribution we

Journal of Physics: Conference Series **282** (2011) 012018 doi:10.1088/1742-6596/282/1/012018

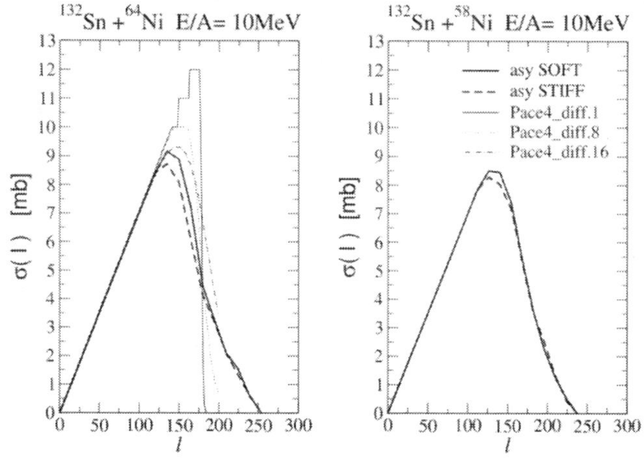

Figure 7. Angular momentum distributions of the fusion cross sections (mb) for the two reactions and the two choices of the symmetry term. For the $^{132}Sn + {}^{64}Ni$ system (left panel), the results of PACE4 calculations are also reported, for different l-diffuseness.

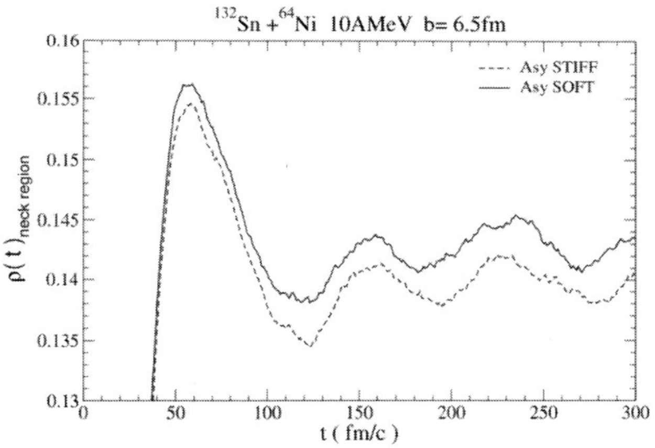

Figure 8. Reaction $^{132}Sn +^{64} Ni$ semiperipheral. Time evolution of the total density in the "neck" region

have to choose rather large diffuseness values, while the suggested standard choice for stable systems is around Δl=4 $\hbar$. This seems to be a nice evidence of the neutron skin effect.

4. Analysis of symmetry energy effects

The larger fusion probability obtained with the Asysoft choice, especially in the more n-rich system, seems to indicate that the reaction mechanism is regulated by the symmetry term at suprasaturation density, where the Asysoft choice is less repulsive for the neutrons [3, 14]. In order to check this point we have performed a detailed study of the density evolution in the region of overlap of the two nuclei, named *neck* in the following. We present results obtained for the system $^{132}Sn +^{64} Ni$ at impact parameter b = 6.5 fm.

The time evolution of the total density in this "neck" region is reported in Fig.8 for the two choices of the symmetry energy . We note that in the time interval of interest we have densities

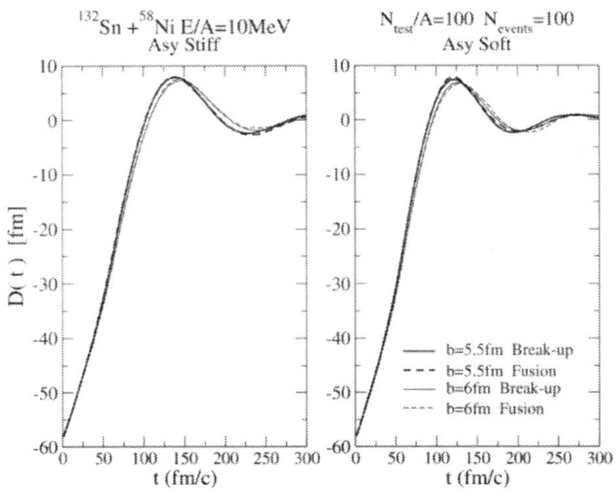

Figure 9. Reaction $^{132}Sn +^{58} Ni$ semiperipheral. Prompt Dipole oscillations in the composite system for break-up (solid lines) and fusion (dashed lines) events. Left Panel: Asystiff. Right Panel: Asysoft.

above or around the normal density and so a less repulsive symmetry term within the Asysoft choice, corresponding to larger fusion probabilities.

This also explains why larger fusion cross sections are seen for the neutron rich system, mainly in the Asysoft case. In fact, the neutron excess pushes the formed hot compound nucleus closer to the stability valley, especially when the symmetry energy is smaller.

Other nice features are: i) the density values found in the Asysoft case are always above the Asystiff ones, to confirm the expectation of a smaller equilibrium density for a stiffer symmetry term [3]; ii) collective monopole oscillations are present after 100 fm/c, showing that also at these low energies we can have some compression energy.

Break-up Events
Within the same transport approach, a first analysis of symmetry energy effects on break-up events in semiperipheral collisions of $^{132}Sn +^{64} Ni$ at 10 $AMeV$ has been reported in ref.[22]. Consistently with the more accurate study presented here, smaller break-up probabilities have been seen in the Asysoft choice. Moreover the neck dynamics on the way to separation is found also influenced by the symmetry energy below saturation. This can be observed in the different deformation pattern of the Projectile-Like and Target-Like Fragments (PLF/TLF), as shown in Fig.1 of [22]. Except for the most peripheral selections, larger deformations are seen in the Asystiff case, corresponding to a smaller symmetry repulsion at the low densities probed in the separation region. The neutron-rich neck connecting the two partners can then survive a longer time producing very deformed primary PLF/TLF. Even small clusters can be eventually dynamically emitted leading to ternary/quaternary fragmentation events [23, 24].

In conclusion not only the break-up probability but also a detailed study of fragment deformations in deep-inelastic (and fast-fission) processes, as well as of the yield of 3-4 body events, will give independent information on the symmetry term around saturation.

5. The Prompt Dipole Mode in Fusion and Break-up Events
From the time evolution of the nucleon phase space occupation, see Eq.(1), it is possible to extract at each time step the isovector dipole moment of the composite system. This is given

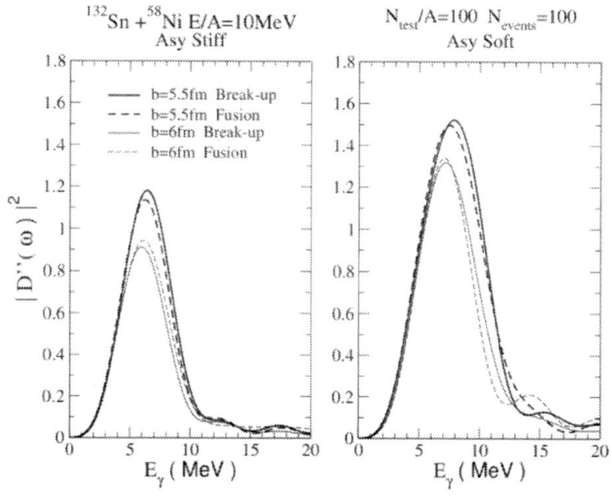

Figure 10. Reaction $^{132}Sn +^{58} Ni$ semiperipheral. Prompt Dipole strengths (in c^2 units), see text, for break-up (solid lines) and fusion (dashed lines) events. Left Panel: Asystiff. Right Panel: Asysoft.

by $D(t) = \frac{NZ}{A}X(t)$, where $A = N + Z$, and $N = N_1 + N_2$, $Z = Z_1 + Z_2$, are the total number of participating nucleons, while $X(t)$ is the distance between the centers of mass of protons and neutrons. It has been clearly shown, in theory as well as in experiments, that at these beam energies the charge equilibration in fusion reactions proceeds through such prompt collective mode. In our study we have focused the attention on the system with larger initial charge asymmetry, the ^{132}Sn on ^{58}Ni case,

In Fig.9 we present the prompt dipole oscillations obtained for semicentral impact parameters, in the transition zone. We nicely see that in both classes of events, ending in fusion or deep-inelastic channels, the dipole mode is present almost with the same strength. We note that such fast dipole radiation was actually observed even in the most dissipative deep-inelastic events in stable ion collisions [25, 26, 28].

The corresponding emission rates can be evaluated, through a "bremsstrahlung" mechanism, in a consistent transport approach to the rection dynamics, which can account for the whole contribution along the dissipative non-equilibrium path, in fusion or deep-inelastic processes [27].

In fact from the dipole evolution $D(t)$ we can directly estimate the photon emission probability ($E_\gamma = \hbar\omega$):

$$\frac{dP}{dE_\gamma} = \frac{2e^2}{3\pi\hbar c^3 E_\gamma}|D''(\omega)|^2, \tag{4}$$

where $D''(\omega)$ is the Fourier transform of the dipole acceleration $D''(t)$. We remark that in this way it is possible to evaluate, in *absolute* values, the corresponding pre-equilibrium photon emission yields. In Fig.10 we report the prompt dipole strengths $|D''(\omega)|^2$ for the same event selections of Fig.9.

The dipole strength distributions are very similar in the fusion and break-up selections in this centrality region where we have a strong competition between the two mechanisms. In any case there is a smaller strength in the less central collisions (b=6.0fm), with a centroid slightly shifted to lower values, corresponding to more deformed shapes of the dinuclear composite system.

In the Asysoft choice we have a systematic increase of the yields, roughly given by the area of the strength distribution, of about 40% more than in the Asystiff case, for both centralities and

Journal of Physics: Conference Series **282** (2011) 012018 doi:10.1088/1742-6596/282/1/012018

selections. In fact from Eq.(4) we can directly evaluate the total γ-multiplicities, integrated over the dynamical dipole region. For centrality b=5.5fm we get 2.3 10^{-3} (1.6 10^{-3}) in the Asysoft (Asystiff) choice, and for b=6.0fm respectively 1.9 10^{-3} (1.3 10^{-3}), with almost no difference betwen fusion and break-up events.

From Fig.1 we see that Asysoft corresponds to a larger symmetry energy below saturation. Since the symmetry term gives the restoring force of the dipole mode, our result is a good indication that the prompt dipole oscillation is taking place in a deformed dinuclear composite system, where low density surface contributions are important, as already observed in ref.[5].

Previously we have shown that the Asysoft choice leads to a large fusion probability since it gives a smaller repulsion at the suprasaturation densities of the first stage of the reaction. Here we see that for the dipole oscillation it gives a larger restoring force corresponding to mean densities below saturation. This apparent contradictory conclusion can be easily understood comparing Figs.8 and 9. We note that the onset of the collective dipole mode is delayed with respect to the first high density stage of the neck region since the composite system needs some time to develop a collective response of the dinuclear mean field.

In this way fusion and dynamical dipole data can be directly used to probe the isovector part of the in medium effective interaction *below and above* saturation density.

Anisotropy

Aside the total gamma spectrum the corresponding angular distribution can be a sensitive probe to explore the properties of preequilibrium dipole mode and the early stages of fusion dynamics. In fact a clear anisotropy vs. the beam axis has been recently observed [29]. For a dipole oscillation just along the beam axis we expect an angular distribution of the emitted photons like $W(\theta) \sim \sin^2\theta \sim 1 + a_2 P_2(cos\theta)$ with $a_2 = -1$, where θ is the polar angle between the photon direction and the beam axis. Such extreme anisotropy will be never observed since in the collision the prompt dipole axis will rotate during the radiative emission. In fact the deviation from the $\sin^2\theta$ behavior will give a measure of the time interval of the fast dipole emission. In the case of a large rotation one can even observe a minimum at 90 degrees [16].

We clearly see that the dominant emission region is the initial one, just after the onset of the collective mode between 80 and 150 fm/c, while the emitting dinuclear system has a large rotation. We can predict a dependence of the angular distribution on the symmetry energy. With a weaker symmetry term at low densities (Asystiff case), the emission probability is a little delayed and presents a smoother behavior. If meanwhile the system has a fast rotation we can expect a clear symmetry term effect on the $W(\theta)$, in particular for the large angular momenta in the transition region.

Hence, from accurate measurements of the angular distribution of the emitted γ's, in the range of impact parameters where the system rotation is significant, one can extract independent information on the density behavior of the symmetry energy.

6. Mass symmetry effects

It is well known that the mass symmetry of the colliding partners is also affecting the fusion probability. Mass symmetric cases are expected to lead to a reduced fusion events for the combined effect of larger coulomb barrier and smaller dissipation (in a macroscopic "window" model). Since the Dynamical Dipole mode is also related to the fusion dynamics we have analysed both the fusion-break-up competition and the DDR emission for n-rich systems selecting entrance channels with roughly the same charge asymmetry but with rather different mass asymmetry.

We report here some preliminary results for the reactions $^{124}Sn(N/Z = 1.48) +^{56} Fe(N/Z = 1.15)$ vs. $^{90}Kr(N/Z = 1.5) +^{90} Zr(N/Z = 1.25)$ at 10 $AMeV$, i.e. with same average $< N/Z >= 1.37$, but with close charge asymmetry and rather different mass symmetry in the entrance channel.

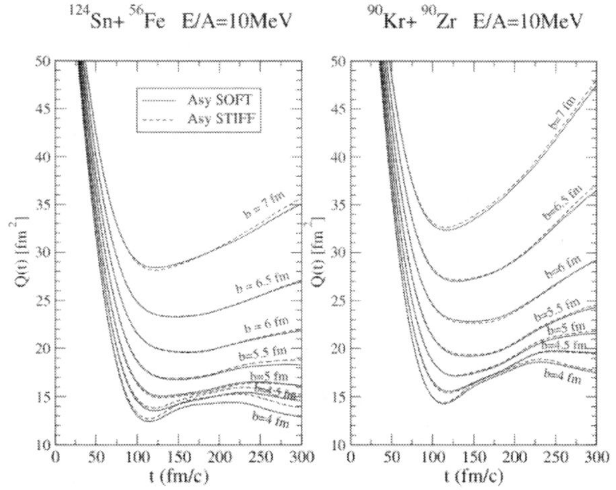

Figure 11. Reactions $^{124}Sn+^{56}Fe$ vs. $^{90}Kr+^{90}Zr$ at 10 $AMeV$. Time evolution of the space quadrupole moments in the angular momentum transition region, between b=4.0 and 7.0 fm. Solid line: Asysoft. Dashed line: Asystiff.

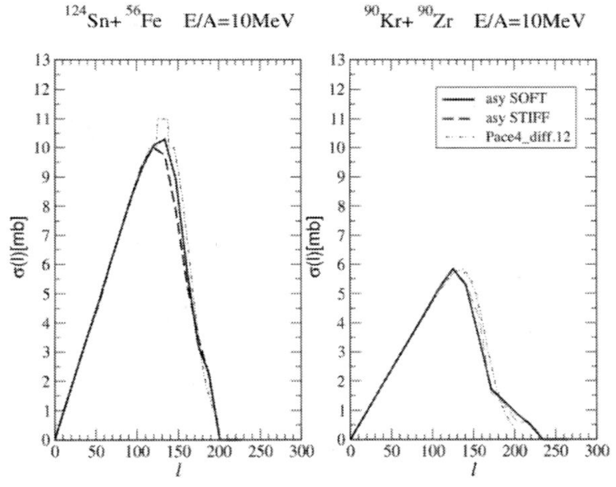

Figure 12. Reactions $^{124}Sn+^{56}Fe$ vs. $^{90}Kr+^{90}Zr$ at 10 $AMeV$. Angular momentum distributions of the fusion cross sections (mb) for the two choices of the symmetry term. The results of PACE4 calculations are also reported, l-diffuseness 12 $\hbar$.

In Fig.11 we show the time evolution of the mean space quadrupole moments (like Fig.2) for the two reactions and the two choices of the symmetry term. We see that for the mass asymmetric case Sn/Fe we have more definite oscillations in the centrality transition region, clear indication of a larger fusion probability. We note also that in mass symmetric Kr/Zr case the fusion process is slower and at the same 300 fm/c time step we do not see clear asymptotic trends for the space quadrupoles in the transition region (and the same is observed for the momentum quadrupoles QKs). It is then difficult to distinguish symmetry energy effects but we can clearly evaluate the impact of mass symmetry on the fusion cross sections. This is shown in Fig.12. The total cross sections are 1115 mb and 630 mb respectively for the mass asymmetric and symmetric cases (Asysoft choice). We are performing an extension of the calculation to later

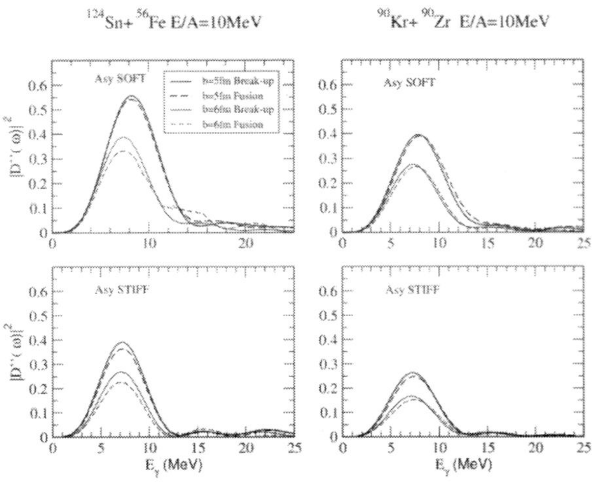

Figure 13. Reactions $^{124}Sn +^{56} Fe$ vs. $^{90}Kr +^{90} Zr$ at 10 $AMeV$. Prompt Dipole strengths (in c^2 units) for break-up (solid lines) and fusion (dashed lines) events for impact parameters in the transition region. Left Panel: Asystiff. Right Panel: Asysoft.

times in order to evaluate the dependence on the stiffness of the symmetry terms. Even in this n-rich case we find that in order to have similar $\sigma(l)$ distributions for the PACE4 simulations we have to use a rather large l-diffuseness value.

It is interesting to look at the DDR excitation. In Fig.13 we report the dipole strength distribution $|D''(\omega)|^2$ for the fusion/break-up event selection in the impact parameter transition region. As in the previous $^{132}Sn +^{58} Ni$ system the DDR emission is larger in the Asysoft choice and not much different for the two reaction mechanism selections. An important point to remark is the larger emission probability for the mass-asymmetric Sn/Fe case. However this effects is almost exactly scaling with the square of the initial Dipole Moment $D(t = 0)$, 26.3 fm for Sn/Fe vs. 22.4 fm for Kr/Fe. This is a clear indication that only the initial charge asymmetry really matters for the prompt dipole emission, which then is not much affected from the different fusion dynamics. All that is fully consistent with the very fast nature of the Dynamical Dipole mode.

7. Conclusions and perspectives

We have undertaken an analysis of the reaction path followed in collisions involving exotic systems at beam energies around 10 AMeV. In this energy regime, the main reaction mechanisms range from fusion to dissipative binary processes, together with the excitation of collective modes of the nuclear shape. Moreover, in charge asymmetric systems, isovector dipole oscillations can be excited at the early dynamical stage, also sensitive to the behavior of the symmetry energy. We have shown that, in neutron-rich systems, fusion vs. break-up probabilities are influenced by the neutron repulsion during the approaching phase, where densities just above the normal value are observed. Hence larger fusion cross sections are obtained in the Asysoft case, associated with a smaller value of the symmetry energy at supra-saturation densities. On the other hand, the isovector collective response, that takes place in the deformed dinuclear configuration with large surface contributions, is sensitive to the symmetry energy below saturation. The relevant point of our analysis is that it is based on the study of the fluctuations that develop during the early dynamics, when the transport calculations are reliable. Fluctuations of the quadrupole moments, in phase space, essentially determine the final reaction path.

In conclusion considerable isospin effects are revealed just selecting the impact parameter window corresponding to semi-peripheral reactions. Interesting perspectives are opening for new experiments on low energy collisions with exotic beams focused to the study of the symmetry term below and above saturation density. We suggest some sensitive observables:

i) Fusion vs. Break-up probabilities in the centrality transition region;

ii) Fragment deformations in break-up processes and probability of ternary/quaternary events.

iii) γ-multiplicity and anisotropy of the Prompt Dipole Radiation, for dissipative collisions in charge asymmetric entrance channels.

iv) Analysis of the same observables vs. the entrance channel mass asymmetry.

Aknowledgements
We warmly thank Alessia Di Pietro for the discussions about the the use of Pace4 fusion simulations. One of authors, V. B. thanks for hospitality at the Laboratori Nazionali del Sud, INFN. This work was supported in part by the Romanian Ministery for Education and Research under the contract PNII, No.ID-1038/2008.

References

[1] Li B A and Schröder W U, Eds., 2001 *Isospin Physics in Heavy Heavy Ion Collisions at Intermediate Energies*, Nova Science Publishers, Inc, New York.
[2] Steiner A W, Prakash M, Lattimer J M, Ellis P J 2005 *Phys. Rep.* **411** 325
[3] Baran V, Colonna M, Greco V, Di Toro M 2005 *Phys. Rep.* **410** 335
[4] Li B A, Chen L W, Ko C M 2008 *Phys. Rep.* **464** 113
[5] Baran V, Rizzo C, Colonna M, Di Toro M, Pierroutsakou D 2009 *Phys. Rev.*C **79** 021603(R)
[6] Chomaz P, Di Toro M, Smerzi A 1993 *Nucl. Phys.* A **563** 509
[7] Baran V et al. 1996 *Nucl. Phys.*A **600** 111
[8] Simenel C, Chomaz P, de France G 2001 *Phys. Rev. Lett.* **86** 2971
[9] Simenel C, Chomaz P, de France G 2007 *Phys. Rev.*C **76** 024609
[10] Rizzo J, Chomaz P, Colonna M 2008 *Nucl. Phys.*A **806** 40
[11] Colonna M et al. 1998 *Nucl. Phys.*A **642** 449
[12] Li G Q, Machleidt R 1993 *Phys. Rev.*C **48** 1702 Phys. Rev. C49, 566 (1994)
[13] Li G Q, Machleidt R 1994 *Phys. Rev.*C **49** 566
[14] Colonna M et al. 1998 *Phys. Rev.*C **57** 1410
[15] Baran V et al. 2002 *Nucl. Phys.*A **703** 603
[16] Rizzo C, Baran V, Colonna C, Corsi A, Di Toro M, 2010 *Symmetry energy effects on fusion cross sections*, arXiv:1010.2927
[17] Marini P et al. (Indra-Vamos Collab) 2010 *IWM2009 Int.Workshop* SIF Conf.Proceedings Vol.101, pp.189-196, Bologna
[18] Amorini F et al. 2009 *Phys. Rev. Lett.* **102** 112701
[19] Liang J F et al. 2007 *Phys. Rev.*C **75** 054607
[20] Gavron A 1979 *Phys. Rev.* **21** 230
[21] Tarasov O B, Bazin D 2003 *Nucl. Inst. Methods* B **204** 174
[22] Di Toro M et al. 2007 *Nucl. Phys.*A **787** 585c
[23] Skwira-Chalot I et al. (Chimera Collab.) 2008 *Phys. Rev. Lett.* **101** 262701
[24] Wilczynski J et al. (Chimera Collab.) 2010 *Phys. Rev.*C **81** 024605
[25] Pierroutsakou D et al. 2003 *Eur. Phys. Jour.*A **16** 423, *Nucl. Phys.*A **687** 245c
[26] Pierroutsakou D et al. 2005 *Phys. Rev.*C **71** 054605
[27] Baran V, Brink D M, Colonna M, Di Toro M 2001 *Phys. Rev. Lett.* **87** 182501
[28] Amorini F et al. 2004 *Phys. Rev.*C **69** 014608
[29] Martin B, Pierroutsakou D et al. (Medea Collab.) 2008 *Phys. Lett.*B **664** 47

Journal of Physics: Conference Series **282** (2011) 012019 doi:10.1088/1742-6596/282/1/012019

Incomplete vs. Complete Fusion at E/A $\approx$ 4-7 MeV

Pushpendra P. Singh[1], Abhishek Yadav[2], Vijay R. Sharma[2], Devendra P. Singh[2], Unnati Gupta[2], Manoj K. Sharma[3], R. Kumar[4], K. S. Golda[4], R. P. Singh[4], S. Muralithar[4], B. P. Singh[2], R. K. Bhowmik[4], R. Prasad[2]

[1]INFN - Laboratori Nazionali di Legnaro, I-35020 Legnaro, Italy
[2]Accelerator Laboratory, Department of Physics, A. M. University, Aligarh-202 002, India
[3]Physics Department, S. V. College, Aligarh-202 001, India
[4]NP-Group, Inter-University Accelerator Center, New Delhi-110 067, India

E-mail: pushpendrapsingh@gmail.com

Abstract. With a view to study onset and strength of incomplete fusion at low projectile energies (i.e., $\approx$ 4-7 MeV/nucleon) three sets of experiments have been performed in ^{12}C,^{16}O+^{169}Tm systems. In first set of experiments, spin-distributions and feeding intensity profiles for xn,αxn/2αxn-channels have been measured to figure out associated ℓ-values. The spin-distributions for direct-α-emitting channels (associated with incomplete fusion) have been found to be distinctly different than that observed for fusion-evaporation (complete fusion) channels. The mean value of driving input angular momenta associated with direct-α-emitting-channels have been found to be higher than that observed for fusion-evaporation xn/α-emitting-channels, and increases with direct-α-multiplicity in forward cone. The second set of experiments has been performed to understand influence of incomplete fusion on complete fusion at these energies. Incomplete fusion strength function has been deduced from the analysis of experimental excitation functions. The third set of experiments deals with the validation of data reduction procedure used to deduce incomplete fusion fraction, and to confirm the fusion incompleteness at slightly above barrier energies. Forward-recoil-ranges of heavy reaction products have been measured and analysed on the basis of break-up fusion model. More than one linear-momentum-transfer components associated with full- and/or partial-fusion of projectile with target nucleus have been observed. Experimental ranges of forward-recoils are found to be in good agreement with that estimated using range-energy formulation. The relative strengths of complete and incomplete fusion components deduced from the analysis of forward-recoil-ranges and excitation functions complement each other. Result presented in this paper conclusively demonstrate substantial incomplete fusion contribution at energy as low as 7% above the barrier.

1. Introduction

Dynamics of incomplete fusion has been intensively investigated in recent years using light heavy ion (A$\leq$16) beams at low projectile energies (i.e., $\approx$ 4-7 MeV/nucleon), where complete fusion (CF) supposed to be the sole contributor [1, 2, 3, 4, 5, 6, 7, 8, 9, 10, 11, 12, 13, 14]. At these energies, the interaction trajectories lying within the target dimensions prominently lead to the reaction modes namely: (*i*) CF: defined as the capture of entire projectile by target nucleus, and (*ii*) IF: where only a part of projectile fuses with target nucleus. In a qualitative way, both reaction modes can be disentangled on the basis of driving angular momenta at different interaction trajectories (ℓ-bins) [15, 16, 17]. Central and/or near-central trajectories ($0 \leq \ell \leq$

$\ell crit$) lead to CF, where an excited composite system forms after intimate contact and transient amalgamation of projectile and target nucleus. In this case, the attractive nuclear potential influences the sum of repulsive Coulomb and centrifugal potentials. Eventually, the projectile's kinetic energy and driving angular momenta are equally distributed among all the accessible internal degrees of freedom of composite system to form an equilibrated compound nucleus (CN). However, at relatively higher ℓ-values ($\geq \ell_{crit}$: imparted in the entrance channel due to non-central interactions) centrifugal potential overwhelms attractive nuclear potential, therefore, the pocket in the entrance channel potential disappears. As a consequence, fusion of entire projectile (CF) hinders and gives way to fusion incompleteness. In this case, a part of projectile emits as a spectator (P^s) to release excess driving angular momenta. After such an emission, the remnant (participant: P^p) supposed to have effective driving angular momenta less than or equal to its own critical limit ($\ell_{eff} \leq \ell_{crit}^{P^p+T}$) for fusion to occur, which leads to an incompletely fused composite (IFC), and direct projectile-like-fragments (PLFs) centred in forward cone. The CN formed via CF is expected to have pre-determined mass/charge, excitation energy and angular-momenta. However, the IFC system (reduced CN) forms with relatively less mass/charge and excitation energy (due to partial fusion of projectile), but at high angular-momenta (imparted due to non-central interactions) as compared to the CN formed via CF.

After first experimental observation of direct PLFs (in massive transfer reactions) by Britt and Quinton [18] and Galin et al.[19], several dynamical models have been proposed to explain ICF dynamics. In SUMRULE model, Wilczynski *et al.*[16, 17], suggested that ICF is mainly confined to the ℓ-space above ℓ_{crit} for CF, and originates from peripheral interactions or non-central collisions. The non-central nature of ICF has been emphasized by Trautmann *et al.*[20], and Inamura *et al.*[21]. The Break-up Fusion (BF)-model [22] of Udagawa and Tamura is based on the Distorted Wave Born Approximation ($DWBA$), in which the projectile is assumed to break-up into constituent α-clusters (e.g., ^{16}O may break-up into ^{12}C+α and/or ^{8}Be+^{8}Be) within the nuclear field of target nucleus. One of the fragments may get fuse with target nucleus and remnant behaves like a spectator (ejected in the forward cone). According to Promptly Emitted Particles (PEP)-model [23], the nucleons transferred from projectile to the target nucleus may get accelerated in the nuclear field of target nucleus and consequently acquire extra velocity to escape before equilibration. Further, Morgenstern *et al.* [24], correlated the onset of ICF with relative velocity (v_{rel}) where ICF found to be contributed significantly above $v_{rel} \approx 0.06$ (6 % speed of light). In a review, Gerschel[25] inferred that the localization of ℓ-window also depends on the target deformation. Despite a variety of existing studies (see ref.[9, 10, 14] for detail), the dynamics of ICF at these energies is still not fairly understood, thus continues to be an active area of investigations. The most debated issues related to ICF are; (i) the localization of ℓ-values, and (ii) the onset and influence on complete fusion at low projectile energies. In order to have an insight into the aforementioned issues and to better understand ICF dynamics, three sets of experiments have been performed [6, 7, 8, 9, 10, 11]. The experimental details, obtained results and their interpretations are discussed in section-2. The summary and conclusions of the present work are given in the last section of this paper.

2. Experimental details, results and interpretations

Experiments have been carried out at the Inter-University Accelerator Center (IUAC), New Delhi, using different setups and techniques. Some of the interesting results along with a short account of experimental conditions are given in the following sub-sections. In first set of experiments, particle-γ-coincidence technique has been employed for spin-distribution measurement [6, 7, 8]. A brief account of experimental conditions related to spin-distribution measurement are given in section-2.1, where the localization of ℓ-values in different reaction products is discussed. For second and third sets of experiments, the novel activation technique has been used where ICF strength function [9, 11], and forward-recoil-ranges (FRRs) [10] of heavy recoils have been measured. An approach to deduce ICF strength function from the analysis of experimental excitation functions is given in section 2.2. While, the validation of this approach through the analysis of experimental FRRs is discussed in section 2.3. Detailed discussion on the results obtained from these experiments can be found in refs.[6, 7, 8, 9, 10, 11].

2.1. spin-distributions: a sensitive probe for incomplete fusion dynamics
Aiming to probe the role of high ℓ-values in the onset of ICF at low projectile energies, the spin-distributions (SDs) of xn/pxn/αxn/2αxn-channels have been measured in ^{12}C,^{16}O+^{169}Tm systems at energies $\approx$ 4-7 MeV/nucleon [6, 7, 8]. In this work, particle-γ coincidence technique has been employed for reaction channel selection. The particle-γ-coincidences have been recorded using Gamma Detector Array (GDA) and Charged Particle Detector Array (CPDA) setup. The GDA is an assembly of 12 Compton suppressed, high resolution HPGe-detectors installed at angles 45°, 99°, 153° with respect to the beam axis and there are 4 detectors at each of these angles. The CPDA is a set of 14-phoswich detectors housed in a 14-cm diameter scattering chamber, covering nearly 90% of the total solid angle. All 14 detectors of CPDA have been divided into the angular rings; (*i*) Forward angle (F) 10° – 60°, (*ii*) Sideways (S) 60° – 120° and (*iii*) Backward angle (B) 120° – 170°. Depending on the fast and slow components of the CPDA, proton and α-particles in each angular ring can be identified. Isotopically pure, self-supporting ^{169}Tm (100%) target of thickness $\approx$ 1 mg/cm^2 (prepared by rolling technique) has been bombarded with $^{12}C^{+5}$(E$\approx$54-90MeV) and $^{16}O^{+7}$(E$\approx$90MeV) beams delivered from 15UD-Pelletron Accelerator. CPD's were covered by Al-absorbers of appropriate thicknesses to remove the scattered beam. At forward angles (F) 10° – 60°, the detectors are supposed to detect both; (*i*) fusion-evaporation (CF) α-component of average energy i.e., E_α (evaporation) $\approx$ 18 MeV, and (*ii*) ICF 'fast' α-component belonging to the same velocity as that of incident projectile, e.g., E_α = 22.5 MeV in case of 90MeV ^{16}O beam. In order to cut-off fusion-evaporation α-particles from forward cone, an additional Al-absorber of appropriate thickness has been kept on forward angle (F) 10°–60° CPD's so that only ICF 'fast' α-component (E$\geq$18 MeV) can be detected in forward cone. In-beam prompt γ-ray spectra have been recorded in event-by-event multi parameter mode, which includes different coincidences like; α and 2α detected in backward (B), forward (F) and 90° (S)-angles. Singles data have also been collected to identify xn-channels (produced via CF). Data analysis has been performed in two steps. In first step, spin distributions of xn-channels have been measured by looking into singles spectra. For the identification of pxn-channels, backward(B)-α-gated spectra has been subtracted from backward(B)-particles(Z=1,2)-gated spectra to achieve proton-gated spectra. However, αxn/2αxn(CN-α)-channels produced via CF have been identified from the backward(B)-α-gated spectra. Further, as per the definition of ICF, the fast-α-particles (particles of the order of projectile velocity) are expected to be emitted only in forward cone (F). As such, αxn/2αxn(Direct-α)-channels produced via ICF have been identified from forward(F)-α-gated spectra. The intensity and area under the photo-peak (efficiency corrected) of the characteristic prompt γ-transitions assigned to the particular reaction products were used to determine the relative production yield.

As a matter of fact, the relative number of statistical and 'yrast'-like transitions depend on entry state angular momenta and available excitation energy (E*). The CF reaction products are formed at high E* and low angular momenta leading to more statistical transitions, where 'yrast' states are expected to fed by statistical γ-transitions. However, the ICF reaction products achieve low E* (due to the involvement of partial degrees of excitations) and high angular momenta (relatively higher values of impact parameters contribute to the high spin states) at a given projectile energy. In such a case, number of 'yrast'-like transitions are much larger than that of statistical ones, where less or no feeding is expected. Therefore, the spin-distribution of CF and ICF products are expected to be entirely different in nature and can be used as a sensitive tool to probe reaction dynamics by looking into entry state spin population. As such, in order to generate experimental spin-distributions of various CF and ICF reaction products, relative production yields have been plotted as a function of experimentally observed spin (J_{obs}^{exp}) corresponding to prompt γ-transitions. For better comparison of different reaction channels (xn, αxn and 2αxn) in a panel, relative production yields have been normalised with their own highest experimentally measured values (Y_{obs}^{max}) at lowest J_{obs}^{min}.

Further, as an analytical representation of data, the experimentally measured spin-distributions obtained as mentioned above have been fitted by a function of following type;

$$Y = Y_\circ/[1 + exp(J - J_\circ)/\Delta] \quad (1)$$

Where; Δ is related to the width of mean input angular momenta ($J_\circ$) and $Y_\circ$ is the normalisation constant. Here, $J_\circ$ is a sensitive parameter, which provides the qualitative information about the driving input angular momenta associated with various reaction channels.

As a representative case, experimental spin-distributions of xn/αxn/2αxn-channels populated via CF and ICF in ^{12}C,^{16}O+^{169}Tm systems are plotted in Fig.1. The errors have not been shown in the figures as they have been estimated to be $\leq$ 10 %, and the inclusion of these errors is not likely to modify the present analysis. The nomenclature used in this figure indicate the involved reaction dynamics i.e., 'B' and 'F' respectively indicate CF-channel: identified from backward(B)-α-gated spectra, and ICF-channel: identified from forward(F)-α-gated spectra. As can be observed from this figure, in general, there is a striking difference in the spin-distribution for different channels which indicates the involvement of entirely different reaction dynamics in the production of these reaction products. The intensity of xn(singles)/αxn-B-channels (only populated via CF) falls off rather quickly with observed spin (J_{obs}), indicating strong feeding and/or broad spin population during the de-excitation of CN. However, for αxn-F-channels (expected to be populated via ICF), the intensity appears to be almost constant up to a certain value of J_{obs}, and then decreases towards entry side. This indicates the absence of feeding to the lowest members of the 'yrast' band and/or the population of low spin states are strongly hindered in ICF channels. Further, the feeding intensity profiles (FIPs) for different reaction products have been generated from experimentally measured spin-distributions, and are plotted in Fig.1(c,f,h). As indicated in Fig.1c, the feeding intensity for CF-channels (xn/pxn-B) is showing sharp exponential rise towards low spin states, which indicates regular population with strong feeding contribution for each γ-transition up to J_{obs}^{min}. For better comparison of direct-α-emitting channels (ICF) and fusion-evaporation α-emitting channels (CF), the feeding intensity profiles for CF and ICF products have been plotted in a single panel (see Figs.1f). As can be observed from this figure, the feeding intensity of αxn-channels (^{174}Ta-CF-B) identified from backward(B)-α-gated spectra shows similar trend as that has been observed for xn/pxn-B-channels, where the band is fed over a broad spin range. While, the feeding intensity for forward(F)-αxn/2αxn-channels (identified from forward α-gated spectra) is found to be increasing up to a certain values of J_{obs} and then decreases gradually towards band head, see Fig.1(f,h). This trend indicates that the high spin states are strongly fed even in case of ICF channels. However, as the residual nucleus de-excites, the feeding intensity decreases gradually with available excitation energy and/or angular momenta, which indicates the absence of feeding to the lowest members of the 'yrast' band or the low spin states are less populated in ICF-αxn/2αxn-F channels. Such feeding intensity pattern is expected to arise from narrow ℓ-window, localized near and/or above to the critical angular momentum for CF. Similar trends have also been observed in other xn/αxn/2αxn-channels (see ref.[6, 7, 8] for detail).

In order to understand the multitude of mean driving angular momenta (ℓ-values), and to examine the possibility to populate high spin states via ICF, mean driving angular momenta involved in CF and ICF channels have been deduced from the analysis of spin-distributions. The ℓ-values involved in various modes of reactions are plotted in Fig.1(i). As indicated in this figure, the value of $< \ell >$ involved in the production of CF-xn/pxn-B/αxn-B, and ICF-αxn-F,

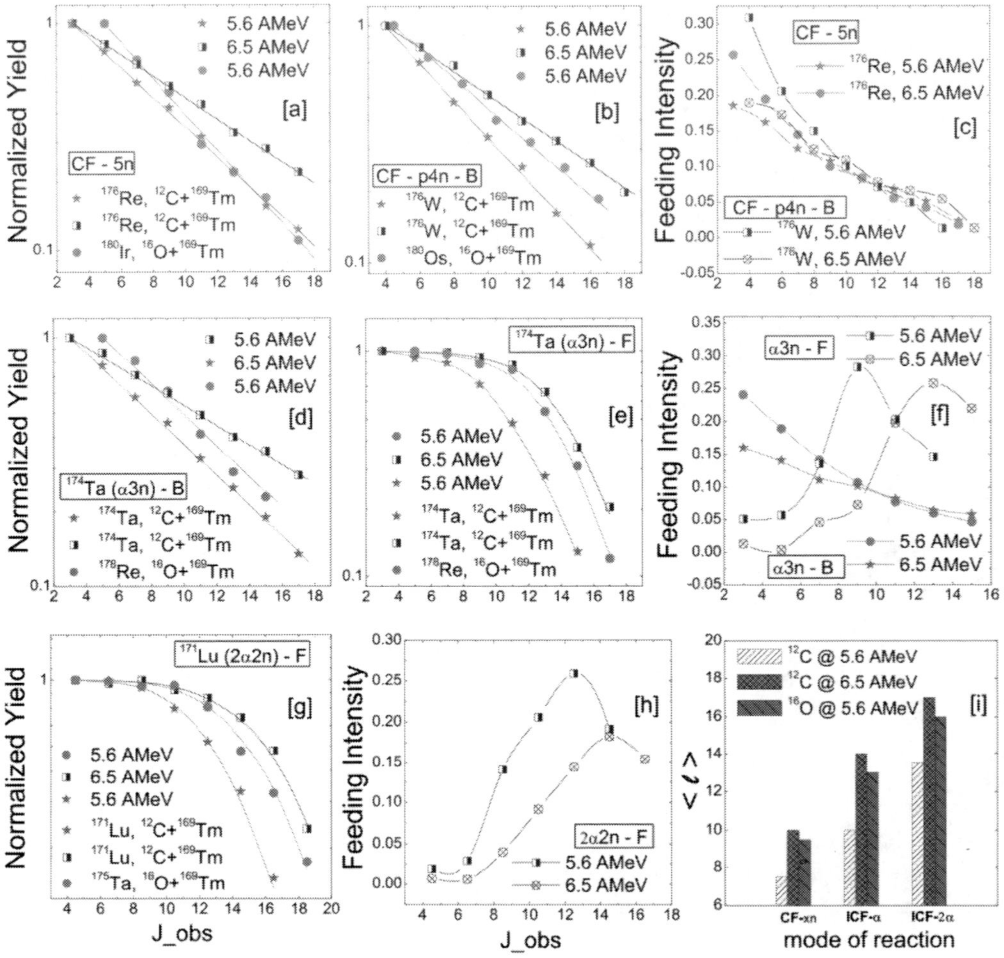

Figure 1. Experimentally measured spin-distributions and feeding-intesity-profiles for xn/pxn/αxn/2αxn channels identified from singles and particle(α)-gated spectra) are plotted. Reaction products have been labelled by self-explanatory notations and emission cascades. The nomenclature shows that the exist channels are composed by the one given residual nucleus. Lines through the data points are the result of best fit procedure explained in the text. Figure (i) shows, the value of mean driving angular momenta (i.e., the value of J_o deduced from the best fitting procedure of spin-distributions as explained in text) involved in various CF and ICF channels as a function of modes of reaction.

and ICF-2αxn-F channels are found to be $\approx 7.5\ \hbar$, $\approx 10\ \hbar$ and $\approx 13.5\ \hbar$, respectively, at projectile energy $\approx$ 5.6 AMeV. However, at projectile energy $\approx$ 6.5 AMeV, the value of $< \ell >$ for CF-xn/pxn-B/αxn-B, and ICF-αxn-F, and ICF-2αxn-F channels are found to be $\approx 10\hbar$, $\approx 14\hbar$ and $\approx 17\hbar$, respectively. The enhancement in the value of $< \ell >$ in case of direct-α-emitting channels (ICF products) indicates their origin from high ℓ-values as compared to CF-channels.

A very useful correlation between the value of $< \ell >$ and the successively opened ICF channels can be obtained from the data presented in Fig.8. Following the approach presented in ref.[?], the value of $< \ell >$ associated with ICF in contrast with CF can be represented as;

(i) for ^{12}C+^{169}Tm system at $E_{lab} \approx 5.6$ AMeV;
$\ell_{(ICF-\alpha xn)} \approx 1.33\ell_{(CF-xn/pxn/\alpha xn)}$,
$\ell_{(ICF-2\alpha xn)} \approx 1.35\ell_{(ICF-\alpha xn)} \approx 1.8\ell_{(CF-xn/\alpha xn)}$, and

(ii) at $E_{lab} \approx 6.5$ AMeV;
$\ell_{(ICF-\alpha xn)} \approx 1.4\ell_{(CF-xn/pxn/\alpha xn)}$,
$\ell_{(ICF-2\alpha xn)} \approx 1.2\ell_{(ICF-\alpha xn)} \approx 1.7\ell_{(CF-xn/\alpha xn)}$.

From the correlations presented above, it is interesting to note that the values of $< \ell >$ involved in the production of various ICF-αxn/2αxn-channels are found to be $\approx$ 30 to 70 % higher as compared to the CF-xn/pxn/αxn-channels at both the energies. This clearly indicates the involvement of high ℓ-values in the production of ICF products at a constant projectile energy, essentially due to non-central interactions, where a significant amount of orbital angular momentum between projectile and target nucleus transformed into high spin states of final reaction products. It can also be observed from Fig.1(i), the involved ℓ-values in different reaction channels are found to increase linearly with the projectile energy, and almost same (within ≤ 0.5 $\hbar$) for each set of reaction channels at a given projectile energy. It is also clear from Fig.1(i), the CF is not able to approach the value of angular momenta even at $\approx$ 6.5 AMeV which has been populated via ICF for the same residue at relatively low projectile energy. The above striking features strongly support the possibility to populate high spin states via ICF in final reaction products even at low projectile energy. The smallness of $J_\circ$ indicates the involvement of less input angular momenta in CF reactions. It may also be seen from the deduced values of $J_\circ$ that the multiplicity of α-particles associated with ICF increases with the driving input angular momenta, indicating the variation of different ℓ-values (impact parameter dependent) at a given projectile energy. This indicates that the partial waves of lower ℓ-values do not contribute to the ICF reaction dynamics significantly. Therefore, it may not be out of order to state that ICF reactions occur in the peripheral interactions, probably at finite values of impact parameters, where the centrifugal force field overtakes the sum of nuclear and Coulomb potentials.

2.2. Incomplete fusion strength function: measure of strength and onset

The measurements presented in previous section are relative and don't give information on how ICF show up with various entrance channel parameters. As such, in order to understand influence of ICF on CF at these energies, absolute cross-sections for individual reaction residues have been measured for for ^{12}C,^{16}O+^{169}Tm systems at energies near and above the fusion barrier [9, 11, 14]. Experiments have been performed using GPSC-setup at IUAC. Experimental methodology and setup have been detailed in refs.[9, 11, 14]. However, a brief account of experimental conditions is given here for ready reference. Isotopically pure (abundance = 100%) ^{169}Tm Targets of thickness $t_m \approx$ 1-1.8 mg-cm^{-2} were prepared by uniform pressure rolling technique. Uniformity of each target foil was verified by multi-point thickness measurement using α-transmission method. The error in thickness was estimated to be $\leq 1\%$. For ^{16}O+^{169}Tm system, energy degradation technique has been employed to cover a wide energy range in single run, and corrections were made for energy spread and beam intensity variations. while, in case of ^{12}C+^{169}Tm system, the energy spread and beam intensity variations have been avoided by adopting single target irradiation methodology for each projectile energy. Irradiations were carried out using ^{12}C,^{16}O-beams (E_{beam} = $\approx$ 4-7 MeV/nucleon with beam current $\approx$20-30 nA) delivered from 15UD Pelletron accelerator. The targets were backed by Al-catchers of

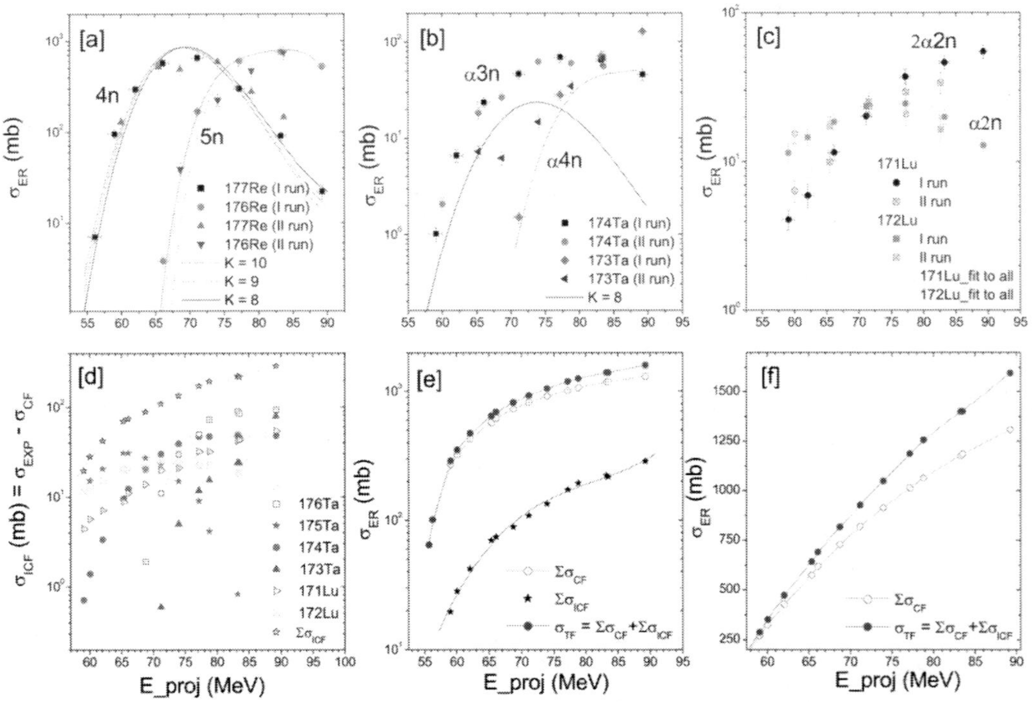

Figure 2. $(a)-(c)$ The experimental excitation functions of xn/αxn/2αxn-channels populated via CF and/or ICF in ^{12}C+^{169}Tm system. Lines and symbols are used for self explanatory notations. In case of ^{173,174}Ta(α4n,α3n) and ^{171,172}Lu(2α2n,2αn) residues, the red dotted lines through the data points represent best fit to the both experimental data sets (I run and II run). The two data sets show good agreement. (d) Experimentally measured and systematically deduced (see text for deduction procedure) EFs of ICF products along with the sum of all ICF-channels ($\Sigma\sigma_{ICF}$), (e) Total fusion cross-section ($\sigma_{TF} = \Sigma\sigma_{CF} + \Sigma\sigma_{ICF}$) along with the sum of all CF-channels ($\Sigma\sigma_{CF}$) and ICF-channels ($\Sigma\sigma_{ICF}$) as a function of projectile energy, and (f) the zoom of $\Sigma\sigma_{CF}$ and σ_{TF} comparison on linear scale for an easy visualization of increasing ICF strength with projectile energy. Solid curves represent best fit to the data points.

appropriate thicknesses, so that the recoiling nuclei can be trapped in the catcher foil thickness. Proper care was taken to maintain the constant beam current during the irradiations so that the uncertainty in the production cross-sections due to beam current fluctuations can be minimised. In order to detect the residues of small half-lives ($t_{1/2} \approx$ 5-10 minutes), in-situ measurements of γ-activities produced during the irradiations were performed off-line using two pre-calibrated HPGe detectors. The uncertainty in geometry dependent efficiency of detectors is estimated to be $\leq 2\%$. Prompt-γ-rays spectra have been recorded in singles mode for an additional degree of freedom in residue identification. Reaction products have been identified by their characteristic prompt and decay γ-lines. In order to gain high confidence in reaction products identification, the decay-curve analysis for each residue has been performed using decay γ-lines. The most intense decay γ-lines have been used for decay-curve analysis and for the production cross-section (σ_{ER}) measurement [7]. The projectile energy dependent reaction cross-sections ($\sigma_r(E)$) for different radio-nuclides have been determined using standard formulation [?]. The overall error in σ_{ER} is estimated to be $\leq$ 13 %, excluding the uncertainty in branching ratio, decay constant etc. Details of error estimation and causing factors are given in ref.[14].

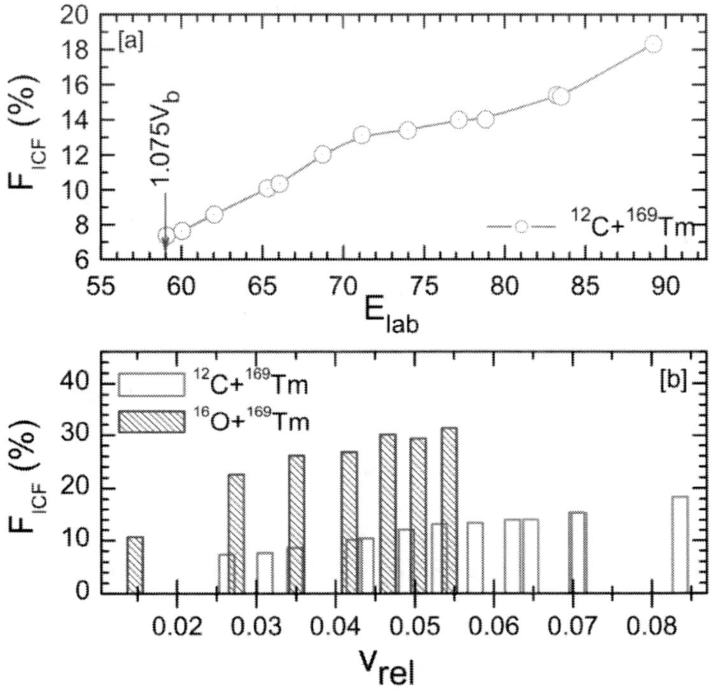

Figure 3. (*a*) the ICF strength function for $^{12}C+^{169}Tm$ system (see text for description), (*b*) the value of F_{ICF} as a function of relative velocity (v_{rel}) for $^{12}C,^{16}O+^{169}Tm$ systems.

In order to deduce ICF fraction (F_{ICF}), experimental EFs have been analysed in the framework of theoretical model code PACE4 based on equilibrated CN-decay. Details of this code can be found elsewhere [9, 14]. In this code, level density parameter (a = A/K) is an important input parameter which affects the equilibrium component. In order to know the suitable value of 'a' for studied systems at given energy range, different values of 'a' from A/10 - A/8 MeV^{-1} have been tested. As a representative case, the predictions of PACE4 for different 'a' values from a = A/10 to a = A/8 MeV^{-1} are plotted in Figs.2(a). Experimental EFs of ^{176,177}Re(5n,4n) residues are reasonably well reproduced for a = A/8 MeV^{-1}. This confirms the production of these residues (^{176,177}Re) through the equilibrated CN-decay via emission of x-neutrons from the excited ^{185}Ir* formed via CF. As such, the value of a = A/8 MeV^{-1} can be used as default parameter for further analysis within the tested energy range. As shown in Fig.2(b), the EFs of α-emitting channels are significantly enhanced than theoretical predictions. The enhanced cross-section in case of αxn-channels points towards some physical effect which is not included in this code. It may be pointed out that PACE4 do not take ICF into account.

Journal of Physics: Conference Series **282** (2011) 012019 doi:10.1088/1742-6596/282/1/012019

Therefore, the enhancement in case of α-emitting channels (α3n,α4n) may be attributed to the onset of ICF. Assuming the enhancement as ICF fraction in case of α3n,α4n-channels, the ICF contribution has been accounted using data reduction procedure suggested by Gomes *et al.*[3]. For some α-emitting channels (as shown in Fig.2c), PACE4 gives negligible cross-section for 2α2n and α2n-channels. This indicates the population of these residues only via CF. Experimental EFs of α-emitting channels are subtracted from the theoretical ones ($\sigma_{ICF} = \sigma_{exp}$ - σ_{pace4}) to get ICF contribution, and are plotted in Fig.1(d). To show how does ICF contribute to the total fusion cross-section ($\sigma_{TF} = \Sigma\sigma_{CF} + \Sigma\sigma_{ICF}$) ? the overall ICF cross-section ($\Sigma\sigma_{ICF}$) is plotted with the sum CF-channels ($\Sigma\sigma_{CF}$) and σ_{TF} in Fig.2(e). For an easy visualisation of increasing ICF contribution with energy, the zoom of $\Sigma\sigma_{CF}$ and σ_{TF} comparison is plotted on linear scale in Fig.2(f). As shown in this figure, the increasing separation between $\Sigma\sigma_{CF}$ and σ_{TF} with projectile energy indicates strong energy dependence on ICF strength. For better insight into the projectile energy effect on ICF strength, the percentage fraction of ICF (F_{ICF}) has been deduced and plotted as a function of projectile energy in Fig.3(a), i.e., termed as ICF strength function. The ICF strength function defines empirical probability of ICF at different projectile energies. As shown in this figure, the value of F_{ICF} is found to be ≈7 % at ≈ 59 MeV, i.e., 1.075Vb (7.5 % above the barrier), and increases smoothly upto ≈18 % at highest measured energy i.e., 1.64 Vb. The existence of ICF at such a low energy has been justified as a consequence of high input ℓ-values imparted into the system in non-central interactions [6, 7, 8, 16, 17].

Further, the values of F_{ICF} obtained in ^{12}C,^{16}O+^{169}Tm systems are plotted as a function of v_{rel} in Fig.3(b). According to Morgenstern's systematics [24], ICF contributes significantly above v_{rel} ≈0.06 (6 % speed of light). As indicated in this figure, the values of v_{rel} are in the range from ≈0.027 (2.7 % of c) to ≈0.084 (8.4 % of c) for ^{12}C, and from ≈0.014 (1.4 % of c) to ≈0.053 (5.3 % of c) for ^{16}O. Therefore, no significant ICF contribution is expected at the given values of v_{rel}. However, the results presented in Fig.3(b) clearly demonstrate the onset of ICF at relatively lower value of v_{rel} i.e, ≈0.027 (F_{ICF} ≈7 %) in ^{12}C+^{169}Tm system, and at ≈0.014 (F_{ICF} ≈10 %) in ^{16}O+^{169}Tm system. In both cases, the observed value of F_{ICF} is significant at well below the proposed onset value of v_{rel} (i.e., 6 % of c). As such, it can be inferred that the ICF starts competing with CF even at slightly above barrier energies.

2.3. Forward-recoil-ranges: proof of fusion incompleteness

In order to validate data reduction procedure to deduce F_{ICF}, and to proof fusion-incompleteness at slightly above barrier energies, forward-recoil-ranges (FRRs) of heavy reaction residues produced in ^{12}C,^{16}O+^{169}Tm systems have been measured at different energies [10, 13]. Kinematically, the FRRs depend upon degree of linear momentum transfer (ρ_{LMT}) from projectile to target nucleus. For ICF, where only a part of projectile fuses with target nucleus, the ρ_{LMT} may be given as;

$$\rho_{LMT} = \frac{P_{frac}}{P_{proj}} \tag{2}$$

where; P_{frac} is the linear momentum of fused fraction of projectile, and P_{proj} is the full linear momentum of projectile. As per kinematical range-energy calculations, entire LMT (i.e., ρ_{LMT} = 1) from projectile to target nucleus (in case of CF) supposed to give maximum recoil velocity to the composite (A_{target}+A_{proj}) system, and relatively less recoil velocity to the incompletely fused composite (ICF: A_T+$A_{proj-fract}$). Consequently, the radio-nuclides populated via small ρ_{LMT} are expected to show relatively smaller penetration depth in the stopping medium as compared to the entire LMT populations. On the basis of aforementioned kinematics, CF and ICF products can be disentangled from the analysis of FRRs of heavy reaction residues.

For the measurement of FRRs, Natural Thulium (^{169}Tm) targets of thickness $\approx$ 38 - 56 μg/cm^2, and Aluminium catcher foils (of thickness $\approx$ 10 - 55 μg/cm^2) were prepared by UH-vacuum deposition technique. The irradiations have been carried out in GPSC-setup and followed by off-line gamma spectroscopy. In order to stop the recoiling reaction products, a stack of thin Al-catcher foils of the thicknesses varying from $\approx$ 10 - 55 μg/cm^2 (sufficient to stop entire LMT events) has been placed just after the target. The recoiling products populated via CF and/or ICF have been stopped at various cumulative depths in the Al-catcher foils. Keeping in mind the half-lives of interest, the irradiations were carried out for $\approx$ 10 hrs duration for each incident energy. After the irradiation, the Al-catcher foils along with the target foil were taken out from the scattering chamber for post-irradiation analysis. The radio-activities produced in catcher foils were counted separately using two pre-calibrated, high resolution, HPGe-detector of 100 c.c. active volume. The HPGe-detectors were pre-calibrated for energy using ^{60}Co, ^{133}Ba and ^{152}Eu γ-sources. The resolution of γ-detector was found to be $\approx$ 2.4 keV, for 1.33 MeV γ-ray of ^{60}Co. The residues were identified by their characteristic γ-lines and confirmed by the decay-curve analysis. The induced activities can be used to measure the production probability of the reaction products. The measured intensities of the characteristic γ-lines have been used to determine the production cross-sections of the residues in each catcher foil using standard formulation. The overall errors from all possible factors (for detail see ref. [10, 13]) including the statistical error are estimated to be $\leq$ 14 %.

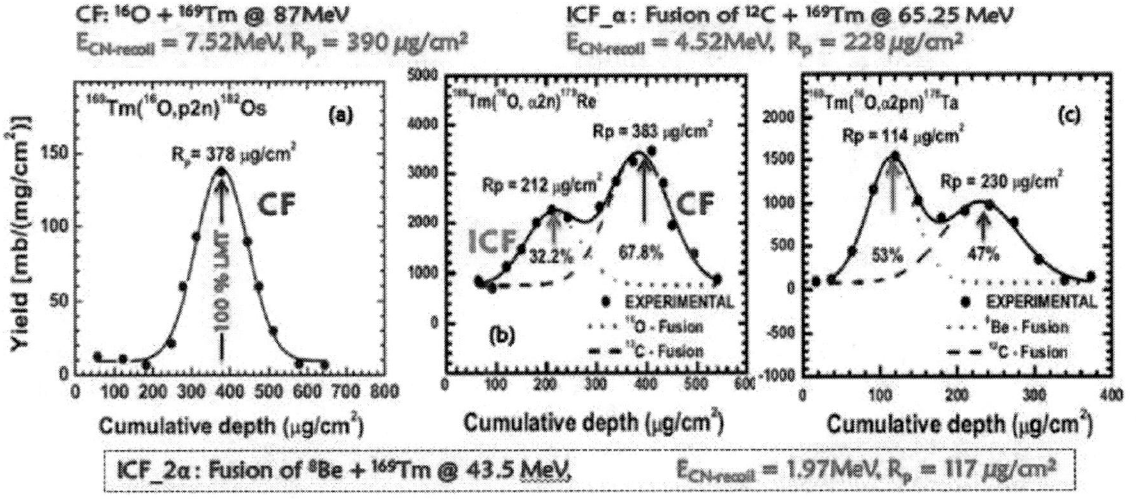

Figure 4. Experimentally measured profile of FRRs for; (*a*) ^{182}Os(p2n), (*b*) ^{179}Re(α2n), and (*c*) ^{178}Ta(α2pn) residues populated via CF and/or ICF of ^{16}O with ^{169}Tm at energies $\approx$ 87 MeV. Relative strengths of different LMT transfer components are indicated in percentage.

In order to generate experimental FRRs, the normalised yield of reaction products have been plotted as a function of cumulative catcher foil thicknesses. The production yield of different reaction products have been deduced by the normalisation of experimentally measured production cross-sections with the respective catcher foil thicknesses. The FRRs have been fitted with Gaussian distribution for better analytical representation of data, and to deduce the value of most probable recoil ranges ($R_{p(exp)}$) in the stopping medium, similar to that given in ref.[10, 13].

Fig.4(a) shows the profile of FRRs for ^{182}Os(p2n), populated in ^{16}O+^{169}Tm system at $\approx$ 87 MeV. As can be seen from this figures, the data can be fitted by a single Gaussian peak indicating

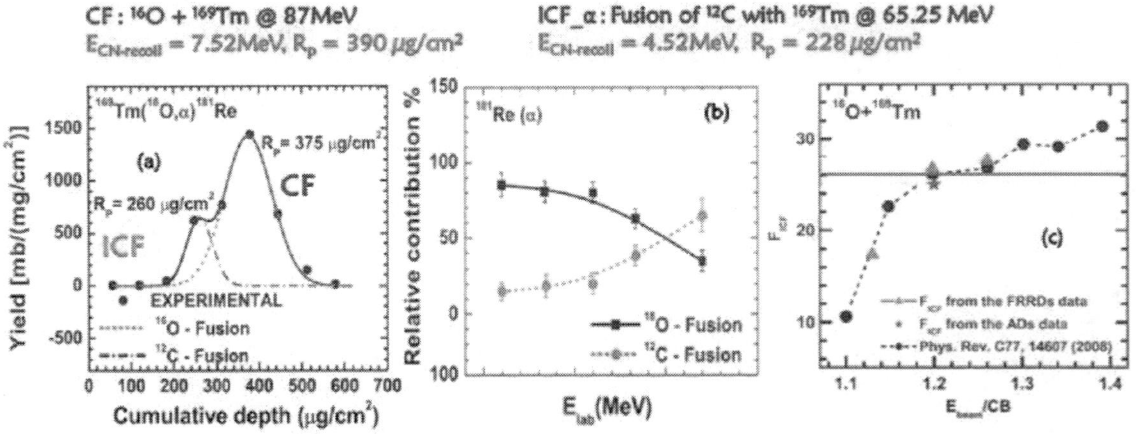

Figure 5. (*a*) Experimentally measured profile of FRRs for ^{181}Re(α/2p2n), (*b*) Relative strengths of CF and ICF of ^{16}O with ^{169}Tm at projectile different energies for ^{181}Re (α or 2p2n - channel). (*c*) The value of F_{ICF} as a function of reduced projectile energy. The the values of F_{ICF} given in different colors have been obtained from the analysis of three different sets of data. As indicated, the value of F_{ICF} deduced from three different techniques at is in good agreement within the experimental uncertainties. Lines through the data points are drawn to guide the eyes.

single LMT-component involved in the production of this residue. However, Fig.4(b-c) show the FRRs for ^{179}Re(α2n) and ^{178}Ta(α2pn) radio-isotopes. As shown in these figures, the FRRs can be resolved into two Gaussian peaks, revealing the presence of more than one LMT-components associated with the fusion of ^{16}O, ^{12}C and/or ^{8}Be with the target nucleus. As an example, the FRR of residue ^{179}Re(α2n) is showing two LMT components at $\approx$383 μg/cm^2 (due to ^{16}O-fusion) and at $\approx$ 212 μg/cm^2 (^{12}C-fusion, and an α as spectator) at projectile energy $\approx$ 87 MeV. As such, on the basis of observed $R_{P(exp)}$, it can be inferred that the residues ^{179}Re(α2n) is populated via both CF and ICF processes. This implies fractional momentum transfer in ICF. The value of R_P^{exp} for different residues have also been compared with the theoretically estimated most probable FRRs have been estimated by adopting break-up-fusion description of massive transfer reactions given by Udagawa and Tamura [22], and are found to be in good agreement with theoretical ones (see ref.[10, 13]).

Further, an attempt has been made to find out the relative strengths of CF and ICF contributions to understand the energy dependence of full (CF products) and partial (ICF products) linear-momentum-transfer components. Fig.5(a) shows the FRRs for ^{181}Re(α or 2p2n) which can be resolved into two Gaussian peaks. Peak at $\approx$ 260 μg/cm^2 indicating contribution from ICF of ^{12}C-fusion. For this case, the relative contributions of CF and ICF obtained from the analysis of FRRs at different energies are plotted in Fig.5(b). As indicated in this figure, the CF (fusion of ^{16}O) contribution decreases at higher projectile energies, while the ICF (fusion of ^{12}C) contribution is found to increase with projectile energy. The observed trend reveals that the influence of ICF over CF increases with projectile energy, as expected.

In order to validate data reduction procedure given in section-2.2, the value of F_{ICF} has been deduced from the analysis of FRRs using following relation,

$$F_{ICF} = \frac{\Sigma\sigma_{ICF}}{\Sigma\sigma_{CF} + \Sigma\sigma_{ICF}} \times 100$$

Where, $\Sigma\sigma_{CF}$ and $\Sigma\sigma_{ICF}$ are the sum of cross-sections (obtained from the analysis of FRRs) of CF and ICF processes, respectively. The F_{ICF} has been deduced as a function of normalized projectile energy (E_{beam}/CB), where E_{beam} is the beam energy and CB is the Coulomb barrier for the $^{16}O+^{169}Tm$ system. For better comparison and to have confidence in the measured values, the values of F_{ICF} deduced from three different techniques, namely: (*i*) from the analysis of FRRs, (*ii*) angular distributions, and (*iii*) from the analysis of EFs, have been plotted as a function of E_{beam}/CB in Fig.5(c). As shown in this figure, the value of F_{ICF} deduced from three different methods at a constant normalized projectile energy (E_{beam}/CB = 1.2) is found to be in good mutual agreement within the experimental uncertainties, which puts faith in our measurements and data reduction procedure. Lines through the data points are drawn to guide the eyes. The value of F_{ICF} deduced from the analysis of FRRs is found to be $\approx$ 15 % at energy $\approx$76 MeV ($\approx$ 10 % above the CB), while, at energy $\approx$ 81 MeV ($\approx$ 17 % above the CB), the value of F_{ICF} approaches up to $\approx$ 29 %.

3. Summary and Conclusions

This paper deals with recent experimental results from three set of experiments performed to study ICF at energies $\approx$ 4-7 MeV/nucleon. In first set of experiments, the spin-distributions and feeding intensity profiles have been measured at near and above barrier energies. The spin-distributions of ICF-αxn/2αxn-channels are found be distinctly different than that observed for CF-xn/pxn/αxn-channels, which indicates entirely different de-excitation patterns in CF and ICF products. In case of CF products, strong feeding through broad range of spin population has been observed towards the band head. However, the spin-distribution(s) associated with ICF are found to be arrised from the narrow spin population, localised near and/or above to the critical angular momentum for CF, where a given PLF is emitted to release excess driving angular momenta. This indicates the competition from successively opened ICF channels for each value of ℓ-value above ℓ_{crit} for normal fusion (CF). The population of low spin states are hindered and/or less fed in case of ICF. This reveals the occurrence of ICF due to the influence of centrifugal potential in peripheral interactions, where driving angular momentum limit do not allow CF. It is also shown that the direct α-multiplicity increases in forward cone with the value of $< \ell >$ at a particular projectile energy. For an example, at $\approx$ 5.6 AMeV the value of $< \ell >$ for CF-xn/pxn-B/αxn-B, and ICF-αxn-F, and ICF-2αxn-F channels are found to be $\approx$ 7.5 $\hbar$, $\approx$ 10 $\hbar$ and $\approx$ 13.5 $\hbar$, respectively. On the basis of above results, it can be concluded that the high ℓ-values associated with non-central interactions essentially contribute to open up direct-α-emitting channels. It may, further, be pointed out that the value of $< \ell >$ associated with 2α-emitting channel likely to be originated from higher impact parameters as that associated with the production of single direct-α-emitting channel. Further, from the comparison of ℓ-values involved in the production of direct-α-emitting (ICF products) and normal-α-emitting channels (CF-products), we presented direct evidence that ICF can populate high spin states in final reaction products which is not possible to achieve via CF at a given projectile energy and/or even higher energy.

In second set of experiments, ICF strength function has been measured using some systematics. Existence of ICF at energy as low as $\approx$ 7.5 % above the barrier is conclusively demonstrated. The value of F_{ICF} increases from $\approx$7 % to $\approx$18 % with in the studied energy range (i.e., 1.02V_b to 1.64V_b). The results presented in section-2.2 clearly demonstrate the onset of ICF at relatively lower value of v_{rel} i.e, $\approx$0.027 (F_{ICF} $\approx$7 %) in $^{12}C+^{169}Tm$ system, and at $\approx$0.014 (F_{ICF} $\approx$10 %) in $^{16}O+^{169}Tm$ system. In both cases, the observed value of F_{ICF} is significant at well below the proposed onset value of v_{rel} (i.e., 6 % of c). Further, a comparison of F_{ICF} for $^{12}C+^{169}Tm$ and $^{16}O+^{169}Tm$ systems displays higher ICF probability for $^{16}O+^{169}Tm$ system for the entire measured energy range. This indicates strong projectile dependence on F_{ICF}. Presence of significant ICF fraction at well below the proposed onset value of v_{rel} (i.e., 6

% of c) supplements Morgenstern's systematics.

Moreover, the results presented in section-2.3 are from the third set of experiments where FRRs of heavy reaction products have been measured to confirm the findings of first two set of experiments. From the analysis of FRRs, it has been observed that the α-emitting channels have significant contribution from both CF and ICF processes. Different LMT components attributed to the full (fusion of ^{16}O) and partial fusion (^{12}C and/or ^{8}Be fusion from ^{16}O) of projectile with target nucleus have been observed. The $R_{P(exp)}$ are found to consistent with the theoretical predictions, within the experimental errors. The value of F_{ICF} obtained from the analysis of FRRs is found to be in good agreement with that estimated by the measurement and analysis of EFs. The results presented in section-2.3 support BF-model description of massive transfer reaction given by Udagawa and Tamura [22]. As such, it can be concluded that the ICF processes contribute significantly to the reaction cross-section at these energies. The extension of this work using ^{13}C, ^{14}N and ^{18}O beams with ^{169}Tm target would be interesting, and will be helpful for further refinement of present systematics.

Acknowledgments
We thank Prof. Amit Roy, Director of IUAC for extending experimental facilities to perform these experiments. One of the authors (P.P.S.) thanks to INFN-Laboratori Nazionali di Legnaro (LNL), and R. P. & B. P. S. thank to DST and UGC for financial support.

References

[1] A. Diaz-Torres, J. Phys. G: Nucl. Part. Phys. **37**, 075109 (2010), Phys. Rev. Lett. **98**, 152701 (2007), and the references therein.
[2] D. J. Hinde and M. Dasgupta, Phys. Rev. C **81**, 064611 (2010), and the references therein.
[3] P. R. S. Gomes *et al.*, Phys. Lett. **B601**, 20 (2004); Phys. Rev. C**73**, 064606 (2006); Phys. Rev. Lett. **B53**, 1630 (1984).
[4] E. Z. Buthelezi *et al.*, Nucl. Phys. **A734**, 553 (2004).
[5] L. R. Gasques *et al.*, Phys. Rev. C**79**, 034605 (2009); Phys. Rev. C**74**, 064615 (2006).
[6] Pushpendra P. Singh *et al.*, Phys. Lett. **B671**, 20 (2009), and the references therein.
[7] Pushpendra P. Singh *et al.*, Phys. Rev. C**80**, 064603 (2009)
[8] Pushpendra P. Singh *et al.*, Phys. Rev. C**78**, 017602 (2008), and the references therein.
[9] Pushpendra P. Singh *et al.*, Phys. Rev. C**77**, 014607 (2008).
[10] Pushpendra P. Singh *et al.*, Eur. Phys. J. A **34**, 29 (2007), and the refences there in.
[11] Pushpendra P. Singh *et al.*, Phys. Rev. C (submitted)
[12] Devendra P. Singh *et al.*, Phys. Rev. C**81**, 054607 (2010).
[13] Unnati Gupta *et al.*, Phys. Rev. C**80**, 024613 (2009).
[14] Unnati Gupta *et al.*, Nucl. Phys. **A811**, 77 (2008).
[15] P. E. Hodgson, E. Gadioli and E. Gadioli Erba, Introductory Nuclear Physics, Chapter **23**, Clarendon Press, Oxford (1997).
[16] J. Wilczynski, et al., Phys. Rev. Lett. 45, 606 (1980); Nucl. Phys. A **373**, 109 (1982).
[17] K. Siwek-Wilczynska et al., Phys. Rev. Lett. 42, 1599 (1979).
[18] H. C. Britt and A. R. Quinton, Phys. Rev. **124**, 877 (1961).
[19] J. Galin et. al., Phys. Rev. C **9**, 1126 (1974).
[20] W. Trautmann, et al., Phys. Rev. Lett. B **53**, 1630 (1984).
[21] T. Inamura, et al., Phys. Lett. **68B**, 51 (1977), Phys. Lett. **84B**, 71 (1982); Phys. Rev. C**32**, 1539 (1985).
[22] T. Udagawa and T. Tamura, Phys. Rev. Lett. **45**, 1311 (1980).
[23] J. P. Bondrof, J. N. De, G. Fai, A. O. T. Karvinen, and J. Randrup, Nucl. Phys. A **333**, 285 (1980).
[24] H. Morgenstern *et al.*, Z. Phys. **A313**, 39 (1983); Phys. Rev. Lett. **52**, 1104 (1984); Z. Phys. **A324**, 443 (1986).
[25] C. Gerschel, Nucl. Phys. A **387**, 297 (1982), and refences there in.

Structure effects and dynamics in fusion reactions of light weakly bound nuclei

E Strano[a,b], A DiPietro[a], P Figuera[a], M Fisichella[c], M Lattuada[a,b], C Maiolino[a], A Musumarra[a,b], M G Pellegriti[a,b], D Santonocito[a], V Scuderi[a], D Torresi[a,b], M Zadro[d]

[a]INFN Laboratori Nazionali del Sud, Catania, Italy
[b]Università degli studi di Catania, Catania, Italy
[c]Università di Messina, Messina, Italy
[d]Ruđer Bošković Institute, Zagreb, Croatia

E-mail: strano@lns.infn.it

Abstract. The study of fusion in collision around the Coulomb barrier, induced by radioactive or stable weakly bound nuclei, has been the subject of many experiments in the last years. From a semiclassical point of view, direct processes like break-up or transfer may be favoured by the low binding energies and one might also expect suppressed fusion cross section due to the competition with the break-up. However, according to the Coupled Channel calculations (CC), it is well known that the presence of strong open reaction channels can be responsible for a fusion cross-section enhancement with respect to the single barrier penetration calculations and the break-up should be included in such calculations. In order to further investigate on this topic, the $^{6}Li+^{64}Zn$ collision has been studied at several energies around the Coulomb barrier, to extract the total fusion and the total reaction cross sections and to study the energy dependence of the optical model potential.

1. Introduction and motivations

In the last few years, several attempts have been made to understand the reaction mechanisms in collisions of weakly bound nuclei around the Coulomb barrier [e.g. 1].

A weakly bound nucleus is a nuclear system with the ground state energy very close to the particle emission threshold. This means that a little excitation is sufficient to dissociate the nucleus into two or more parts. These peculiar nuclei can be either stable or unstable and may show a cluster structure. The low binding energy leads to a diffused mass distribution and, in extreme cases, to the halo structure [2].

In collisions involving a weakly bound nucleus direct processes like break-up and transfer may be favoured by the low binding energy and/or cluster structure.

Two kinds of effects on fusion cross section may be expected: from a static point of view, a diffused mass distribution is affecting the projectile-target potential lowering the Coulomb barrier and increasing the fusion cross section; from a dynamical point of view, it is known that the strong coupling of the entrance channel with inelastic excitation or other reaction channels like transfer leads to a modulation of the barrier resulting in an enhancement of the fusion cross section with respect to single barrier penetration model. Due to the low break-up thresholds, in weakly bound nuclei the coupling to inelastic excitations may correspond to the coupling with the break-up channel, therefore we need to understand if the break-up is simply removing flux from fusion suppressing it or, if the coupling to break-up will produce enhancement of the fusion cross section around and below the barrier.

In order to investigate on the previously discussed topic, different authors have performed many experiments. Recently, to consistently compare all existing results, it has been developed a dimensionless universal fusion function [1, 3, 4] to be compared with the experimental data in order to look for systematic effects of the break-up on fusion cross-section. In such comparison the experimental data have been scaled to eliminate the differences due to the coupling to excited bound states, different charges, different barrier shapes and positions [1, 3, 4 and references therein].

From this analysis the authors concluded that reactions on heavy targets show a complete fusion cross-section enhancement below the Coulomb barrier and complete fusion suppression above the barrier [4]. In the case of light and medium mass targets the experimental results apparently show no effects on fusion cross section above the barrier but data are missing at energies below the barrier [3]. It has been noticed in [3] that although experimental data for different systems clearly show a consistent systematic behaviour. A few of them, such as the ^{6}Li+^{64}Zn data on ref [6], deviate from the systematic and possible problems on the experimental data have been suggested for these systems.

In order to investigate on the previously discussed topic, the goal of the present experiment is to study the collisions induced by ^{6}Li and ^{7}Li on a ^{64}Zn target. ^{6}Li and ^{7}Li are both weakly bound nuclei but there are differences in their structure: ^{6}Li does not have discrete excited states and its binding energy is 1.4 MeV; ^{7}Li has a bound excited state at 0.5 MeV and its binding energy is 2.5 MeV. Possible effects of these differences on reaction dynamics are discussed in [5], where are investigated the collisions induced by ^{6}Li and ^{7}Li on a ^{59}Co target. At sub-barrier energies, it has been observed a ^{6}Li total fusion cross-section enhancement with respect to the ^{7}Li total fusion cross-section. The authors concluded that this behaviour could be explained only taking into account the coupling with the break-up and the different binding energy of the two nuclei. For this reason we want to measure the ^{6,7}Li+^{64}Zn total fusion excitation functions down to energies far below the Coulomb barrier. Moreover, as noticed before, the ^{6}Li+^{64}Zn collision is one of the few systems where the total fusion cross-section deviates from the systematics [3] and the presence of problems in the ^{6}Li+^{64}Zn experimental data of [6] have been suggested in [3]. In addition, the comparison between the total fusion cross-section and the total reaction cross-section [6] shows, in this case, an unusually large elastic break-up cross-section never seen in other similar systems.

For this reason we propose a new measure of the total fusion cross-section with a different technique.

Complementary information on the study of the ^{6,7}Li+^{64}Zn collision has also been obtained by measuring precise elastic scattering angular distributions at several energies around the Coulomb barrier. Such angular distributions have been used to investigate on the possible absence of usual threshold anomaly on optical potentials due to the presence of break-up as suggested by several studies [8-14].

Although the complete goal of the present experiment is a comparison of the results for the systems ^{6,7}Li+^{64}Zn, here we will discuss only some of the results for the ^{6}Li+^{64}Zn system since the analysis is not yet complete.

2. Total fusion cross section

In order to measure the total fusion cross section around and below the barrier we cannot detect directly the evaporation residues because a great number of them could remain in the target or be below the detection energy threshold due to the low incident energy. For this reason we used an activation technique detecting off-line the X-rays following the electron capture decay of the evaporation residues. The experimental setup is shown in figure 1. The beam passes through a thin Au foil in order to extract the beam current profile by detecting the Rutherford scattered particles with two monitor detectors. Then, the beam impinges on the target followed by a catcher, which stops all the evaporation residues coming out from the ^{64}Zn target. The target thicknesses varied between 550 $\mu g/cm^2$ and 250 $\mu g/cm^2$ decreasing as the beam energy reduced. At low energies the target is thinner because below the barrier the fusion cross section changes rapidly so, a little energy loss in the target may lead to a large variation of the fusion cross section.

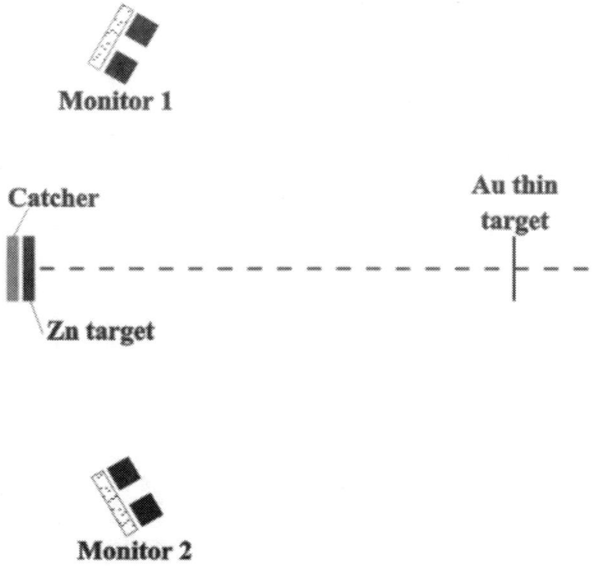

Figure 1. Experimental setup for the ^{64}Zn activation with the ^{6}Li beam. The beam passes through a thin Au foil in order to extract the beam current profile detecting the Rutherford scattered particles. Then, the beam impinges on the target followed by a catcher to collect the recoiling evaporation residues.

After the activation, the target and the catcher are placed in front of a silicon-lithium detector in order to measure the activity curve. This activation technique can be applied when the largest fraction of the evaporation residues decay by electron capture. In the activated target can be found, at the same time, several isotopes of different elements. After the electron-capture decay of the parent nucleus in its daughter, there is emission of a characteristic atomic X-ray having the typical energy of the daughter atom. Since each element has a different characteristic X-ray energy, the X-ray spectrum is the sum of different Gaussian peaks. With a multi-Gaussian fit, the different evaporation residues are identified in charge localizing the peak of their daughters but the isotopes of the same element cannot be discriminated by the X-ray energy. Each energy peak is therefore the sum of the contribution of the different isotopes of the same element. In figure 2, as an example spectrum, can be clearly seen the primary peaks (Kα) of Zn, Ga and Ge and the secondary peak (Kβ) of Ge on the right. The secondary peaks of Zn and Ga are also considered in the fit.

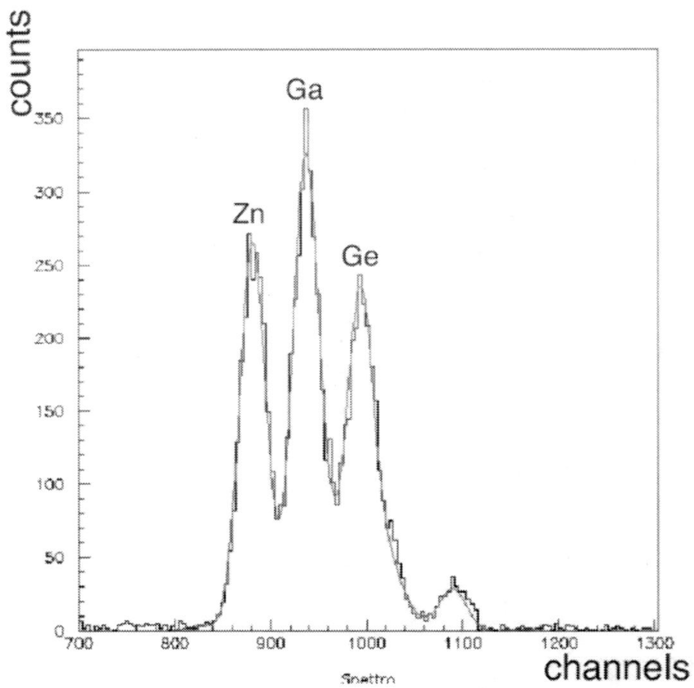

Figure 2. X-ray spectrum from the activated target. Since in the target can be found several different evaporation residues, the spectrum is the sum of different Gaussian peaks.

Several spectra from the same activated target have been measured at different times. Integrating the peak of a specific element and dividing by the duration of the spectrum acquisition run, we obtain the average activity of that element during the run. The activity curve is built by measuring at different times the X-ray spectra from the same activated target and obtaining the activity for each peak. Since different isotopes of the same element have different half-lives, the identification in mass can be done by fitting the activity curve as the sum of different half lives contribution as showed in figure 3.
As a parameter of the fit, we obtain the time zero activity for each evaporation residue that is the activity of the residue at the end of the activation.
Normalizing the time zero activity with the geometric and intrinsic efficiency of the detector, considering the beam current profile during the activation, the target thickness and the (K-alfa) fluorescence probability one can obtain the production cross section for each evaporation residue.
Adding the contribution of all evaporation residues we obtain the total fusion cross section.

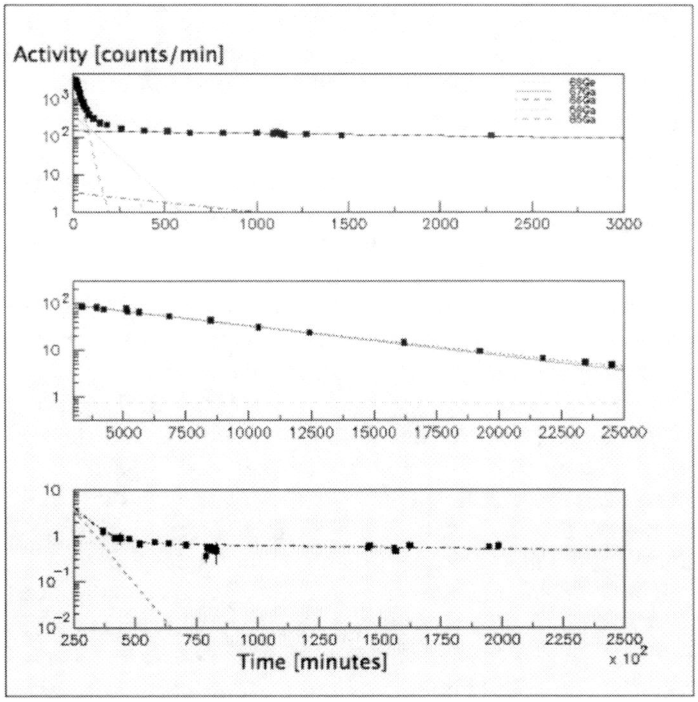

Figure 3. Activity curve fit to identify different Ga isotopes by their half-live.

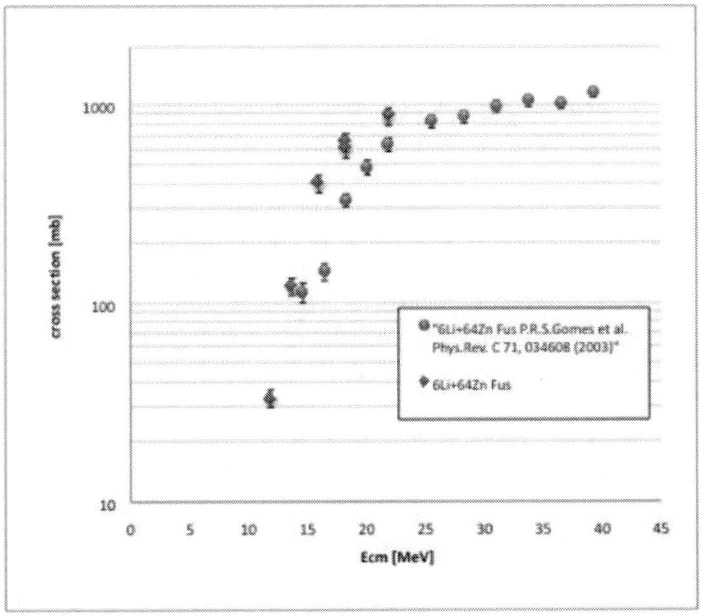

Figure 4. Total fusion excitation functions for the $^{6}Li+^{64}Zn$ collision. The present data are plotted with the green diamonds and are compared to the previous data from the Brazilian group [6]. The disagreement between the two cross-section confirms experimental problems in the previous data.

In figure 4 our total fusion cross section with green diamonds is compared with the previous data of [6] in red dots. The error bar in our data is of the order of 10%. There is a disagreement between our data and the existing ones most probably due to detection threshold problems in the previous experiment as suggested by the authors [6]. With the activation technique, we do not have the threshold problems connected with the evaporation residues energies.

3. Elastic scattering angular distributions, total reaction cross-section and threshold anomaly.

As anticipated, we measured precise elastic scattering angular distributions at several energies around the Coulomb barrier in order to extract the total reaction cross-section and to investigate on the possible absence of usual threshold anomaly on optical potentials [8-14].

The usual behaviour of the optical potential as a function of energy is described in [7]. As the incident energy decreases below the barrier, the imaginary part is reducing and the real part shows a "bump". This behaviour is called threshold anomaly. The real and the imaginary part are linked together by the dispersion relation. The physics interpretation of the imaginary part decrease is the reaction channels closure. The bump in the real part is due to the coupling effects that introduce an attractive polarization potential. For the weakly bound nuclei it has been suggested that the coupling to the break-up may produce a repulsive polarization potential destroying the usual threshold anomaly. In some cases, we may observe the opposite behaviour with an increase of the imaginary part around the barrier [8]. Some authors called this behaviour "break-up threshold anomaly".

The available experimental data do not allow one to draw general conclusions regarding the behaviour of the optical potential around the Coulomb barrier for the weakly bound ^{6,7}Li nuclei. The results for different systems are often not conclusive because there are not enough data available at energies near and below the barrier. Thus, it would be important to perform measurements with small energy steps at energies in the vicinity of the Coulomb barrier. The measure of the elastic scattering angular distribution at low energy is a difficult task because, in this case, the Coulomb scattering is dominant. Therefore, to see variation with respect to the Coulomb scattering it is necessary to measure with an accuracy of the order of 1% and to go to very backward angles. The systematic error can be reduced by paying special attention to the normalization and to the mechanical alignment. For this reason, at the lowest beam energies, we measured at symmetrical angles with respect to the beam direction. The experimental setup is shown in figure 5. The scattered particles are collected by an array of five ΔE(10-μm thick)-E(200-μm thick) silicon telescopes mounted on a rotating plate to cover a wide angular range. Absolute values of the cross sections were obtained by normalizing the data with the assumption that the scattering at small angles, where monitor detectors were placed, is purely of the Rutherford type. The ratio between the monitors solid angles and telescopes ones has been determined with a Rutherford scattering of ^{6}Li on a ~200 μg/cm^2 thick ^{197}Au target.

The ^{64}Zn target thickness of ~400 μg/cm^2 is small enough to allow the discrimination of the target first excited state. Using telescopes one can easily separate the elastic events from the reaction ones. This is particularly important for very backward angles since, in this case, the low intense elastic peak is in the same energy range of the continuum alpha spectrum. In figure 6 are shown the measured elastic scattering angular distributions.

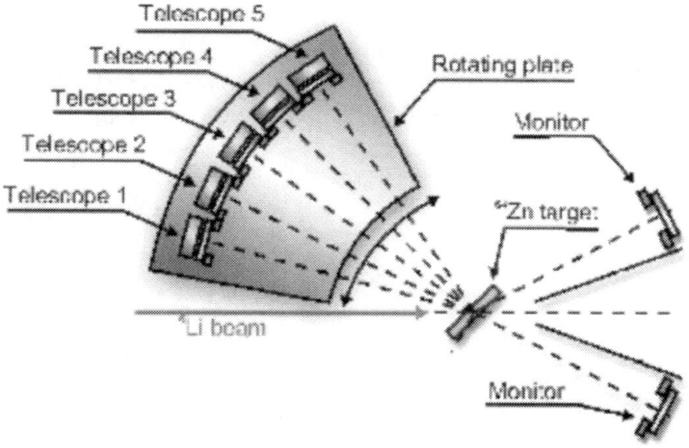

Figure 5. Experimental setup to measure the elastic scattering angular distributions. The scattered particles are collected by an array of five silicon telescopes mounted on a rotating plate to cover a wide angular range.

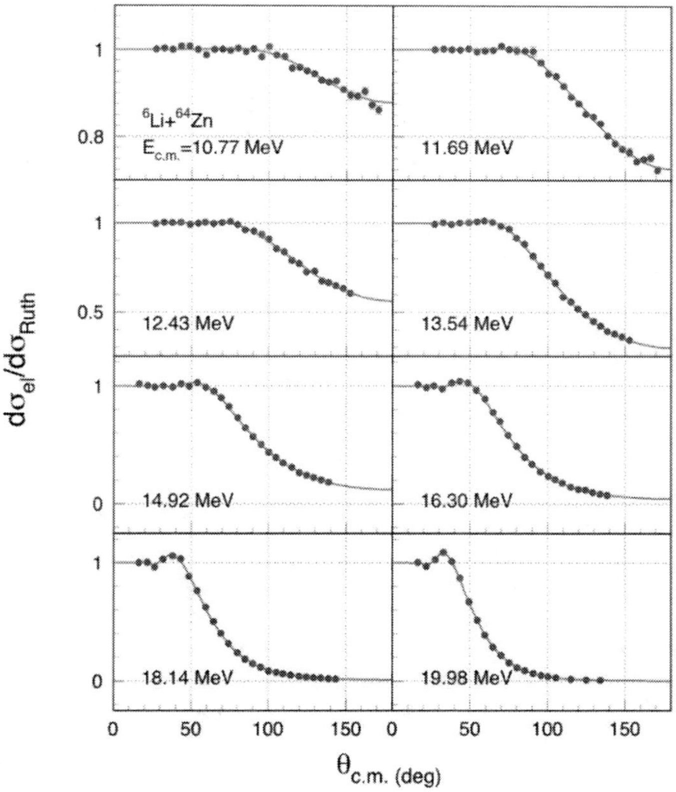

Figure 6. Elastic scattering angular distributions at different energies for the $^{6}Li+^{64}Zn$ system.

3.1. Optical model analysis and results

The angular distributions have been fitted using different optical model potentials [9]: renormalized double folding potential for both real and imaginary part (DF1); renormalized double folding potential for the real part and Woods-Saxon potential for the imaginary part (DFWS). In this last case two different density distributions have been used for the double folding potential (DF1WS & DF2WS). The results are displayed in figure 7.

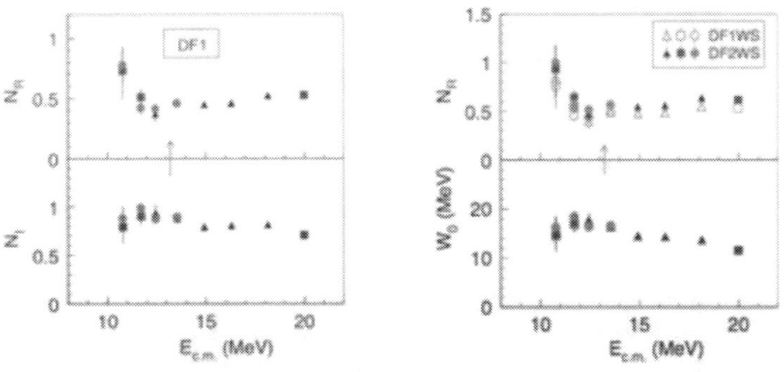

Figure 7. Trend of the optical model potential obtained by fitting the elastic scattering angular distributions with two different potentials. Renormalized double folding potential for both real and imaginary part (DF1); renormalized double folding potential for the real part and Woods-Saxon potential for the imaginary part (DF1WS, DF2WS). See the text for details.

In figure 7-left we see the normalization factors from the fit using the double folding potential for both real and imaginary part (DF1). In figure 7-right we see the other fit with the Woods-Saxon potential for the imaginary part and the two different density distributions for the double folding real potential (DF1WS and DF2WS). Solid symbols, in the right picture, correspond to the DF2WS fit; empty symbols, in the same picture, correspond to the DF1WS fit. Results of independent measurements at a fixed energy, which are plotted in different colours, agree always together.

As we observe in both pictures, the optical potential trend is the same independently by the potential used for the fit. As it can be seen from figure 7, around the Coulomb barrier, indicated by the arrow, there is no presence of usual threshold anomaly.

3.2. Total reaction cross-section

Fitting the elastic scattering angular distribution with the optical model we also extracted the total reaction cross sections. These cross sections are plotted in figure 8 (red triangles) together with the corresponding ones measured by the Brazilian group [6] (yellow squares) and the total fusion cross section data already shown before. Our total reaction cross-section agrees with the previous data of [6]. The difference between the total reaction cross-section and the total fusion cross section is the break-up contribution. In the present data the deduced break-up contribution is smaller then the one extracted from the previous data [6] due to the larger fusion cross-section.

We can also observe that the relative contribution of the break-up is more important at sub barrier energies.

Journal of Physics: Conference Series **282** (2011) 012020 doi:10.1088/1742-6596/282/1/012020

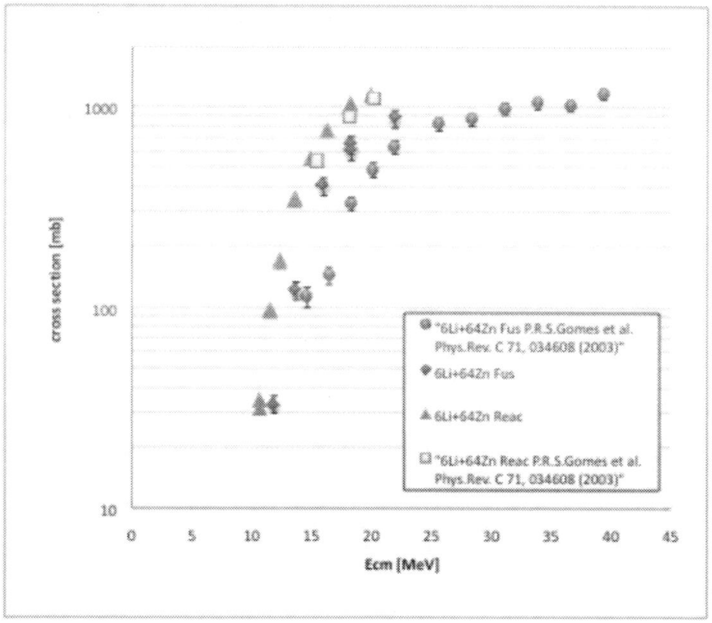

Figure 8. Present total reaction cross-section (red triangles) and total fusion cross-section (green diamonds) are compared with the previous data from the Brazilian group (red dots for the total fusion cross-section; yellow squares for the total reaction cross-section).

4. Summary and conclusions

By using an activation technique we measured the total fusion excitation function for the $^{6}Li+^{64}Zn$ system. Our excitation function appears to be systematically larger than the one reported in [6]. This result seems to confirm the presence of experimental problems in the $^{6}Li+^{64}Zn$ fusion data of [6] as suggested by the same authors of [6] in [3]. These problems are probably due to the energy threshold in the direct detection of the evaporation residues. The activation technique is not affected by this kind of problems. The difference between the reaction cross-section and the fusion cross-section is now smaller than in [6], leaving less space for elastic break-up cross-section. Moreover, as expected, this comparison shows that the relative contribution of the break-up to the total reaction cross-section become more important when decreasing the beam energy below the barrier.

From the analysis of the elastic scattering angular distributions within the optical model, the energy dependence of the optical potential around the Coulomb barrier has been studied showing absence of the usual threshold anomaly.

5. References

[1] L. F. Canto, P. R. S. Gomes, R. Donangelo, and M. S. Hussein, Phys. Rep. 424, 1 (2006)
[2] P. G. Hansen, A. S. Jensen, and B. Jonson, Annu. Rev. Nucl. Part. Sci. 45, 591 (1995)
[3] P.R.S. Gomez et al Phys.Rev. C 79, 027606 (2009)
[4] L.F. Canto et al Nucl.Phys. A 821 (2009) 51
[5] C.Beck et al Phys.Rev. C67, 054602 (2003)
[6] P.R.S. Gomez et al Phys.Rev. C 71, 034608 (2003)
[7] G.R.Satchler Phys.Rep. 199 (1991) 147
[8] J. M. Figueira Phys.Rev. C73, 054603 (2006)
[9] M.Zadro et al., Phys. Rev. C80, 064610 (2009)

[10] N.Keeley, S.J.Bennett, N.M.Clarke, B.R.Fulton, G.Tungate et al. Nucl. Phys. A 571, 326 (1994)
[11] A. M. M. Maciel, P. R. S. Gomes, J. Lubian, R. M. Anjos et al. Phys. Rev. C 59, 2103 (1999)
[12] A. Pakou, N. Alamanos, A. Lagoyannis, A. Gillibert et al. Phys. Lett. B 556, 21 (2003)
[13] A. Pakou, N. Alamanos, G. Doukelis, A. Gillibert et al. Phys. Rev. C 69 , 054602 (2004)
[14] J. M. Figueira, J. O. Fern ´andezNiello, D. Abriola et al. Phys. Rev. C 75, 017602 (2007)
[15] F.A. Souza, L.A.S. Leal, N. Carlin, M.G. Munhoz et al. Phys. Rev. C 75, 044601 (2007)

Quasi-elastic reactions : an interplay of reaction dynamics and nuclear structure

S. Szilner[1], L. Corradi[2], G. Pollarolo[3], S. Beghini[5], S. Courtin[4], E. Farnea[5], E. Fioretto[2], A. Gadea[2], F. Haas[4], D. Jelavić-Malenica[1], D. Lebhertz[4], S. Lunardi[5], D. Mengoni[2], T. Mijatović[1], G. Montagnoli[5], F. Scarlassara[5], N. Soić[1], A. M. Stefanini[2], C. A. Ur[5], J. J. Valiente-Dobón[2]

[1] Ruđer Bošković Institute and University of Zagreb, Zagreb, Croatia.

E-mail: szilner@irb.hr

[2] INFN - Laboratori Nazionali di Legnaro, Legnaro, Italy

[3] INFN and Università di Torino, Italy

[4] IPHC, CNRS/IN2P3 and Université de Strasbourg, Strasbourg, France

[5] INFN and Università di Padova, Padova, Italy

Abstract.
The revival of transfer reaction studies benefited from the construction of the new generation large solid angle spectrometers based on trajectory reconstruction that reached an unprecedented efficiency and selectivity. The coupling of these spectrometers with large γ arrays allowed the identification of individual excited states, their population pattern and decay modes via particle-γ coincidences.

In the present paper aspects of fragment-γ coincidence studies measured with the Prisma-Clara set up in ^{40}Ca+^{96}Zr and ^{40}Ar+^{208}Pb are discussed. In particular, we report about states of particle-phonon character, supporting the idea that the relevant degrees of freedom acting in the reaction dynamics define the final yield distributions.

1. Introduction

Transfer reactions play an essential role in the study of the structure of nuclei, in particular with light ions they have been instrumental for the construction of the shell model and they provided important data to establish the properties of particle-particle correlations in the nuclear medium [1]. With heavy ions, the study of transfer reactions was important in the definition of the reaction mechanism [2, 3] that describes the evolution of the reaction from the quasi-elastic to the more complex deep-inelastic and fusion regime [4, 5]. For energies in the vicinity of the Coulomb barrier, transfer of many nucleons constitutes the largest part of the total reaction cross section thus providing the main source for the loss of flux from the elastic channel and for the energy dissipation from the relative motion to the intrinsic excitation.

In the last decade, the renewed interest in transfer reactions has been mainly due to the realization that multinucleon transfer reactions could be used to populate nuclei moderately rich in neutrons. This renewed interest benefited from the construction of the new generation large solid angle spectrometers based on trajectory reconstruction. The coupling of these spectrometers with large γ arrays allowed the identification of γ rays coming from the decay of the excited states, providing information on the contributions of surface vibrations (bosons), single particles (fermions) and their coupling.

2. Detection techniques for heavy-ion transfer products

In the past, different techniques have been employed to identify nuclei produced in transfer reactions. Most of these techniques make use of magnetic spectrographs or spectrometers for a complete identification of final reaction products in nuclear charge, mass and energy.

The need to distinguish excited states populated in light ion transfer reactions was persuaded by combining magnetic elements of different complexity to focus momenta at definite positions on the focal plane, as for example in Q3D spectrometers. In order to reach an energy resolution of a few tenths keV, the careful corrections for ion optical aberrations were employed.

In order to keep a good resolution and detection efficiency for heavy ions with the large energy dynamic range, solutions with emphasis on the complexity of magnetic elements and/or on the detector systems have been adopted. Dealing with the complex magnetic elements implied that the distribution of atomic charge states of these heavy ions has to be carefully taken into account. To partially overcome this problem, time-of-flight (ToF) spectrometers have been designed with magnetic quadrupole elements which focus ions of different atomic charge states to a relatively small focal plane. All these devices have solid angles in the range $3 - 10$ msr. Beyond these values, it becomes unfeasible to use complex magnetic elements to correct for the ion optical aberrations, thus the presently adopted solution is to simplify the magnetic element configuration and to apply the concept of trajectory reconstruction. This is, at the present, done by using a detector system which, besides nuclear charge, energy and timing, provides the necessary position information along the ion path. The reconstruction of the ion trajectory inside magnetic elements in order to identify the transfer reaction products, has been successfully employed in the very large solid angle (80 msr) spectrometer Prisma [6].

2.1. The Prisma-Clara set-up

Prisma consists of a magnetic quadrupole singlet, placed at 50 cm from the target, and a magnetic dipole. Its main characteristics are a large dispersion and a large momentum acceptance, well adapted for the detection of the kinematics of grazing collisions, where a wealth of nuclei are produced in a wide energy and angular range and with cross sections spanning several orders of magnitude. A position-sensitive micro-channel plate (MCP) detector [7] is placed at the entrance of the spectrometer, providing a start signal for time-of-flight measurements and position signals. Ions pass through the optical elements of the spectrometer and enter a focal plane [8] which is made of a parallel plate of multiwire type (MWPPAC), providing timing and position signals with resolutions similar to the MCP ones. Behind the MWPPAC is an array of a transverse field multiparametric ionization chambers (IC), providing nuclear charge (ΔE) and total energy (E). The IC due to its sub-division can be optimized for the best Z and energy resolutions and to stop at once the ions that differ by more than 20% in kinetic energy.

The γ-ray array Clara [9] consists of 24 HP-Ge clover-type detectors placed as a hemisphere close to the target and opposite to Prisma. The total photopeak efficiency of Clara is of the order of 3% for $E_\gamma = 1.33$ MeV. Typical γ-ray energy resolutions obtained after Doppler correction are 0.6% to 0.9% over the whole velocity distribution of the fragments detected in Prisma.

In Prisma, the mass identification of the reaction products is obtained via an event-by-event reconstruction of the ion trajectory inside the magnetic elements [10]. The identification of nuclear charge Z is obtained through the measurement of energy loss E in the IC, which also provides the total energy E. The charge resolution of Prisma is illustrated in Fig. 1, where we plotted the two-dimensional spectrum of the total energy (IC) vs. total range (the combination

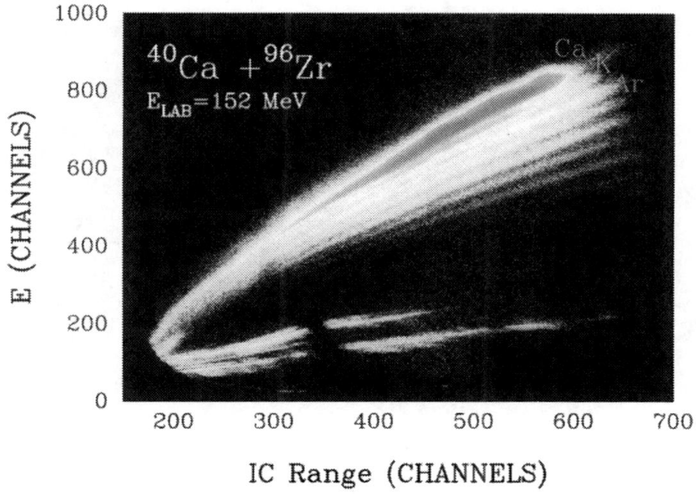

Figure 1. The two-dimensional spectrum of the total energy (IC) vs. total range (the combination of different ΔE sections of IC) for the ^{40}Ca+^{96}Zr reaction.

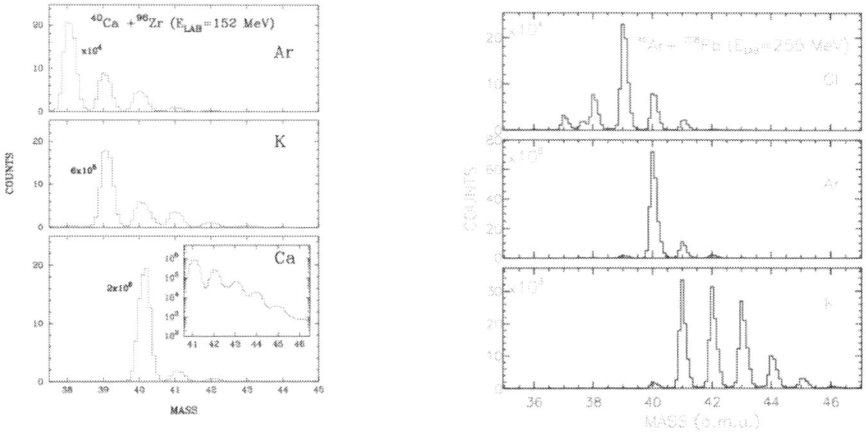

Figure 2. Left panels: Mass spectra for argon, potassium, and calcium isotopes populated in the ^{40}Ca+^{96}Zr reaction. The inset in the bottom panel displays the neutron pick-up channels on a logarithmic scale. Right panels: Mass spectra for potassium, argon, and chlorine isotopes populated in the ^{40}Ar+^{208}Pb reaction. (Note the different scaling factors in each panel.)

of different ΔE sections of IC) for the ^{40}Ca+^{96}Zr reaction measured at $E_{\rm lab}$ =152 MeV and at the grazing angle.

The reconstructed mass spectra for calcium (pure neutron pick-up channels), potassium (one-proton stripping channels), and argon (two-proton stripping channels) isotopes of the ^{40}Ca+^{96}Zr (left panels) and for chlorine (one-proton pick-up channels), argon (neutron transfer channels) and potassium (one-proton stripping channels) isotopes of the ^{40}Ar+^{208}Pb (right panels) reactions are shown in Fig. 2.

The fragment-γ coincidence obtained from the coupling of Clara with Prisma allows us to attribute to each specific reaction product its characteristic γ rays. Since the γ rays are emitted in flight it is mandatory to perform Doppler correction. This is done from the knowledge of the

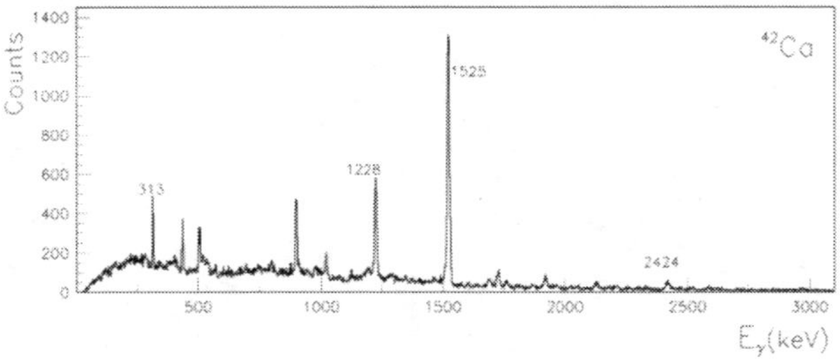

Figure 3. The γ spectrum of ^{42}Ca, the two-neutron pick-up channel of the ^{40}Ca+^{96}Zr reaction.

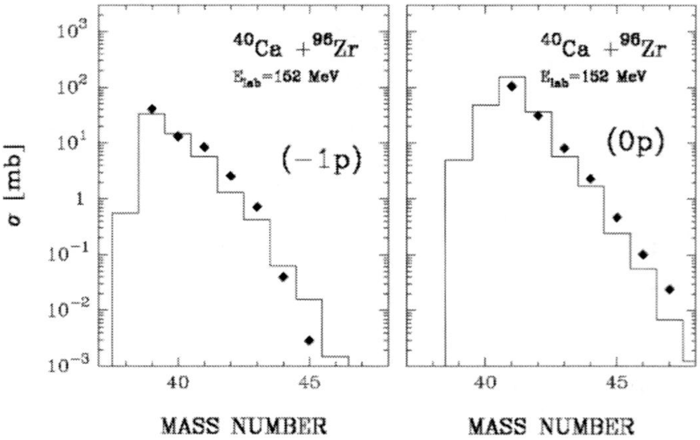

Figure 4. Total cross sections for pure neutron pick-up (right panel) and one-proton stripping (left panel) channels in the ^{40}Ca+^{96}Zr reaction. The points are the experimental data and the histograms are the calculation performed with the code GRAZING.

trajectory reconstructed in Prisma, which provides the velocity vector of the emitting nuclei. The γ spectrum plotted in Fig. 3 demonstrates the reached resolution with this procedure.

3. Phenomenology of MNT

The presently achieved quality of data and of the agreement between theory and measurement is demonstrated in Fig. 4. The figure shows a comparison of the experimental integrated cross sections for one-proton stripping and pure neutron pick-up channels of the ^{40}Ca+^{96}Zr reaction as a function of the number of transfered nucleons with semi-classical calculations [11, 12]. The figure reflects some of the main characteristics of quasi-elastic processes, which we discuss here in more details.

The dependence of the cross sections on the number of transferred neutrons is very similar to the one observed in other studied systems [13, 14]. Multinucleon transfer reactions are governed by optimum Q-value consideration, and using stable ions they dominantly populate neutron pick-up and proton stripping channels. The charge and mass distribution of produced nuclei indicates the dominance of a direct mechanism in the population of different fragments. The neutron pick-up drops by almost a constant factor for each transfered neutron as an independent mechanism would suggest. The pure proton transfer cross sections behave differently, with the population of the $2p$ channel as strong as the $1p$ channel. This suggests the contribution of processes involving the transfer of correlated nucleons in addition to the successive transfer. For the massive proton transfer channels the isotopic distributions drift toward lower masses, a clear indication that these large proton transfer channels are affected by the secondary processes, as evaporation. The calculated cross sections have been obtained by using the semiclassical model GRAZING [11, 12]. This model calculates the evolution of the reaction by taking into account, besides the relative motion variables, the intrinsic degrees of freedom of projectile and target. These are the surface modes and the single-nucleon transfer channels. The multinucleon transfer channels are described via a multistep mechanism. The relative motion of the system is calculated in a nuclear plus Coulomb field. The excitation of the intrinsic degrees of freedom is obtained by employing the well-known form factors for the collective surface vibrations and the one-particle transfer channels. The model takes into account in a simple way the effect of neutron evaporation. The good agreement between experiment and theory gives us confidence on the correct treatment of the quasi-elastic regime.

4. The character of states populated in transfer reactions

The experimental yields, shown in the previous section, have been interpreted with a model that explicitly treats the internal degrees of freedom of the two ions in terms of elementary modes, surface vibration and single particles. Following this description heavy-ion collisions provide a suitable tool for the studies of the particle-vibration coupling scheme. Here we concentrate on the one-neutron transfer channels and spectra of ^{95}Zr populated in the ^{40}Ca+^{96}Zr, and of ^{41}Ar populated in the ^{40}Ar+^{208}Pb reactions. In both cases, the spectra of the nuclei populated in the reaction can be mostly explained by single-particle or single-hole states and states that involve combinations of a single particle or hole with a collective boson.

A well studied laboratory for the weak particle-phonon coupling is the vicinity of the ^{208}Pb nucleus, since its first excited state is the collective 3^- state [15]. The collectivity of 3^- is not dominated by a few particle-hole components but derives from the cooperative action of many configurations. The ^{96}Zr and ^{40}Ar nuclei present a more complicated situation. The low-energy spectrum of ^{96}Zr is dominated by a 2^+ state at 1.75 MeV and by the 3^- at 1.90 MeV. This last state is very collective [B(E3;$3^- \rightarrow 0^+ = 51$ W.u.] and decays via an E1 transition to the 2^+ and via an E3 transition to the ground state. The ground state of ^{95}Zr is well described by a neutron hole in the $d_{5/2}$ orbital. By coupling this hole state with the 3^- one expects a sextuplet of states ($1/2^-$,$3/2^-$, ..., $11/2^-$) at an energy close to the one of the 3^-. Similarly, by coupling the same hole state to the first 2^+ one expects a quintuplet ($1/2^+$, $3/2^+$,. . ., $9/2^+$) at an energy close to one of the 2^+ state in ^{96}Zr. This situation is illustrated in Fig. 5. The reaction mechanism does not populate the components of the two multiplets uniformly but favors the stretched configurations $11/2^-$ and $9/2^+$ since the transfer probability has its maximum at the largest angular momentum transfer. From the adopted levels of ^{95}Zr, we recognized transitions $11/2^- \rightarrow (9/2^+)$, $(9/2^+) \rightarrow (7/2^+)$, and $(7/2^+) \rightarrow 5/2^+_{g.s.}$, as well as the new transition at 2022 keV which we naturally interpret as the E3 decay of $11/2^-$ to the ground state. The intensity of this transition, relative to the E1, is very similar to the one observed in ^{96}Zr, thus reinforcing our interpretation that the $11/2^-$ is a member of a boson-hole $|3^-,(d_{5/2})^{-1} >$ multiplet [10].

A very similar situation seams to emerge in the Ar isotopic chain populated via neutron pick-up channels in the ^{40}Ar+^{208}Pb reaction. Looking at the spectra of the even argon isotopes we notice the strong population of the yrast states, and in particular of the 2^+ state. As already mentioned, in the reaction model the excitation and transfer processes are mediated by the well known single-particle form factors for the fermion degrees of freedom and by the collective form factors, for the vibrational modes. These states act as core excitation in the odd isotopes when a neutron is added. In fact, in the odd Ar isotopes, ^{39,41,43}Ar, besides the transition between the

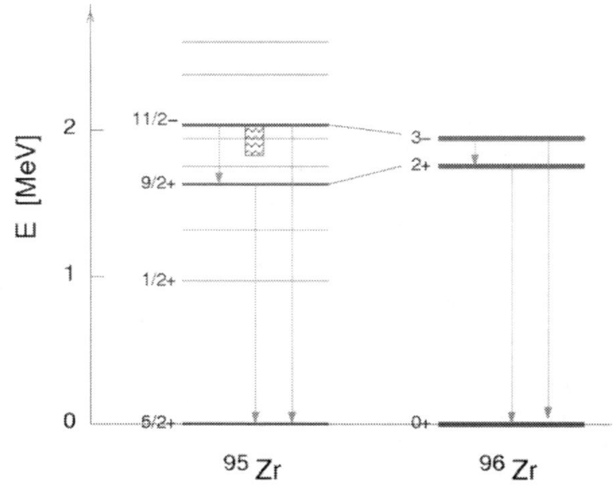

Figure 5. An illustration of particle-phonon coupling for the Zr isotopes.

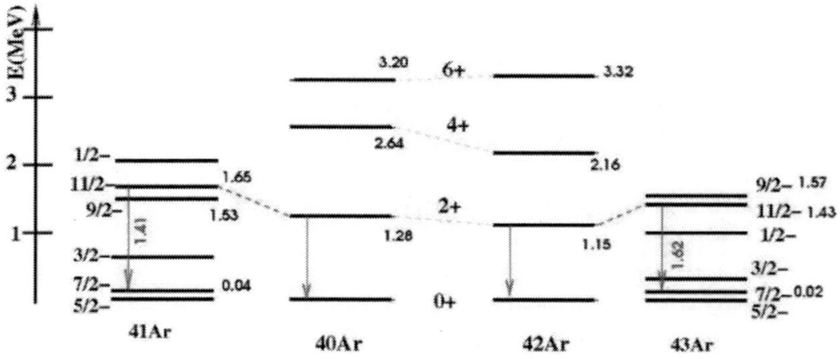

Figure 6. Energy levels calculated in the SDPF-U shell model [16] are plotted as an illustration of particle-vibration coupling scheme. Only relevant levels of ^{40,41,42,43}Ar have been chosen from the SM calculation. Red arrows indicate the E2 transitions.

lowest lying levels, a strong population of the $11/2^-$ states has been observed. The $11/2^-$ state can be understood as a coupling of collective boson to single-particle states (i.e. $|2^+, (f_{7/2})^1>$ giving an $11/2^-$ stretched configuration). The energies of these $11/2^-$ states were compared with the $sd - pf$ large-scale-shell-model calculations, and the coupling schemes are illustrated in Fig. 6.

This significant population of $11/2^-$ states, reached via neutron transfer, demonstrates the importance of excitation of the states whose structure can be explained with the same degrees of freedom which are needed in the reaction model, i.e. coupling of the valence neutron to the vibration quanta.

5. Summary

We have presented some of the recent results of reaction mechanism studies carried out at LNL, Legnaro laboratory. Multi-nucleon transfer reactions have been studied with the large solid angle magnetic spectrometer Prisma, where ions identification is achieved by reconstructing event-by-event the trajectory of the ions in the magnetic elements. Experimental yields have been compared with the semi-classical model GRAZING that includes surface vibrations and single particle transfer modes and taking into account the effect of neutron evaporation. By coupling the Prisma spectrometer with the high resolution Clara γ-array it has been shown how the transfer process populates reaction products in regions of angular momentum and excitation energy quite different from fusion-evaporation processes. This has been exploited to gain information on states that can be interpreted along the particle-vibration coupling scheme.

5.1. Acknowledgments

The authors are grateful to the LNL Tandem-ALPI staff for providing us with good quality beams and to the target laboratory for the excellent target preparation. This work was partly supported by the European Community FP7 REGPOT-2007-3 Contract CLUNA. This work was also supported by the Croatian Ministry of Science, Education and Sports, Grant No. 0098-1191005-2890, and program Cogito.

6. References

[1] Broglia R A, Hansen O and Riedel C 1973 *Advances in Nuclear Physics* vol 6, ed M Baranger and E Vogt (New York: Plenum) p 287
[2] Broglia R A and Winther A 1991 *Heavy Ion Reactions* (Redwood City, CA: Addison-Wesley)
[3] Corradi L, Pollarolo G and Szilner S 2009 *J. Phys. G: Nucl. Part. Phys.* **36** 113101
[4] Reisdorf W 1994 *J. Phys. G: Nucl. Part. Phys.* **20** 1297
[5] Rehm K E 1991 *Annu. Rev. Nucl. Part. Sci.* **41** 429
[6] Stefanini A M et al 2002 *Nucl. Phys.* A **701** 217c
[7] Montagnoli G et al 2005 *Nucl. Instrum. Methods Phys. Res.* A **547** 455
[8] Beghini S et al 2005 *Nucl. Instrum. Methods Phys. Res.* A **551** 364
[9] Gadea A et al 2004 *Eur. Phys. J.* A **20** 193
[10] Szilner S et al 2007 *Phys. Rev.* C **76** 024604
[11] Winther A program GRAZING, htpp:/www.to.infn.it/~nanni/grazing (unpublished)
[12] Winther A 1994 *Nucl. Phys.* A **572** 191; 1995 *Nucl. Phys.* A **594** 203
[13] Corradi L et al 2002 *Phys. Rev.* C **66** 024606
[14] Szilner S et al 2005 *Phys. Rev.* C **71** 044610
[15] Rejmund M et al 2000 *Eur. Phys. J.* A **8** 161 and references therein
[16] Caurier E, Martínez-Pinedo G, Nowacki F, Poves A and Zuker A P 2005 *Rev. Mod. Phys.* **77** 427

Impact parameter dependence of the isospin effects and mass dependence of balance energy

Sakshi Gautam

Department of Physics, Panjab University, Chandigarh -160 014, India.

E-mail: sakshigautm@gmail.com

Abstract. We study the effect of isospin degree of freedom on the balance energy (E_{bal}) as well as its mass dependence throughout the mass range for two different sets of isobaric systems with N/Z = 1 and 1.4 at different colliding geometries ranging from the central to peripheral ones. Our findings reveal the dominance of Coulomb repulsion in isospin effects on E_{bal} as well as its mass dependence throughout the range of the colliding geometry. Our results also indicate that the effect of symmetry energy on the energy of vanishing flow is uniform throughout the mass range and throughout the colliding geometry.

1. Introduction

The existing and upcoming radioactive ion beam (RIB) facilities [1, 2, 3, 4] around the world led to the increased interest in isospin degree of freedom and equation of asymmetric nuclear matter. The above study helps to isolate the isospin-dependent part of nuclear equation of state which is vital in understanding the astrophysical phenomena. Isospin degree of freedom plays an important role through both the nuclear matter equation of state and as well as via in-medium cross section. Therefore, heavy-ion collisions induced by neutron-rich nuclei provide a good opportunity to explore the isospin dependence of in-medium nuclear interactions.

After about three decades of intensive efforts in both nuclear experiments and theoretical calculations, equation of state for isospin symmetric matter is now relatively well understood. The effect of isospin degree of freedom on the collective transverse in-plane flow as well as on its disappearance [5, 6, 7] (there exists a particular incident energy called *balance energy* (E_{bal}) or *energy of vanishing flow* (EVF) at which transverse in-plane flow disappears) has been reported in the literature [8, 9, 10], where it was found that neutron-rich systems have higher E_{bal} compared to neutron-deficient systems at all colliding geometries varying from central to peripheral ones. The effect of isospin degree of freedom on E_{bal} was found to be much more pronounced at peripheral colliding geometries compared to central ones. As reported in the literature, the isospin dependence of collective flow as well as its disappearance has been explained as a competition among various reaction mechanisms, such as nucleon-nucleon collisions, symmetry energy, surface property of the colliding nuclei, and Coulomb force. The relative importance among these mechanisms is not yet clear [8]. In recent study, we [11] confronted theoretical calculations (using isospin-dependent quantum molecular dynamics (IQMD) model [12]) with the data at all colliding geometries and were able to reproduce the data within 5% on the average at all colliding geometries. Motivated by this good agreement of the calculations with data, we [11] studied the isospin effects on the E_{bal} throughout the mass range 48-350 for two sets of isobaric systems with N/Z = 1.0 and 1.4 at semi central colliding geometry. These results showed that the difference between the E_{bal} for two isobaric systems is mainly due to the Coulomb repulsion. It was also shown that Coulomb repulsion dominates over symmetry energy. These findings also indicated towards the dominance of the Coulomb repulsion in larger magnitude of isospin effects in E_{bal} at peripheral collisions. Here we aim to extend the study over full range of colliding geometry varying from central to peripheral ones. Section 2 describes the model in brief. Section 3 explains the results and discussion and Sec. 4 summarizes the results.

2. The model

The present study is carried out within the framework of the IQMD model [12] which is an extension of the QMD model [13, 14, 15]. The IQMD model treats different charge states of nucleons, deltas, and pions explicitly, as inherited from the Vlasov-Uehling-Uhlenbeck (VUU) model. The IQMD model has been used successfully for the analysis of a large number of observables from low to relativistic energies. The isospin degree of freedom enters into the calculations via symmetry potential, cross sections, and Coulomb interaction.

In this model, baryons are represented by Gaussian-shaped density distributions

$$f_i(\vec{r},\vec{p},t) = \frac{1}{\pi^2\hbar^2}\exp(-[\vec{r}-\vec{r_i}(t)]^2\frac{1}{2L}) \times \exp(-[\vec{p}-\vec{p_i}(t)]^2\frac{2L}{\hbar^2}) \quad (1)$$

Nucleons are initialized in a sphere with radius R = 1.12 $A^{1/3}$ fm, in accordance with liquid-drop model. Each nucleon occupies a volume of h^3, so that phase space is uniformly filled. The initial momenta are randomly chosen between 0 and Fermi momentum ($\vec{p}_F$). The nucleons of the target and projectile interact by two- and three-body Skyrme forces, Yukawa potential, Coulomb interactions, and momentum-dependent interactions. In addition to the use of explicit charge states of all baryons and mesons, a symmetry potential between protons and neutrons corresponding to the Bethe-Weizsacker mass formula has been included. The hadrons propagate using Hamilton equations of motion:

$$\frac{d\vec{r_i}}{dt} = \frac{d\langle H\rangle}{d\vec{p_i}}; \qquad \frac{d\vec{p_i}}{dt} = -\frac{d\langle H\rangle}{d\vec{r_i}} \quad (2)$$

with

$$\begin{aligned}\langle H\rangle &= \langle T\rangle + \langle V\rangle \\ &= \sum_i \frac{p_i^2}{2m_i} + \sum_i\sum_{j>i}\int f_i(\vec{r},\vec{p},t)V^{ij}(\vec{r}\,',\vec{r}) \\ &\quad \times f_j(\vec{r}\,',\vec{p}\,',t)d\vec{r}\,d\vec{r}\,'\,d\vec{p}\,d\vec{p}\,'. \end{aligned} \quad (3)$$

The baryon potential V^{ij}, in the above relation, reads as

$$\begin{aligned} V^{ij}(\vec{r}\,'-\vec{r}) &= V^{ij}_{Sky} + V^{ij}_{Yuk} + V^{ij}_{Coul} + V^{ij}_{mdi} + V^{ij}_{sym} \\ &= [t_1\delta(\vec{r}\,'-\vec{r}) + t_2\delta(\vec{r}\,'-\vec{r})\rho^{\gamma-1}(\frac{\vec{r}\,'+\vec{r}}{2})] \\ &\quad + t_3\frac{\exp(|(\vec{r}\,'-\vec{r})|/\mu)}{(|(\vec{r}\,'-\vec{r})|/\mu)} + \frac{Z_iZ_je^2}{|(\vec{r}\,'-\vec{r})|} \\ &\quad + t_4\ln^2[t_5(\vec{p}\,'-\vec{p})^2+1]\delta(\vec{r}\,'-\vec{r}) \\ &\quad + t_6\frac{1}{\varrho_0}T_{3i}T_{3j}\delta(\vec{r_i}\,'-\vec{r_j}). \end{aligned} \quad (4)$$

Here Z_i and Z_j denote the charges of *ith* and *jth* baryon, and T_{3i} and T_{3j} are their respective T_3 components (i.e., 1/2 for protons and −1/2 for neutrons). The parameters μ and t_1,....,t_6 are adjusted to the real part of the nucleonic optical potential. For the density dependence of the nucleon optical potential, standard Skyrme type parametrization is employed. Note that such Skyrme forces and related potential are essential tool to understand low energy phenomena such as fusion, fission and cluster radioactivity [16]. The momentum-dependence V_{mdi}^{ij} of the nn interactions, which may optionally be used in IQMD, is fitted to the experimental data in the real part of the nucleon optical potential.

3. Results and discussion

For the present study, we simulate several thousands events of each reaction at incident energies around E_{bal} in small steps of 10 MeV/nucleon. In particular, we simulate the reactions ^{24}Mg+^{24}Mg, ^{58}Cu+^{58}Cu, ^{72}Kr+^{72}Kr, ^{96}Cd+^{96}Cd, ^{120}Nd+^{120}Nd, ^{135}Ho+^{135}Ho, having N/Z = 1.0 and reactions ^{24}Ne+^{24}Ne, ^{58}Cr+^{58}Cr, ^{72}Zn+^{72}Zn, ^{96}Zr+^{96}Zr, ^{120}Sn+^{120}Sn, and ^{135}Ba+^{135}Ba, having N/Z = 1.4, respectively, in the whole range of colliding geometry. The colliding geometry is divided into four impact parameter bins of $0.15 < \hat{b} < 0.25$ (BIN 1), $0.35 < \hat{b} < 0.45$ (BIN 2), $0.55 < \hat{b} < 0.65$ (BIN 3), and $0.75 < \hat{b} < 0.85$ (BIN 4), where $\hat{b} = b/b_{max}$.

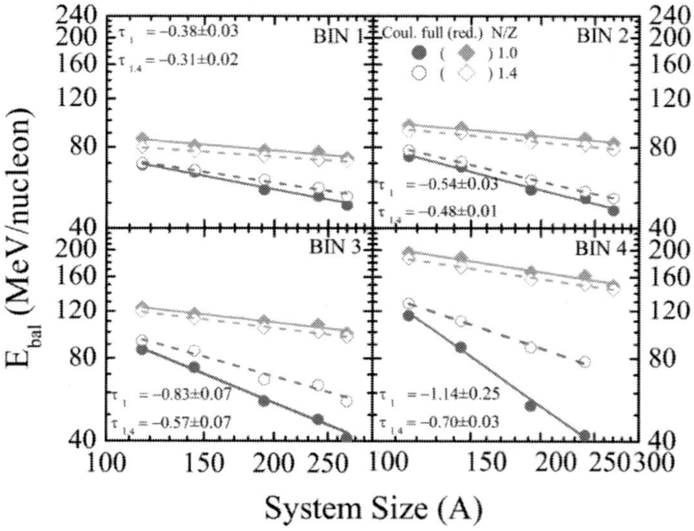

Figure 1. (Color online) E_{bal} as a function of combined mass of system for different impact parameter bins. Solid (open) symbols are for systems having N/Z = 1.0 (1.4). The solid (dashed) lines are power law fit of A^{τ}. τ values for full Coulomb calculations are displayed in figure. The values of τ for reduced Coulomb calculations are given in the text.

Here N/Z is changed by keeping the mass fixed. We use anisotropic standard isospin- and energy-dependent nn cross section $\sigma = 0.8\ \sigma_{NN}{}^{free}$. The details about the elastic and inelastic cross sections for proton-proton and proton-neutron collisions can be found in Ref. [12]. The cross sections for neutron-neutron collisions are assumed to be equal to the proton-proton cross sections. We also use soft equation of state along with momentum-dependent interactions (MDI). The results with the above choice of equation of state and cross section were in good agreement with the data. The reactions are followed until the transverse flow saturates. The saturation time varies form 100 fm/c for lighter masses to 300 fm/c for heavier masses. For transverse flow, we use the quantity "*directed transverse momentum* $\langle p_x^{dir} \rangle$" which is defined as [17]

$$\langle p_x^{dir} \rangle = \frac{1}{A} \sum_{i=1}^{A} sign\{y(i)\} p_x(i), \tag{5}$$

where $y(i)$ is the rapidity and p_x(i) is the momentum of ith particle. The rapidity is defined as

$$Y(i) = \frac{1}{2} \ln \frac{\mathbf{E}(i) + \mathbf{p}_z(i)}{\mathbf{E}(i) - \mathbf{p}_z(i)}, \tag{6}$$

where $\mathbf{E}(i)$ and $\mathbf{p}_z(i)$ are, respectively, the energy and longitudinal momentum of ith particle. In this definition, all the rapidity bins are taken into account. A straight line interpolation is used to calculate E_{bal}.

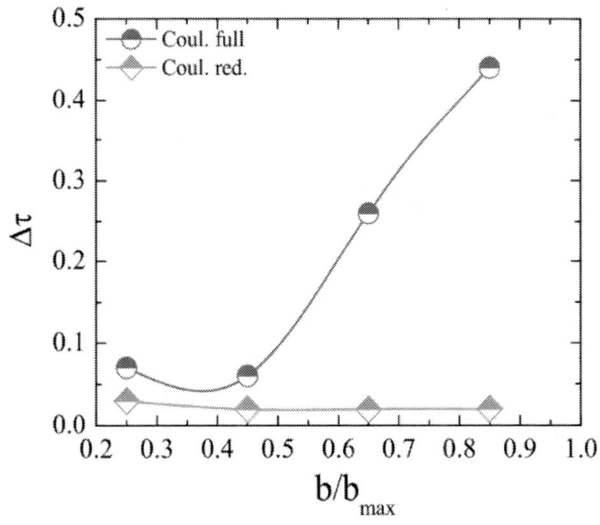

Figure 2. (Color online) τ as a function of reduced impact parameter. Lines are only to guide the eye.

Figure 1 displays the mass dependence of E_{bal} for four impact parameter bins. The solid (open) green circles indicate E_{bal} for systems with lower (higher) N/Z. Lines are power law fit of A^τ. E_{bal} follows a power law behavior of A^τ for both N/Z = 1 and 1.4 (τ being labeled as $\tau_{1.0}$ and $\tau_{1.4}$ for systems having N/Z = 1 and 1.4, respectively) at all colliding geometries. Also, for the clarity, we did not include the very light system (A=48) as isospin effects are negligible at all colliding geometries for this system.

From fig. 1 we see that isospin effects are clearly visible for all the four bins as neutron-rich system has higher E_{bal} throughout the mass range in agreement with the previous studies [9, 8, 11]. The magnitude of the isospin effects increases with the increase in the mass of the system at all colliding geometries. The effect is much more pronounced at larger colliding geometries. One can see that the difference between $\tau_{1.0}$ and $\tau_{1.4}$ increases with increase in the impact parameter. In Ref. [18], one of us and coworkers studied the isospin effects on the mass

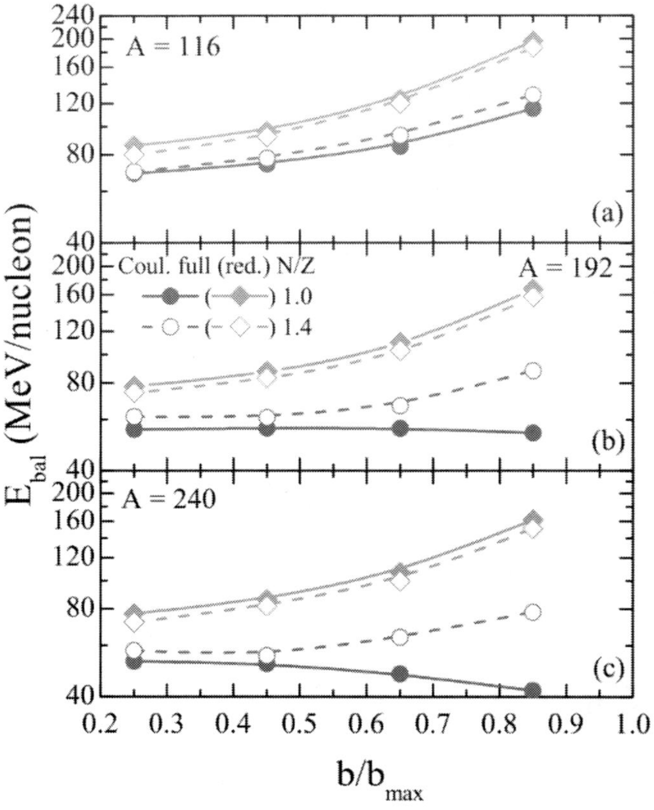

Figure 3. (Color online) E_{bal} as a function of impact parameter for different system masses. Symbols have same meaning as in Fig.1. Lines are only to guide the eye. [18]

dependence of E_{bal} for BIN 2. There the Coulomb potential was reduced by a factor of 100 and was showed that the Coulomb repulsion plays dominant role over symmetry energy in isospin effects on E_{bal} as well as its mass dependence at semi central colliding geometry (BIN 2). Since here we plan to extend that study over a full range of colliding geometry, so here also we reduce the Coulomb potential by a factor of 100 and calculate the E_{bal} throughout the mass range at all colliding geometries. Solid (open) diamonds represent E_{bal} calculated with reduced Coulomb for systems with lower (higher) neutron content. Lines are power law fit of A^{τ}. The values of $\tau_{1.0}$ ($\tau_{1.4}$) are -0.17 (-0.14), -0.17 (-0.19), -0.24 (-0.26), and -0.31 (-0.29) for BIN 1, BIN 2, BIN 3, and BIN 4, respectively. Interestingly, we find that the magnitude of isospin effects (difference in E_{bal} for a given pair) is now nearly same throughout the mass range which indicates that the effect of symmetry energy is uniform throughout the mass range. This is true for all the colliding geometries. This is supported by Ref. [17] where Sood and Puri studied the average density as a function of mass of the system (throughout the periodic table) at incident energies equal to E_{bal} for each given mass. There they found that although both E_{bal} and average density follows a power law behavior of A^{τ}, E_{bal} decreases more sharply with the combined mass of the system (with τ = -0.42), whereas the average density (calculated at incident energy equal to

E_{bal}) is almost independent of the mass of the system with τ = -0.05. It is worth mentioning here that the trend will be different at fixed incident energy in which case density increases with increase in the mass of the system [19, 20]. From fig. 1 one can also see that the enhancement in E_{bal} (by reducing Coulomb) is more in heavier systems as compared to lighter systems for all colliding geometries. The effect is more pronounced at higher colliding geometries. Moreover, throughout the mass range at all colliding geometries, the neutron-rich systems have less E_{bal} as compared to neutron-deficient systems when we reduce the Coulomb. This trend is quite the opposite to the one which we have when we have full Coulomb. This (as explained in Ref. [18] also) is due to the fact that the reduced Coulomb repulsion leads to higher E_{bal}. As a result, the density achieved during the course of the reaction will be more due to which the impact of the repulsive symmetry energy will be more in neutron-rich systems, which in turn leads to less E_{bal} for neutron-rich systems.

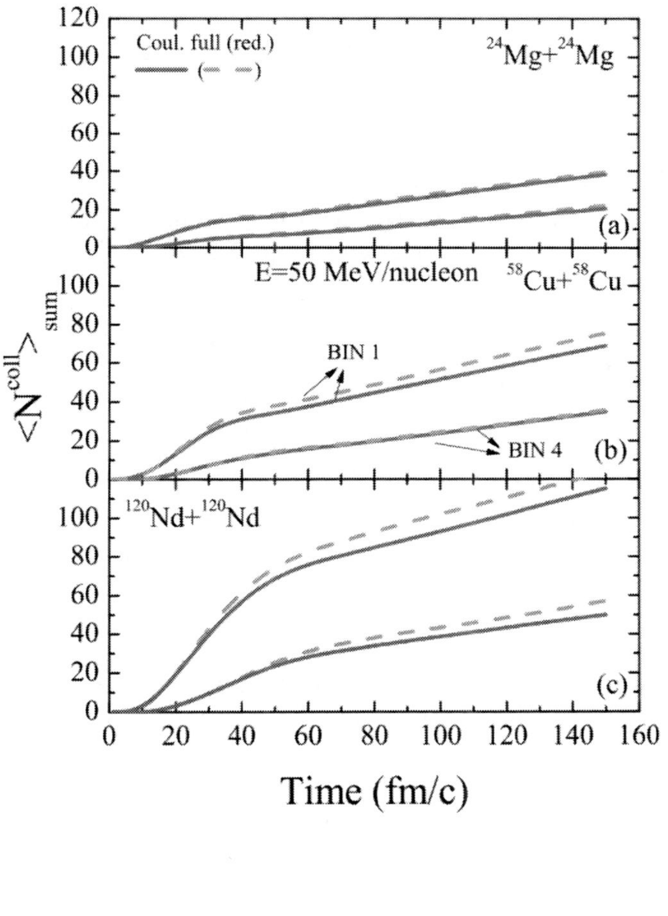

Figure 4. (Color online) The time evolution of number of collisions for various system masses at 50 MeV/nucleon for BIN 1 and BIN 4. Various lines are explained in the text .

In fig. 2, we show the variation in $\Delta\tau$ as a function of colliding geometry for both full and reduced Coulomb where $\Delta\tau = \tau_{1.4} - \tau_{1.0}$. Half filled circles (diamonds) represent calculations with full (reduced) Coulomb. $\Delta\tau$ values are plotted at upper limit of impact parameter for each bin. Lines are only to guide the eye. From figure (green circles), we see that the difference between $\tau_{1.4}$ and $\tau_{1.0}$ increases with increase in impact parameter which is due to the sharp increase in value of $|\tau_1|$, whereas when we reduce the Coulomb (orange diamonds), this difference is negligible, which shows an enhanced effect of Coulomb at higher impact parameters.

In fig. 3a, 3b, and 3c, we display E_{bal} as a function of $\hat{b}$ for masses 116, 192, and 240, respectively, for both full and reduced Coulomb. Symbols have the same meaning as in fig. 1. For full Coulomb (green circles), for all the masses at all colliding geometries, system with higher N/Z has larger E_{bal} in agreement with previous studies [8, 9]. Moreover, the difference between E_{bal} for a given mass pair, increases with increase in colliding geometry. This is more clearly visible in heavier masses. Also for N/Z = 1.4, E_{bal} increases with increase in impact parameter as expected [22, 21]. However for N/Z = 1, this is true only for lighter mass system such as A = 116. For heavier masses E_{bal} infact begins to decrease with increase in impact parameter in contrast to the previous studies [8, 9, 11, 22, 21]. However, when we reduce the Coulomb (by a factor of 100 (diamonds)), we find that:

(i) Neutron-rich systems have smaller E_{bal} as compared to neutron-deficient systems as mentioned previously also. This is true at all the colliding geometries throughout the mass range. This clearly shows the dominance of Coulomb repulsion over symmetry energy in isospin effects throughout the mass range at all colliding geometries.

(ii) The difference between E_{bal} for systems with different N/Z remains almost constant as a function of colliding geometry which indicates that the effect of symmetry energy is uniform throughout the range of $\hat{b}$ as well. This also shows that the large differences in E_{bal} values for a given isobaric pair are due to the Coulomb repulsions.

To see the role of Coulomb repulsion on nn collisions, in fig. 4 we display no. of collisions for A = 48, 116, and 240 for calculations with and without Coulomb potential at central (BIN 1) and peripheral (BIN 4) colliding geometry at incident energy of 50 MeV/nucleon. Solid (dashed) lines represent Coulomb full (reduced) calculations. Higher (lower) lines represent the results for BIN 1 (BIN 4). We find that Coulomb decreases the no. of collisions in medium and heavier mass systems for central bin whereas for peripheral bin, the effect of Coulomb on collisions is only for heavier masses (lower panel).

4. Summary

We have studied the isospin effects in the disappearance of flow as well as its mass dependence throughout the mass range for two sets of isobaric systems with N/Z = 1 and 1.4 in the whole range of colliding geometry. Our results clearly demonstrate the dominance of Coulomb repulsion in isospin effects on E_{bal} as well as its mass dependence throughout the range of colliding geometry. The above study also shows that the effect of symmetry energy on E_{bal} is uniform throughout the mass range and colliding geometry.

This work has been supported by a grant from Indo-French Centre For The Promotion Of Advanced Research (IFCPAR) under project no. 4104-1.

5. References

[1] Zhan W *et al* 2006 *Int J Mod Phys* E **15** 1941
see e.g.http://www.impcas.ac.cn/zhuye/en/htm/247.htm
[2] See e.g. http://www.gsi.de/fair/indexe.html
[3] Yano Y 2007 *Nucl Inst Methods B* **261** 1009
[4] See e.g. White Papers of the 2007 NSAC Long Range Plan town Meeting Jan 2007 Chiacgo http://dnp.aps.org.
[5] Scheid W, Ligensa R and Greiner W 1968 *Phys Rev Lett* **21** 1479
[6] Bonasera A and Csernai L P 1987 *Phys Rev Lett* **59** 630
[7] Krofcheck D *et al* 1989 *Phys Rev Lett* **63** 2028

[8] Li B A, Ren Z, Ko C M and Yennello S J 1996 *Phys Rev Lett* **76** 4492
[9] Pak R *et al* 1997 *Phys Rev Lett* **78** 1022
Pak R *et al* 1997 *Phys Rev Lett* **78** 1026
[10] Daffin F and Bauer W, arxiv: 9809.0241v1.
[11] Gautam S *et al* 2010 *J Phys G: Nucl Part Phys* **37** 085102
[12] Hartnack C *et al* 1998 *Eur Phys J* A **1** 151
[13] Aichelin J 1991 *Phys Rep* **202** 233
Puri R K *et al* 1996 *Phys Rev* C **54** R28
Puri R K *et al* 2000 *J Comput Phys* **162** 245
Vermani Y K *et al* 2009 *Europhys Lett* **85** 62001
Vermani Y K *et al* 2010 *J Phys G: Nucl Part Phys* **37** 015105
Kumar S *et al* 2008 *Phys Rev* C **78** 064602
[14] Kumar S *et al* 1998 *Phys Rev* C **58** 3494
Lehmann E *et al* 1996 *Z Phys* A **355** 55
Sood A D *et al* 2009 *Phys Rev* C **79** 064618
Sood A D *et al* 2004 *Phys Rev* C **70** 034611
Kumar S *et al* 2010 *Phys Rev* C **81** 014611
Lehmann E *et al* 1993 *Prog Part Nucl Phys* **30** 219
Lehmann E *et al* 1995 *Phys Rev* C **51** 2113
Kumar S *et al* 1990 *Phys Rev* C **58** 1618
[15] Singh J *et al* 2000 *Phys Rev* C **62** 044617
Batko G *et al* 1994 *J Phys G: Nucl Part Phys* **20** 461
Huang S W *et al* 1993 *Prog Part Nucl Phys* **30** 105
Puri R K *et al* *Nucl Phys* A **575** 733
Fuchs C *et al* 1996 *J Phys G: Nucl Part Phys* **22** 131
Vermani Y K *et al* 2009 *Phys Rev* C **79** 064613
Vermani Y K *et al* 2009 *J Phys G: Nucl Part Phys* **36** 105103
[16] Dutt I *et al* 2010 *Phys Rev* C **81** 064608
Dutt I *et al* *Phys Rev* C **81** 064609
Dutt I *et al* *Phys Rev* C **81** 044615
Dutt I *et al* *Phys Rev* C **81** 047601
Puri R K *et al* 2005 *Eur Phys J* A **23** 429
Puri R K *et al* 1998 *Eur Phys J* **A3** 277
Puri R K *et al* 1991 *Phys Rev* C **43** 315
Puri R K *et al* 1992 *Phys Rev* C **45** 1837
Puri R K *et al* 1992 *J Phys G: Nucl Part Phys* **18** 903
[17] Sood A D and Puri R K 2004 *Phys Lett* **B594** 260
Sood A D and Puri R K 2006 *Eur Phys J* A**30** 571
Sood A D and Puri R K 2006 *Phys Rev* C **73** 067602
[18] Gautam S and Sood A D 2010 *Phys Rev* C **82** 014604
Gautam S, Sood A D, Puri R K and Aichelin J *Phys Rev* C (submitted)
[19] Blättel B *et al* 1991 *Phys Rev* C **43** 2728
[20] Khoa D T *et al* 1992 *Nucl Phys* **A548** 102
[21] Magestro D J, Bauer W and Westfall G D 2000 *Phys Rev* C **62** 041603(R)
[22] Chugh R and Puri R K 2010 *Phys Rev* C **82** 014603

Effect of mass asymmetry on the mass dependence of balance energy.

Supriya Goyal
Department of Physics, Panjab University, Chandigarh 160 014, India
E-mail: ashuphysics@gmail.com

Abstract. We demonstrate the role of the mass asymmetry on the balance energy (E_{bal}) by studying asymmetric reactions throughout the periodic table and over entire colliding geometry. Our results, which are almost independent of the system size and as well as of the colliding geometries indicate a sizeable effect of the asymmetry of the reaction on the balance energy.

1. Introduction

Reaction dynamics has always captured a central place in nuclear physics mainly due to the wider domain of physics it caters. Right from the low energy (where fusion-fission takes place) to the ultra-relativistic energies (where sub-nucleonic degrees of freedom are dominant) a large number of new phenomena has been observed/predicted. The reaction dynamics in heavy-ion collisions at intermediate energies has been used extensively during the last three decades to understand the nature of nuclear matter at extreme conditions of temperature and density and the nature of nuclear equation of state (EoS) [1, 2, 3, 4, 5, 6, 7]. The phenomena that are mainly observed in this energy range are mainly multifragmentation, collective transverse flow, particle production etc.

In the search of nuclear equation of state as well as of nuclear interactions and forces, collective flow has been found to be of immense importance [1, 2, 3, 4, 5, 6, 7]. Among collective flow, transverse in-plane flow enjoys special status. During the last two decades, much emphasis has been put on the study of collective flow [1, 2, 3, 4, 5, 6, 7]. Lots of experiments have been performed and number of theoretical attempts have also been employed to explain and understand these observations. As reported by [8, 9] for the first time and later on by many others, collective flow is negative at low incident energies whereas it is positive at a reasonable higher incident energies. At a particular incident energy, however, a transition occurs. This transition energy is also known as energy of vanishing flow or balance energy (E_{bal}). This balance energy (E_{bal}) has been subjected to intensive theoretical calculations using variety of equations of state as well as nucleon-nucleon cross-sections [1, 2, 3, 4, 5, 6, 7, 8, 9]. This also includes the mass-dependence of E_{bal} which have been reproduced successfully by various theoretical models [1, 3, 4, 5, 6, 7].

Interestingly, most of these studies take symmetric or nearly symmetric reactions into account. Recently, FOPI group studied the flow for the asymmetric reaction of ^{40}Ca+^{197}Au [10, 11]. They noted that the flow in the asymmetric collisions is a key observable for investigating the reaction dynamics. In contrast to symmetric collisions, where center of mass is one of the nucleus, this quantity is not known *a priori* in asymmetric nuclei experimentally. Later on, FOPI conducted experiment on ^{58}Ni and ^{208}Pb [10, 11]. In other class of studies, the total mass of the system was kept fixed as 96 units whereas charge was varied [12]. Theoretically, recently Kaur and Kumar [13] conducted a complete study of the multifragmentation by varying the asymmetry of the colliding nuclei. Asymmetry parameter η is defined as $(A_T - A_p/A_T + A_P)$; where A_T and A_P are the masses of the target and projectile, respectively. All these attempts point towards a need for the systematic study of the disappearance of flow for asymmetric colliding nuclei. Further as noted, asymmetry of a reaction plays dramatic role in heavy-ion collisions [14, 15, 16, 17]. This happens because excitation energy in symmetric colliding nuclei leads to larger compression while asymmetric reactions lack the compression since large part of the excitation energy is in the form of thermal energy.

Note that some isolated studies with asymmetric nuclei are already done in the literature where the reactions of $^{20}Ne+^{12}C$, $^{20}Ne+^{27}Al$, $^{20}Ne+^{63}Cu$, $^{58}Ni+^{12}C$, $^{64}Zn+^{27}Al$, $^{1}H+^{197}Au$, $^{12}C+^{197}Au$, $^{197}Au+^{12}C$, $^{40}Ar+^{207}Pb$ etc are taken into account [10, 11, 18, 19, 20, 21, 22, 23, 24, 25, 26]. Interestingly, none of the studies focus on the E_{bal} for asymmetric colliding nuclei. To address this, we here present a systematic study of the E_{bal} as a function of asymmetry of the colliding nuclei. While total mass of the reactions remain fixed, the asymmetry is varied by transferring the neutrons/protons from one nucleus to other. The QMD model used for the present analysis is explained briefly in the section 2. Results and discussion are presented in section 3 followed by summary in section 4.

2. Description of the model

The quantum molecular dynamics model (QMD) simulates the reaction on an event by event basis [2]. This is based on a molecular dynamics picture where nucleons interact via two and three-body interactions. The explicit two and three-body interactions preserves the fluctuations and correlations which are important for *N*-body phenomenon such as multifragmentation [2].

In the QMD model, the (successfully) initialized nuclei are boosted towards each other with proper center-of mass velocity using relativistic kinematics. Here each nucleon α is represented by a Gaussian wave packet with a width of $\sqrt{L}$ centered around the mean position $\vec{r}_\alpha$(t) and mean momentum $\vec{p}_\alpha$(t) [2]:

$$\phi_\alpha(\vec{r},\vec{p},t) = \frac{1}{(2\pi L)^{3/4}} e^{\left[-\{\vec{r}-\vec{r}_\alpha(t)\}^2/4L\right]} e^{\left[i\vec{p}_\alpha(t)\cdot\vec{r}/\hbar\right]}. \tag{1}$$

The Wigner distribution of a system with $A_T + A_P$ nucleons is given by

$$f(\vec{r},\vec{p},t) = \sum_{\alpha=1}^{A_T+A_P} \frac{1}{(\pi\hbar)^3} e^{\left[-\{\vec{r}-\vec{r}_\alpha(t)\}^2/2L\right]} e^{\left[-\{\vec{p}-\vec{p}_\alpha(t)\}^2 2L/\hbar^2\right]'}, \tag{2}$$

with L = 1.08 fm^2.

The center of each Gaussian (in the coordinate and momentum space) is chosen by the Monte Carlo procedure. The momentum of nucleons (in each nucleus) is chosen between zero and local Fermi momentum [= $\sqrt{2m_\alpha V_\alpha(\vec{r})}$; $V_\alpha(\vec{r})$ is the potential energy of nucleon α]. Naturally, one has to take care that the nuclei, thus generated, have right binding energy and proper root mean square radii.

The centroid of each wave packet is propagated using the classical equation of motion [2]:

$$\frac{d\vec{r}_\alpha}{dt} = \frac{dH}{d\vec{p}_\alpha}, \tag{3}$$

$$\frac{d\vec{p}_\alpha}{dt} = -\frac{dH}{d\vec{r}_\alpha}, \tag{4}$$

Journal of Physics: Conference Series **282** (2011) 012023 doi:10.1088/1742-6596/282/1/012023

where the Hamiltonian is given by

$$H = \sum_{\alpha} \frac{\vec{p}_{\alpha}^{2}}{2m_{\alpha}} + V^{tot}. \tag{5}$$

Our total interaction potential V^{tot} reads as [2]

$$V^{tot} = V^{Loc} + V^{Yuk} + V^{Coul} + V^{MDI}, \tag{6}$$

with

$$V^{Loc} = t_1\delta(\vec{r}_\alpha - \vec{r}_\beta) + t_2\delta(\vec{r}_\alpha - \vec{r}_\beta)\delta(\vec{r}_\alpha - \vec{r}_\gamma), \tag{7}$$

$$V^{Yuk} = t_3 e^{-|\vec{r}_\alpha - \vec{r}_\beta|/m} / \left(|\vec{r}_\alpha - \vec{r}_\beta|/m\right), \tag{8}$$

with m = 1.5 fm and t_3 = -6.66 MeV.

The static (local) Skyrme interaction can further be parametrized as:

$$U^{Loc} = \alpha\left(\frac{\rho}{\rho_o}\right) + \beta\left(\frac{\rho}{\rho_o}\right)^{\gamma}. \tag{9}$$

Here α, β and γ are the parameters that define equation of state. The momentum dependent interaction is obtained by parameterizing the momentum dependence of the real part of the optical potential. The final form of the potential reads as

$$U^{MDI} \approx t_4 ln^2[t_5(\vec{p}_\alpha - \vec{p}_\beta)^2 + 1]\delta(\vec{r}_\alpha - \vec{r}_\beta). \tag{10}$$

Here t_4 = 1.57 MeV and t_5 = 5 $\times 10^{-4} MeV^{-2}$. A parameterized form of the local plus momentum dependent interaction (MDI) potential (at zero temperature) is given by

$$U = \alpha\left(\frac{\rho}{\rho_0}\right) + \beta\left(\frac{\rho}{\rho_0}\right)^{\gamma} + \delta ln^2[\epsilon(\rho/\rho_0)^{2/3} + 1]\rho/\rho_0. \tag{11}$$

The parameters α, β, and γ in above Eq. (11) must be readjusted in the presence of momentum dependent interactions so as to reproduce the ground state properties of the nuclear matter. The set of parameters corresponding to different equations of state can be found in Ref. [2].

3. Results and Discussion

For the present study, we simulated various reactions for 1000-5000 events in the incident energy range between 90 and 500 MeV/nucleon. In particular, we simulated the reactions of $^{17}_{8}O$+$^{23}_{11}Na$ ($\eta = 0.1$), $^{14}_{7}N$ +$^{26}_{12}Mg$ ($\eta = 0.3$), $^{10}_{5}B$ +$^{30}_{14}Si$ ($\eta = 0.5$), and $^{6}_{3}Li$ +$^{34}_{16}S$ ($\eta = 0.7$) for total mass (A_{TOT}) = 40, $^{36}_{18}Ar$+$^{44}_{20}Ca$ ($\eta = 0.1$), $^{28}_{14}Si$+$^{52}_{24}Cr$ ($\eta = 0.3$), $^{20}_{10}Ne$+$^{60}_{28}Ni$ ($\eta = 0.5$), and $^{10}_{5}B$+$^{70}_{32}Ge$ ($\eta = 0.7$) for total mass (A_{TOT}) = 80, $^{70}_{32}Ge$+$^{90}_{40}Zr$ ($\eta = 0.1$), $^{54}_{26}Fe$+$^{106}_{48}Cd$ ($\eta = 0.3$), $^{40}_{20}Ca$+$^{120}_{52}Te$ ($\eta = 0.5$), and $^{24}_{12}Mg$+$^{136}_{58}Ce$ ($\eta = 0.7$) for total mass (A_{TOT}) = 160, and $^{108}_{48}Cd$+$^{132}_{56}Ba$ ($\eta = 0.1$), $^{84}_{38}Sr$+$^{156}_{66}Dy$ ($\eta = 0.3$), $^{60}_{28}Ni$+$^{180}_{74}W$ ($\eta = 0.5$), and $^{36}_{18}Ar$+$^{204}_{82}Pb$ ($\eta = 0.7$) for total mass (A_{TOT}) = 240. The present study is for peripheral collisions (i.e. b/b$_{max}$ = 0.5). Note that in some cases, slight variation can be seen for charges. The charges are chosen in a way so that colliding nuclei are stable nuclides. A momentum dependent soft equation of state with standard energy dependent cugnon cross-section (labeled as SMD) is used for the present calculations.

The balance energy (E_{bal}) is calculated using the *directed transverse momentum* $< P_x^{dir} >$, which is defined as:

$$\langle P_x^{dir}\rangle = \frac{1}{A}\sum_i \text{sign}\{Y(i)\}\ \mathbf{p}_x(i), \tag{12}$$

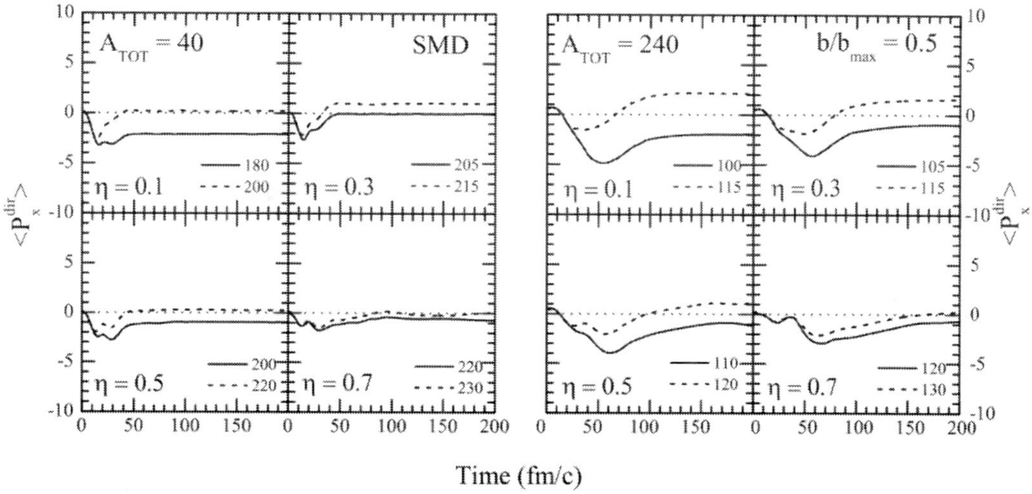

Figure 1. The time evolution of the directed transverse flow $< P_x^{dir} >$ as a function of time for different mass asymmetries. Columns 1 and 2 from left are for system mass $(A_{TOT}) = 40$, whereas columns 3 and 4 are for system mass $(A_{TOT}) = 240$. Here we display the results for peripheral collisions at different incident energies using SMD equation of state.

where $Y(i)$ and $\mathbf{p}_x(i)$ are the rapidity distribution and transverse momentum of ith particle, respectively.

In Fig. 1, we display the time evolution of the directed transverse flow $< P_x^{dir} >$, for various asymmetric reactions at different incident energies with system mass $(A_{TOT}) = 40$ and 240 units. From the figure, it is clear that transverse flow is always negative during the initial phase of the reaction irrespective of incident energy, system mass, and asymmetry of the reaction. This shows that the interactions among nuclei are attractive during initial phase of the reaction. These interactions turn repulsive depending on the incident energy. The transverse in-plane flow in lighter colliding nuclei for all values of η saturates earlier compared to heavy colliding nuclei. A sharp transition from negative to positive flow is seen for lighter nuclei at lower asymmetry. For the heavier nuclei at all asymmetries; the transition is gradual. From the figure, one can see that nearly symmetric reactions ($\eta = 0.1$ and 0.3) respond strongly to the change with incident energy compared to highly asymmetric reactions. The cause behind is that with the increase in the asymmetry of a reaction lesser binary collisions take place resulting in lesser density and therefore, less response occurs for variation in the collective flow. On the other hand, for nearly symmetric reactions, the binary collisions increases linearly with incident energy, therefore, huge difference can be seen.

It would be interesting to see how mass dependence character of the E_{bal} behaves at a fixed asymmetry. In Fig. 2, we display the E_{bal} verses combined system mass by keeping the asymmetry fixed. We notice that in all the cases, a perfect power law dependence (with power factor close to 1/3) can be seen for all asymmetries right from 0.1 to 0.7. The values of power factor are -0.34, -0.36, -0.35, and -0.34, respectively, for $\eta = 0.1$, 0.3, 0.5, and 0.7. Strikingly, the results are similar to the one as shown recently by Chugh and Puri [27]. There it was predicted that the role of MDI for symmetric systems is more dominant at peripheral geometries and the value of power factor is nearly equal to -0.35. From the figure, it is clear that all points are

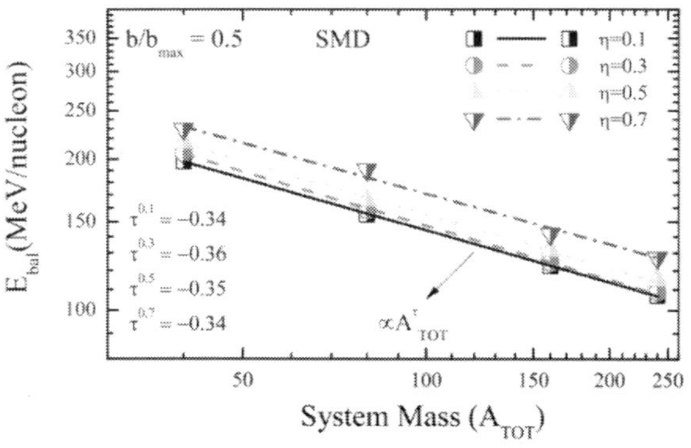

Figure 2. (Color Online) The E_{bal} as a function of system mass (A_{TOT}) of reacting partners for SMD EoS and for peripheral collisions. The results for different asymmetries η = 0.1, 0.3, 0.5, and 0.7 are represented, respectively, by the half filled squares, circles, triangles and inverted triangles. Lines are power law fit $\propto A^{\tau}$.

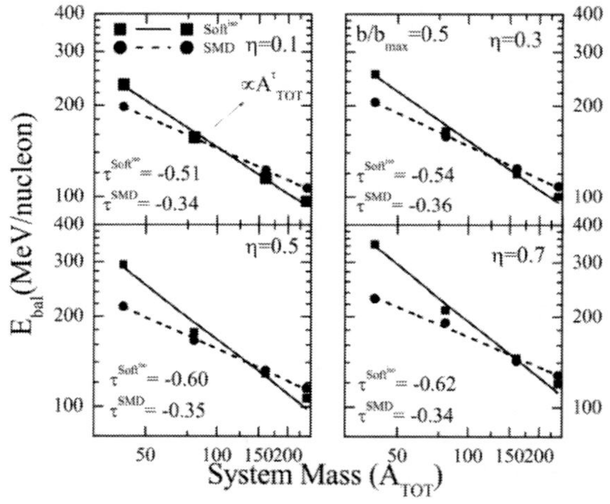

Figure 3. The E_{bal} as a function of system mass. Solid squares(circles) are for Soft^{iso}(SMD) EoS. The top(bottom) left and right panels are for asymmetry parameter of 0.1(0.5) and 0.3(0.7) respectively. Lines are power law fit $\propto A^{\tau}_{TOT}$.

lying on the line, indicating that asymmetry plays a major role in the mass dependence. The variation in E_{bal} for η varying between 0.1 and 0.7 is $\approx$ 32, 35, 19, and 19 for A_{TOT} = 40, 80, 160, and 240, respectively. This explains why in the earlier mass dependence studies [1], though average behavior was a power law, individual E_{bal} were quite far from the average law [1, 3, 4]. There this happened because no control was made for the asymmetry of a colliding pair. Had all reactions been analyzed on a fixed asymmetry, one could have obtain perfect match with average power law.

The effect of MDI on the E_{bal} is shown in Fig. 3, where we display the E_{bal} as a function of combined mass of the system with Soft^{iso} (Soft^{iso} represents the soft equation of state with isotropic standard energy dependent cugnon cross-section) and SMD EoS for η = 0.1, 0.3, 0.5,

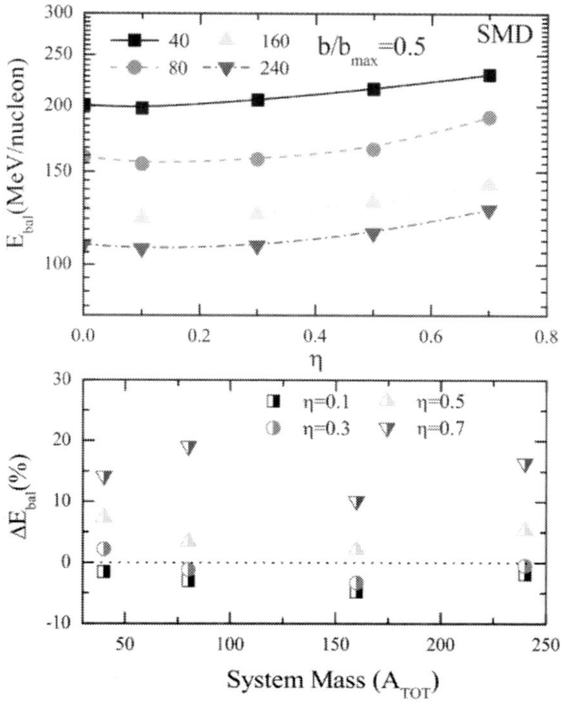

Figure 4. (Color Online) E_{bal} as a function of asymmetry parameter η is displayed in the upper panel, whereas the percentage difference $\Delta E_{bal}(\%)$ as a function of system mass of reacting partners is shown in the lower panel. SMD EoS and an reduced impact parameter of 0.5 is used in both cases. The results for the systems having total mass A_{TOT} of 40, 80, 160, and 240 are represented, respectively, by the solid squares, circles, triangles and inverted triangles. Lines are to guide the eye. The results of the percentage difference for different asymmetries η = 0.1, 0.3, 0.5, and 0.7 are represented, respectively, by the half filled squares, circles, triangles and inverted triangles.

and 0.7. Lines represent the power law fitting and different symbols are explained in the caption of the figure. It is clear from the figure that as η increases, the suppression of E_{bal} for the lighter systems with MDI also increases while the increase in E_{bal} for heavier nuclei remains almost same. Since transverse flow decreases with increase in asymmetry due to the decrease in nn collisions and Coulomb repulsions, therefore, the effect of the repulsive nature of MDI increases with increase in asymmetry mainly for the lighter masses. With the increase of asymmetry of the reaction, the role of MDI on the mass dependence of E_{bal} is similar to its role for the symmetric reactions at semi-peripheral and peripheral geometries.

In Fig. 4, we display E_{bal} as a function of η for a fixed mass equal to 40, 80, 160, and 240 units (top panel). In agrement with all previous studies, E_{bal} decreases with increase in the mass of the system. This decrease has been attributed to the increasing role of Coulomb forces in heavier colliding nuclei. As discussed in earlier figures, a sizeable influence can be seen towards E_{bal} with variation of the asymmetry of a reaction. For lighter masses, the effect of the variation of η can result about 35 MeV change in the E_{bal}. As noted in absolute terms, lighter nuclei are more affected compared to heavier ones. Overall, one sees that the effect of the asymmetry of a reaction is not at all negligible. It can have sizeable effect which goes as power law with power factor close to 1/3. We also display the percentage difference $\Delta E_{bal}(\%)$ defined as $\Delta E_{bal}(\%)$ = $((E_{bal}^{\eta\neq0}$-$E_{bal}^{\eta=0})/E_{bal}^{\eta=0})\times100$ as function of system mass (bottom panel). Very interestingly, we

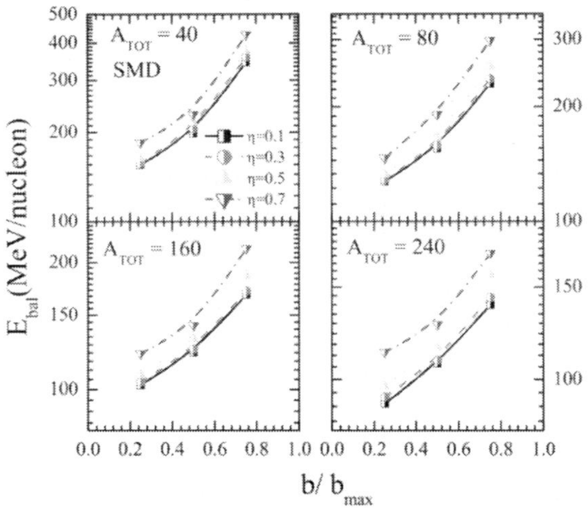

Figure 5. (Color Online) The E_{bal} as a function of reduced impact parameter. The top(bottom) left and right panels are for A_{TOT} = 40(80) and 160(240) respectively. The results for different asymmetries η = 0.1, 0.3, 0.5, and 0.7 are represented, respectively, by the half filled squares, circles, triangles and inverted triangles. Lines are to guide the eye.

see that the effect of the asymmetry variation is almost uniform throughout the periodic table. In other words, asymmetry of a reaction can play significant role in E_{bal} and the deviation from the mean line can be eliminated if proper care is taken for the asymmetry of colliding nuclei.

In Fig. 5, we display the E_{bal} as a function of the impact parameter for different asymmetries. A well known trend i.e. increase in the E_{bal} with impact parameter can be seen for all ranges of system mass. Further, as demonstrated by many authors, the impact parameter variation is less effected in the presence of MDI. The striking result is that the effect of mass asymmetry variation is almost independent of the impact parameter.

4. Summary
Using the quantum molecular dynamics model, we presented a detailed study of the balance energy with reference to mass asymmetry. Almost independent of the system mass as well as impact parameter, an uniform effect of the mass asymmetry can be seen at the energy of vanishing flow. We find that for large asymmetries, (η = 0.7), the effect of asymmetry can be 15% with MDI and in the absence it can be 40%. This also explain the deviation in the individual E_{bal} from the mean values as reported earlier.

5. Acknowledgments
Author is thankful to Council of Scientific and Industrial Research (CSIR) for providing the Junior Research Fellowship.

6. References
[1] Sood A D and Puri R K 2004 *Phys. Lett.* B **594** 260
[2] Aichelin J 1991 *Phys. Rep.* **202** 233
[3] Westfall G D *et al.* 1993 *Phys. Rev. Lett.* **71** 1986
[4] Magestro D J *et al.* 2000 *Phys. Rev.* C **61** 021602(R)
[5] Pan Q and Danielewicz 1993 *Phys. Rev. Lett.* **70** 2062
[6] Pak R *et al.* 1996 *Phys. Rev.* C **54** 2457
[7] Buta A *et al.* 1995 *Nucl. Phys.* A **584** 397
[8] Bonasera A and Csernai L P 1987 *Phys. Rev. Lett.* **59** 630

[9] Krofcheck D *et al.* 1989 *Phys. Rev. Lett.* **63** 2028
[10] Hartmann O 2003 Dissertation TU Darmstadt
[11] Eskola K J 1989 *Nucl. Phys.* B **323** 37
[12] Kim Y Jin 2004 Ph.D. thesis Korea University
[13] Kaur V and Kumar S 2010 *Phys. Rev.* C **81** 064610
[14] Jakobsson B *et al* 1990 *Nucl. Phys.* A **509** 195
[15] Barz H W *et al* 1992 *Nucl. Phys.* A **548** 427
[16] Begemann-Blaich M *et al* 1993 *Phys. Rev.* C **48** 610
[17] Tsang M B *et al* 1993 *Phys. Rev. Lett.* **71** 1502
[18] Li B A, Bauer W and Bertsch G F 1991 *Phys. Rev.* C **44** 2095
[19] Gosset J *et al.* 1989 *Phys. Rev. Lett.* **62** 1251
[20] Stöcker H, Maruhn J A and Greiner W 1980 *Phys. Rev. Lett.* **44** 725
[21] Bowman D R *et al.* 1987 *Phys. Lett.* B **189** 282
[22] Bertsch G F, Lynch W G and Tsang M B 1987 *Phys. Lett.* B **189** 384
[23] Wilson W K *et al.* 1991 *Phys. Rev.* C **43** 2696
[24] Sun R *et al.* 2000 *Phys. Rev. Lett.* **84** 43
[25] Avdeyev S P *et al.* 2001 *Phys. Lett.* B **503** 256
[26] Hauger J A *et al.* 1996 *Phys. Rev. Lett.* **77** 235
[27] Chugh R and Puri R K 2010 *Phys. Rev.* C **82** 014603

AUTHOR INDEX

AUTHOR INDEX

Institute of Physics
Dirac House, Temple Back
Bristol BS1 6BE UK

CURRAN ASSOCIATES INC
proceedings.com